T0141983

Advances in Intelligent Systems and Computing

Volume 945

The series "Advances in Intelligent Systems and Computing" contains publications on theory, applications, and design methods of Intelligent Systems and Intelligent Computing. Virtually all disciplines such as engineering, natural sciences, computer and information science, ICT, economics, business, e-commerce, environment, healthcare, life science are covered. The list of topics spans all the areas of modern intelligent systems and computing such as: computational intelligence, soft computing including neural networks, fuzzy systems, evolutionary computing and the fusion of these paradigms, social intelligence, ambient intelligence, computational neuroscience, artificial life, virtual worlds and society, cognitive science and systems, Perception and Vision, DNA and immune based systems, self-organizing and adaptive systems, e-Learning and teaching, human-centered and human-centric computing, recommender systems, intelligent control, robotics and mechatronics including human-machine teaming, knowledge-based paradigms, learning paradigms, machine ethics, intelligent data analysis, knowledge management, intelligent agents, intelligent decision making and support, intelligent network security, trust management, interactive entertainment, Web intelligence and multimedia.

The publications within "Advances in Intelligent Systems and Computing" are primarily proceedings of important conferences, symposia and congresses. They cover significant recent developments in the field, both of a foundational and applicable character. An important characteristic feature of the series is the short publication time and world-wide distribution. This permits a rapid and broad dissemination of research results.

**** Indexing: The books of this series are submitted to ISI Proceedings, EI-Compendex, DBLP, SCOPUS, Google Scholar and Springerlink ****

More information about this series at http://www.springer.com/series/11156

Piotr Kulczycki · Janusz Kacprzyk ·
László T. Kóczy · Radko Mesiar ·
Rafal Wisniewski
Editors

Information Technology, Systems Research, and Computational Physics

Editors
Piotr Kulczycki
Faculty of Physics and Applied
Computer Science
AGH University of Science and Technology
Krakow, Poland

Systems Research Institute
Polish Academy of Sciences
Warsaw, Poland

László T. Kóczy
Faculty of Engineering Sciences
Széchenyi István University
Győr, Hungary

Faculty of Electrical Engineering
and Informatics
Budapest University of Technology
and Economics
Budapest, Hungary

Rafal Wisniewski
Department of Electronic Systems
Aalborg University
Aalborg, Denmark

Janusz Kacprzyk
Systems Research Institute
Polish Academy of Sciences
Warsaw, Poland

Radko Mesiar
Faculty of Civil Engineering
Slovak University of Technology
Bratislava, Slovakia

ISSN 2194-5357 ISSN 2194-5365 (electronic)
Advances in Intelligent Systems and Computing
ISBN 978-3-030-18057-7 ISBN 978-3-030-18058-4 (eBook)
https://doi.org/10.1007/978-3-030-18058-4

This Springer imprint is published by the registered company Springer Nature Switzerland AG
The registered company address is: Gewerbestrasse 11, 6330 Cham, Switzerland

Preface

Information technology (IT)—highly sophisticated information processing—forms the intellectual basis at the forefront of the current, third scientific-technical revolution. The first, generally placed at the turn of the eighteenth and nineteenth centuries, created the foundations for replacing muscle power with that of steam, and—as a consequence—manual production started to be displaced by industrial mass production. The second, which took place at the turn of the nineteenth and twentieth centuries, was brought about by a large number of groundbreaking concepts and inventions occurring thanks to the new energy carrier—electricity. This third scientific-technical revolution, started after the Second World War, is of a different character from the previous ones in that it is nonmaterial. In essence, it constitutes the collection, treatment, and transmission of data, and so the subject of research and operation is here abstract, unreal objects. The dominant discipline for innovative development and progress became information technology as it is widely understood.

If therefore the crux of the changes does not consist of creating new machines or devices, but of the radical transformation of preexisting essence and character, then the spectrum of research and practical interests is unusually broad, even unlimited in the framework of contemporary science and applicational fields. Thus, such is the differing material of this book. It consists of three parts. In the first part, a specialized calculating apparatus is presented, which originated within the context of and for the needs of information technology: data analysis (especially exploratory) and its main tool—intelligent computing. This part contains 13 chapters. Applicational aspects for problems in the scope of information technology and system research have become the topics of seven texts comprising the second part. In turn, the final ten-chapter third part is devoted to tasks of basic disciplines – mathematics and physics – describing the world's reality, investigated in those presentations with the aid of contemporary information technology methods.

The subject choice of particular parts of this edited book was determined through discussions and reflection arising during the *Third Conference on Information Technology, Systems Research and Computational Physics (ITSRCP'18)*[1], creating together with the sister *Sixth International Symposium CompIMAGE'18—Computational Modeling of Objects Presented in Images: Fundamentals, Methods, and Applications* (CompIMAGE'18)[2] components of the event joining them the *International Multi-Conference of Computer Science (CS'18)*[3], which took place on July 2–5, 2015, at the AGH University of Science and Technology in Kraków, Poland. In total, at the above meeting participants from 21 countries—Austria, Bangladesh, Colombia, Czech Republic, Germany, Hungary, India, Italy, Jordan, Mexico, Morocco, Poland, Portugal, Slovakia, Spain, Sri Lanka, Taiwan, Tunisia, Ukraine, USA, Yemen—from 5 continents, published their papers. The proceedings of this event are available in the e-book [1]. The authors of selected papers presented at the ITSRCP'18 conference were invited to present their research as part of this post-conference edited book.

The above multiconference was the essential continuation of the previous *Congress on Information Technology, Computational and Experimental Physics* (*CITCEP 2015*), December 18–20, 2015, Kraków, Poland, whose materials were published in the edited book [2].

As the organizers of the *ITSRCP'18* conference, we direct special words of recognition to Reneta P. Barneva (State University of New York at Fredonia, USA) being Scientific Chair, as well as Valentin E. Brimkov (SUNY Buffalo State, USA), Joao Manuel R. S. Tavares and Renato M. Natal Jorge (both from University of Porto, Portugal), accompanied together with Piotr Kulczycki (AGH Academy of Science and Technology, Poland), the Steering Committee of the sister *CompIMAGE'18* conference. Its post-proceedings are currently printed in the edited book [3].

We also would hereby like to express our heartfelt thanks to Technical Associate Editors of this book, Dr. Piotr A. Kowalski and Dr. Szymon Łukasik, as well as Co-Organizers of the above conferences, Dr. Joanna Świebocka-Więk, Grzegorz Gołaszewski, and Tomasz Rybotycki, as well as all participants of this very interesting interdisciplinary event.

September 2018

Piotr Kulczycki
Janusz Kacprzyk
László T. Kóczy
Radko Mesiar
Rafal Wisniewski

[1] http://itsrcp18.fis.agh.edu.pl/.
[2] http://isci18.fis.agh.edu.pl/.
[3] http://cs2018.fis.agh.edu.pl/.

References

1. P. Kulczycki, P.A. Kowalski, S. Łukasik (eds.), *Contemporary Computational Science*[4]. AGH-UCT Press, Kraków, 2018. (ISBN: 978-83-66016-22-4).
2. P. Kulczycki, L.T. Kóczy, R. Mesiar, J. Kacprzyk (eds.), *Information Technology and Computational Physics*[5]. *Advances in Intelligent Systems and Computing* series, vol. 462, Springer, Cham, 2017 (ISBN: 978-3-319-44259-4).
3. Barneva R.P., Brimkov V.E., Kulczycki P., Tavares J.M.R.S. (eds.) *Computational Modeling of Objects Presented in Images. Fundamentals, Methods, and Applications. Lecture Notes in Computer Sciences* series, Springer, 2019, in press.

[4] http://itsrcp18.fis.agh.edu.pl/wp-content/uploads/Contemporary_Computational_Science_-_eds_Kulczycki_Kowalski_Lukasik.pdf.

[5] http://www.fis.agh.edu.pl/Conf-ITCEP/.

Associate Editors

Piotr A. Kowalski
Faculty of Physics and Applied Computer Science
AGH University of Science and Technology
Al. Mickiewicza 30
30-059 Krakow
Poland
Email: pkowal@agh.edu.pl

Systems Research Institute
Polish Academy of Sciences
Ul. Newelska 6
31-447 Warsaw
Poland
Email: pakowal@ibspan.waw.pl

Szymon Łukasik
Faculty of Physics and Applied Computer Science
AGH University of Science and Technology
Al. Mickiewicza 30
30-059 Krakow
Poland
Email: slukasik@agh.edu.pl

Systems Research Institute
Polish Academy of Sciences
Ul. Newelska 6
31-447 Warsaw
Poland
Email: slukasik@ibspan.waw.pl

Contents

Data Analysis and Intelligent Computing

Using Random Forest Classifier for Particle Identification in the ALICE Experiment

Tomasz Trzciński[1(✉)], Łukasz Graczykowski[2], and Michał Glinka[1] for the ALICE Collaboration

[1] Institute of Computer Science, Warsaw University of Technology, Warsaw, Poland
t.trzcinski@ii.pw.edu.pl, mglinka2@stud.elka.pw.edu.pl
[2] Faculty of Physics, Warsaw University of Technology, Warsaw, Poland
lukasz.graczykowski@pw.edu.pl

Abstract. Particle identification is very often crucial for providing high quality results in high-energy physics experiments. A proper selection of an accurate subset of particle tracks containing particles of interest for a given analysis requires filtering out the data using appropriate threshold parameters. Those parameters are typically chosen sub-optimally by using the so-called "cuts" – sets of simple linear classifiers that are based on well-known physical parameters of registered tracks. Those classifiers are fast, but they are not robust to various conditions which can often result in lower accuracy, efficiency, or purity in identifying the appropriate particles. Our results show that by using more track parameters than in the standard way, we can create classifiers based on machine learning algorithms that are able to discriminate much more particles correctly while reducing traditional method's error rates. More precisely, we see that by using a standard Random Forest method our approach can already surpass classical methods of cutting tracks.

Keywords: Monte Carlo tracks · Random forest classification

1 Introduction

Particle identification (PID), *i.e.* the identification of the mass and flavour composition of particles produced during a collision, is very often a first preprocessing step of a typical analysis in high energy physics experiments. It is of a particular importance in the case of ALICE (A Large Ion Collider Experiment) [1], one of the experiments of the Large Hadron Collider (LHC) [2], whose goal is to study all aspects of ultra-relativistic heavy-ion collisions (lead–lead (Pb–Pb)) in order to measure the properties of the Quark-Gluon Plasma [3,4]. ALICE is a complex detector composed of 18 different detection systems that employ various methods of interaction with highly energetic particles to measure

P. Kulczycki et al. (Eds.): ITSRCP 2018, AISC 945, pp. 3–17, 2020.
https://doi.org/10.1007/978-3-030-18058-4_1

physical phenomena. Thanks to the complexity of detection systems, a separation of signals left by different particle species is possible in a very broad momentum range, from around 100 MeV/c up to around 10 GeV/c, making ALICE the most powerful of the LHC experiments in terms of PID capabilities.

PID requires filtering out a signal corresponding to a given particle species from signals produced by other particle types. This is typically achieved by applying certain threshold parameters on the deviation of the signal in a given detector from its nominal value. Those parameters are typically chosen by using the so-called "cuts" – sets of simple linear classifiers which remove unwanted data below or above a given threshold value. This method works well when the signals are well separated from each other and information from one detector is sufficient. Yet, when they start to overlap and proper combination of the PID data from two or more detectors is required, setting correct threshold values becomes unintuitive and non-trivial. The sub-optimality of such approach typically results in low accuracy, efficiency, or purity of the sample of selected particles, hence reducing available data and reducing statistical power of physical analyses.

In this paper, we address the above-mentioned shortcoming of the currently used method and we propose to improve the selection of particles using machine learning methods. More precisely, we propose a set of classifiers, called the Random Forest [5], that are trained specifically to improve the discrimination between particle types. The classifiers improve the sensitivity, defined as a ratio of correctly classified signal particles to its total amount, over the currently employed method, while not decreasing the precision of the method. In this paper, we show how a state-of-the-art Random Forest classifier can be used to solve the address the problem of particle identification. We evaluate the quality of the proposed selection method and compare it with the traditional method, proving that our approach can significantly outperform the currently used algorithm both in terms of the number of correctly classified particles and the error rates of their selection.

The preliminary version of this paper was presented at the 3rd Conference on Information Technology, Systems Research and Computational Physics, 2–5 July 2018, Cracow, Poland [6].

2 Particle Identification with TPC and TOF

PID of light-flavour charged hadrons (pions, kaons, protons), the most abundantly produced particles in heavy-ion collisions, is usually performed using two ALICE detectors, that is the Time Projection Chamber (TPC) [7] and the Time-Of-Flight system (TOF) [8]. Understanding the detector response of both of them is crucial for any PID technique.

The TPC – the main tracking device of ALICE – is a cylindrical gaseous detector which provides, for a given particle, measurements of the mean energy loss per unit path length, $\langle \mathrm{d}E/\mathrm{d}x \rangle$, from up to 159 independent $\mathrm{d}E/\mathrm{d}x$ measurements along its trajectory. The $\langle \mathrm{d}E/\mathrm{d}x \rangle$ as a function of particle's momentum p is described by the Bethe-Bloch empirical formula, whose parameters depend on various factors, including the detector intrinsic resolution, track reconstruction details, parameters of the gas, and others [9]. The distribution of measured $\langle dE/dx \rangle$ around the value expected from the Bethe-Bloch parameterisation has a Gaussian profile with a standard deviation σ_{TPC}. Therefore, the best PID estimator for the TPC detector is defined as a distance, in numbers of standard deviations $N_{\sigma_{\mathrm{TPC}}}$, of the measured $\langle dE/dx \rangle$ to the nominal signal from the Bethe-Bloch curve.

The TOF detector provides the measurements of velocity of a charged particle by measuring its time of flight t_{TOF} over a given distance along the particle trajectory l. The arrival time is measured by employing the Multigap Resistive Plate Chamber technology which have an intrinsic resolution of 80 ps. For each particle type the measured arrival time, t_{TOF}, is compared to the nominal (expected) time, t_{exp}. The former is defined as a difference between the arrival time in TOF and the event collision time, evaluated using a sophisticated procedure (for details see [10]), while the latter is the time it would take for a particle of a given mass to travel from the interaction point to the TOF. The distribution of the arrival time measured in TOF around the expected time has a Gaussian profile with a standard deviation $\sigma_{\mathrm{PID}}^{\mathrm{TOF}}$. By analogy, the best PID estimator for the TOF detector is defined as a distance, in numbers of standard deviations $N_{\sigma_{\mathrm{PID}}^{\mathrm{TOF}}}$, of the measured arrival time t_{TOF} to the nominal signal t_{exp}. In practice, t_{TOF} is often expressed as $\beta = v/c$, where $v = l/t_{\mathrm{TOF}}$ is the particle's velocity and c is the speed of light.

Figure 1 shows typical TPC energy loss and TOF velocity measurements as a function of particle's momentum from proton–proton collisions at the center-of-mass energy of $\sqrt{s} = 13$ TeV measured by ALICE.

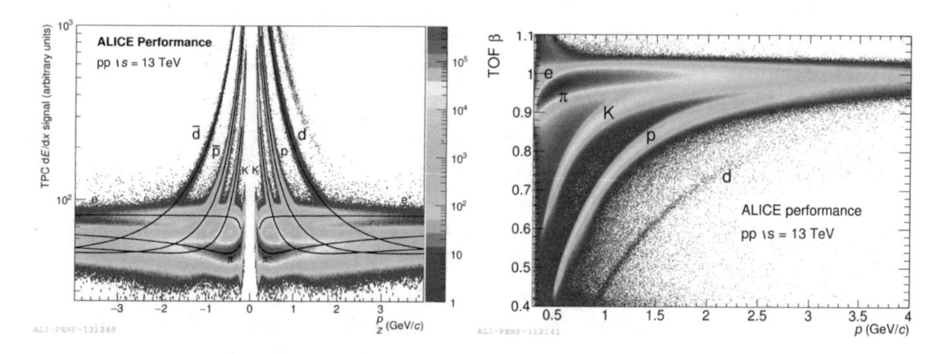

Fig. 1. Results of the measurements of (left) the energy loss in the TPC with lines corresponding to Bethe-Bloch parameterisation and (right) particle's velocity β in the TOF detector as a function of particle's momentum.

3 Baseline PID Method

The typical approach to the PID involves applying simple linear selection criteria where single cut-off values are used to accept or reject tracks with specific

$N\sigma_{\mathrm{PID}}^{\mathrm{TPC}}$ and $N\sigma_{\mathrm{PID}}^{\mathrm{TOF}}$ based on particle's momentum. They are often chosen arbitrarily depending on the goals of the analysis (i.e. whether maintaining the very high purity of the selected sample is required) and experience of a scientist performing the study. In addition, other parameters of the track which may be relevant for PID in a non-trivial way are omitted. Finally, adjusting the cut-off values in order to achieve the desired parameters of the sample requires a lot of time and effort on the physicists side.

Some alternative approaches for PID exist, *e.g.* the Bayesian PID [11], which relies on Bayesian probabilistic approach to the problem of particle identification. Although this method was proven to provide higher signal-to-noise rations, it is not widely used in practice yet and therefore in this work we use the traditional PID as a baseline for our method.

4 Our Approach

In this paper, we propose an approach for machine learning-based PID that is based on the Random Tree method [12]. More precisely, we choose a classifier, called the Random Forest [5], that uses numerous random decision trees to classify given observation. In this section, we describe the method and justify its selection for our application.

4.1 Random Forest Algorithm

Random Forest is an ensemble of decision trees that generate the classification decision based on a set of sub-decisions. In each decision tree, attributes are represented as nodes and classifier's decisions as leaves. Each node is created by selecting best attribute from the fixed size random subset of attributes used to train current tree. Their quality might be assessed via calculating entropy gain for each attribute or its Gini index, which is defined as probability of wrong classification while using only a given attribute. Each node splits a dataset into two subsets, trying to maximize the chances that the samples of the same class end up in the same subset. If one of those subsets has low number of data samples from other classes, it is converted into a leaf and no further splitting is needed. Minimal impurity of a node needed for this to happen equals zero by default, but can be modified as one of classifier's hyper-parameters. The whole process is then repeated to create new nodes. In a classical approach, Random Forest trees are not pruned in any way and those steps are repeated until all leaves are created.

The Random Forest classifier has several hyper-parameters that can be tuned:

- number of decision trees inside the forest
- maximum depth of each decision tree
- minimal impurity of a node for it to be converted into a leaf
- maximum number of attributes used per tree training
- minimal impurity decrease of resulting subdatasets for a node to be created

The final classifier is created by training multiple decision trees. When final classification is to be made, each of the tree processes the data independently. Then, the final decision is chosen as a result of majority voting from all trees inside the forest.

4.2 Unique Properties of Random Forest

In the context of PID, processing high amounts of data at large speed is essential. This is why the Random Forest classifier is a perfect fit. Thanks to its simplicity and ability to scale horizontally[1], the classification decision process can be sped up not only by increasing the computational power, but also by using more machines. Many independent training processes on separate datasets can be parallelized on separate machines and later all resulting decision trees can be simply aggregated into a single classifier. This approach enables to fully exploit the potential of GRID [13], a global collaboration of more than 170 computing centres in 42 countries providing global computing resources to store, distribute and analyse enormous amounts of data as part of CERN's computing infrastructure.

Additionally, the Random Forest classifier is very resistant to *overfitting*. Overfitting is a common phenomenon known in machine learning which defines the situation when the classifier loses its ability to correctly classify samples outside of a training dataset, because its parameters are over-tuned on the training dataset.

Another advantage of the Random Forest over other machine learning methods, such as Support Vector Machines, is its interpretability. The analysis of the parameters used for a given split is straightforward and does not require any additional tools. Thanks to this property, we can create a simple set of 'sanity checks' to see whether particle classification is based on a complimentary set of attribute choices rather than some hidden correlations within our dataset.

Finally, the Random Forest classifier is used in numerous applications across domains, *e.g.* genetics – to identify DNA binding proteins [14] – or medicine – to classify diabetic retinopathy [15]. This wide adoption of the Random Forest method indicates the potential of this method and leads to a significant amount of resources, such as tutorials, publications and libraries, that can be found in the Internet and used to improve the final performance.

4.3 Implementation

In our work, we use a Python implementation of the Random Forest [5] classifier provided by scikit-learn package [16], instead of the ROOT environment, which is widely used at CERN and it is in fact our ultimate production environment.

We use Python for development purposes, as it offers a diverse set of tools used in both machine learning and data engineerings tasks. Thanks to C and Fortran snippets of code integrated in the Python backend, all calculations are efficient

[1] Horizontal scalability is an attribute of a system, which may be expanded by adding new nodes (machines, servers, computers), rather than by only increasing computing power of existing ones.

while a high level of API abstraction enables relatively easy code development. As we envision ROOT as our final production environment, we maintain a full integration between those two distinct systems by using external libraries that allow us to convert CERN datatypes into most commonly used Python one – pandas DataFrame [17]. After the conversion is done we can use Python tools not only to train a classifier, but also to evaluate its performance and tune its parameters.

5 Experiments

In this section, we describe the results of a comparison between our approach and the traditional PID method. Both methods are used to perform a classification of three types of particles. We choose pions, kaons and protons as they are the most abundantly produced particles in a typical pp collisions and, therefore, an excellent use case for our PID algorithm.

The remainder of this section is organized as follows. We first present the dataset generated using Monte Carlo simulations that we use in our evaluation. We then present the results of the initial experiments that aimed at determining the importance of input parameters for the final classification task. We then outline the final results of our evaluation which shows that the proposed Random Forest classifier significantly outperforms currently used PID methods in the task of the particle classification.

5.1 Dataset

As our dataset, we use Monte Carlo proton–proton data at the center of mass energy of $\sqrt{s} = 7$ TeV, generated by PYTHIA 6.4 [18] model, Perugia-0 [19] tune. After generating the particles, we transport (process) them using GEANT3 [20] package, simulating the passage of the particles through the detector medium. We also perform a full simulation of the detector response as well as the reconstruction of full trajectories of the generated particles. The experimental conditions correspond to 2010 data-taking period. This way, we obtain 413,896 particle tracks with the associated label determining its particle class. This information is crucial for classifier training as we can then cast the PID as a classification problem and assess the performance of the Random Forest classifier when solving it.

For completeness, we present here the exact criteria used to select particle trajectories that form our dataset:

- for tracks with $p_{\mathrm{T}} > 0.5$ GeV/c the combined information from both the TPC and TOF was used, ${N_{\sigma,\mathrm{PID}}}^2 = {N_{\sigma,\mathrm{TPC}}}^2 + {N_{\sigma,\mathrm{TOF}}}^2$, resulting in a circular cut in the $N_{\sigma,\mathrm{TPC}}$ and $N^a_{\sigma,\mathrm{TOF}}$ space,
- for tracks with $p_{\mathrm{T}} < 0.5$ GeV/c, where only a few have an associated signal in the TOF and information only from the TPC was used $N_{\sigma,\mathrm{PID}} = N_{\sigma,\mathrm{TPC}}$.

For each of the particle trajectories, our dataset contains 37 attributes that can be used as an input of the classification method:

- $\beta = v/c$ – particle's velocity measured in TOF relative to the speed of light c,
- $p_{\mathrm{x}}, p_{\mathrm{y}}, p_{\mathrm{z}}$ – components of the particle's momentum along x, y, z directions, respectively,
- $p = \sqrt{p_{\mathrm{x}}^2 + p_{\mathrm{y}}^2 + p_{\mathrm{z}}^2}$ – total momentum of a particle,
- $p_{\mathrm{t}} = \sqrt{p_{\mathrm{x}}^2 + p_{\mathrm{y}}^2}$ – transverse momentum of a particle
- No. of TPC clusters – number of TPC clusters belonging to a given particle track,
- TPC signal – mean energy loss signal measured the TPC detector,
- $N_{\sigma^{\pi}_{\mathrm{TPC}}}, N_{\sigma^{\mathrm{K}}_{\mathrm{TPC}}}, N_{\sigma^{\mathrm{P}}_{\mathrm{TPC}}}, N_{\sigma^{\mathrm{e}}_{\mathrm{TPC}}}$ – difference expressed in units of standard deviation σ between the measured and the expected signals for a pion, kaon, proton and electron from the TPC detector, respectively,
- $N_{\sigma^{\pi}_{\mathrm{TOF}}}, N_{\sigma^{\mathrm{K}}_{\mathrm{TOF}}}, N_{\sigma^{\mathrm{P}}_{\mathrm{TOF}}}, N_{\sigma^{\mathrm{e}}_{\mathrm{TOF}}}$ – difference expressed in units of standard deviation σ between the measured and the expected signals for a pion, kaon, proton and electron in the TOF detector, respectively
- cov0-20 – components of a covariance matrix between x,y,z spatial coordinates and the components of the particle's momentum along x,y,z directions.

5.2 Attribute Importance

To understand the impact of each of the 37 attributes on the final classification results, we train our Random Forest classifier using only subset of attributes, *i.e.* excluding the attribute whose importance we evaluate. For this purpose we use All Relevant Feature Selection, which is provided by package Boruta [21]. Main principal behind this algorithm is that an attribute is irrelevant if it contributes to the discrimination task less than a noise. To create an estimate of such *noise* we permute all values inside single columns of a copy of all attributes so that they lose their correlation with class label. Then new classifier is created and the importance of all its attributes is computed. Maximal importance of permuted parameters is compared to the each attribute and those which are significantly lower are rendered as irrelevant. Additionally, those which importance is much higher are rendered relevant. After that, all the steps are repeated until there is no unclassified attribute left.

After our research we estimated that there are 12 attributes that are redundant. Those attribute are: cov0, cov1, cov3, cov4, cov6, cov7, cov8, cov10, cov11, cov12, cov15, cov16. To further validate our choice, we use our training dataset to compare two classifiers: with all attributes and only with *relevant*. We choose OOB-score as an evaluation method, which returns accuracy of classifying all training dataset, but for each observation using only those decision trees, which were not trained with that sample. Using this method we get very similar results: 0.9906 for bigger set of attributes and 0.9907 after reduction, which allows us to think that not only will that decrease computing time, but also heighten our scores. Final attributes importances are shown in Fig. 2

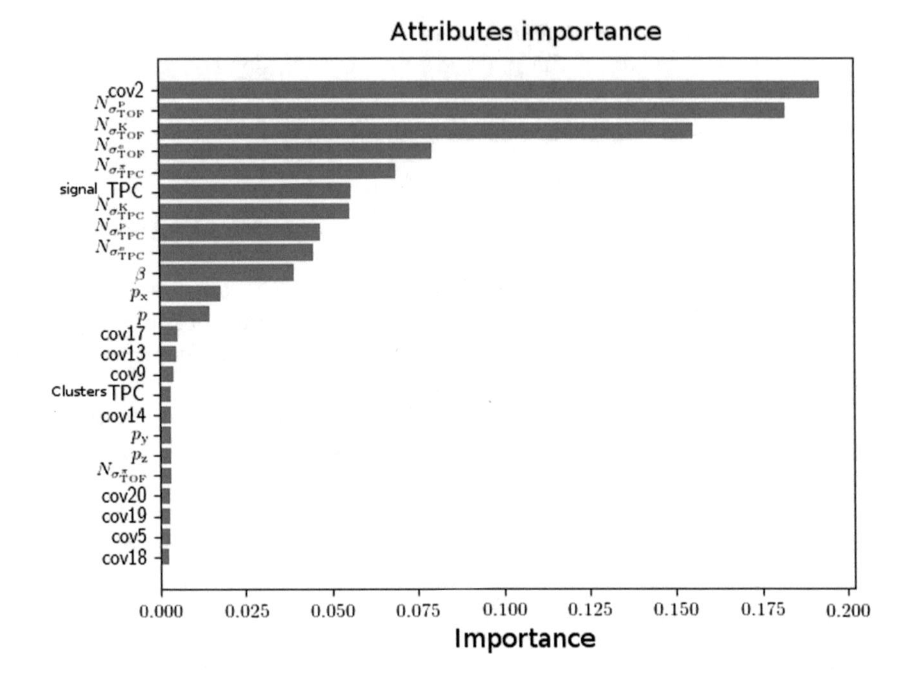

Fig. 2. Contribution of various track parameters to the overall PID from the training of the classifier. The highest importance matches parameters used in traditional PID, which proves correctness of this approach.

5.3 Parameter Tuning

We tune the hyper-parameters of our approach using a mean OOB cross-validation score computed on the classification task. Figure 3 shows the results of a set of experiments performed with different number of decision trees. One can see that the performance of our Random Forest method for PID saturates for more than 75 trees and we use this value in the following experiments.

Additionally we test different maximal depths of decision trees taking by ranging them from none to forty. Figure 4 shows the results of a set of experiments performed. One can see that depth above 20 trees doesn't influence score significantly. We don't restrict it, as it is recommended practice for the Random Forest classifier.

Finally, we also tune maximal number of attributes used to train a single decision tree. Table 1 shows the results of a set of experiments performed. We choose the default value which is square root of all the attributes, as it provides one of the best scores without significantly increasing training time.

The final set of values used in the rest of our work can be found below:

- maximum depth of each decision tree: infinite
- minimal impurity of a node for it to be converted into a leaf: 0.0
- maximum number of attributes used per tree training: 5
- minimal impurity decrease of resulting subdatasets for to create a node: 0.0

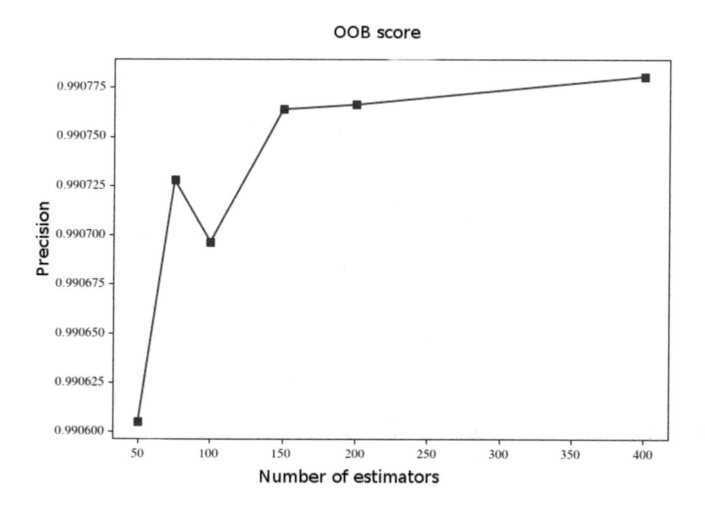

Fig. 3. Mean cross-validation score, using OOB-score function to determine the best number of estimators for Random Forest classifier.

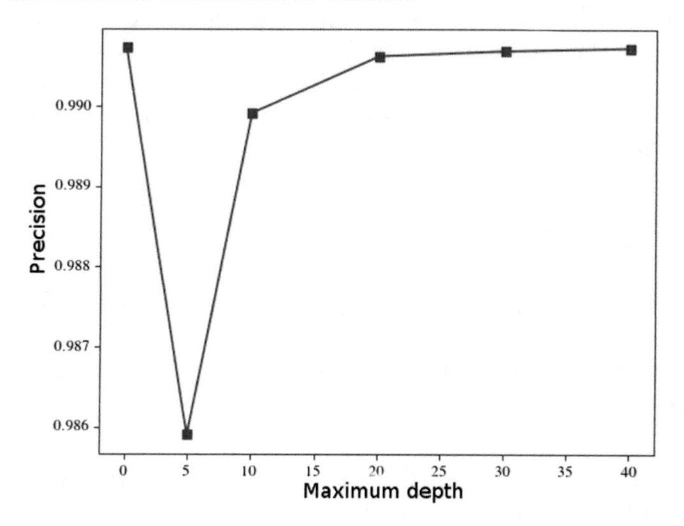

Fig. 4. Mean cross-validation score, using OOB-score function to determine the best maximum depth of a tree for Random Forest classifier.

Table 1. Results of analysis of maximum attributes number used for single decision tree training.

Maximum attributes	OOB score	Training time [s]
25	0.990485	704
12	0.990732	317
5	0.990728	128
4	0.990639	105

5.4 Evaluation Protocol

As our evaluation metrics, we use standard PID quantities: the *PID efficiency* and *purity*. They are defined in Eqs. (1) and (2), respectively:

$$\text{Efficiency} = \frac{\text{number of tracks correctly classified as a given particle specie}}{\text{number of all tracks of a given particle specie available in the sample}}, \tag{1}$$

$$\text{Purity} = \frac{\text{number of tracks correctly classified as a given particle specie}}{\text{number of tracks classified as a given particle specie}}. \tag{2}$$

Those metrics can be closely related to *precision* and *recall* metrics, widely used in the machine learning community.

5.5 Results

Here, we present the final results of the performance of the traditional PID and our method. We show a comparison of the efficiency and purity of the traditional and ML-based PID methods of particles selection as a function of their transverse momentum p_{T}. Overall, our proposed approach significantly outperforms the traditional method, across all metrics.

Kaon Classification. Kaon classification results are shown on Figs. 5 and 6. The achieved classifier qualities are shown below:

- for traditional PID the efficiency is 80.75% and the purity of the kaon sample is 88.28%,
- for the Random Forest the obtained efficiency is 97.57% and the purity of the kaon sample is 99.67%.

Pion Classification. Pion classification results are shown on Figs. 7 and 8. The achieved classifier qualities are shown below:

- for traditional PID the efficiency is 85.51% and the purity of the pion sample is 99.15%,
- for the Random Forest the obtained efficiency is 99.83% and the purity of the pion sample is 99.15%.

Proton Classification. Proton classification results are shown on Figs. 9 and 10. The achieved classifier qualities are shown below:

- for traditional PID the efficiency is 79.66% and the purity of the proton sample is 97.21%,
- for the Random Forest the obtained efficiency is 99.01% and the purity of the proton sample is 98.92%.

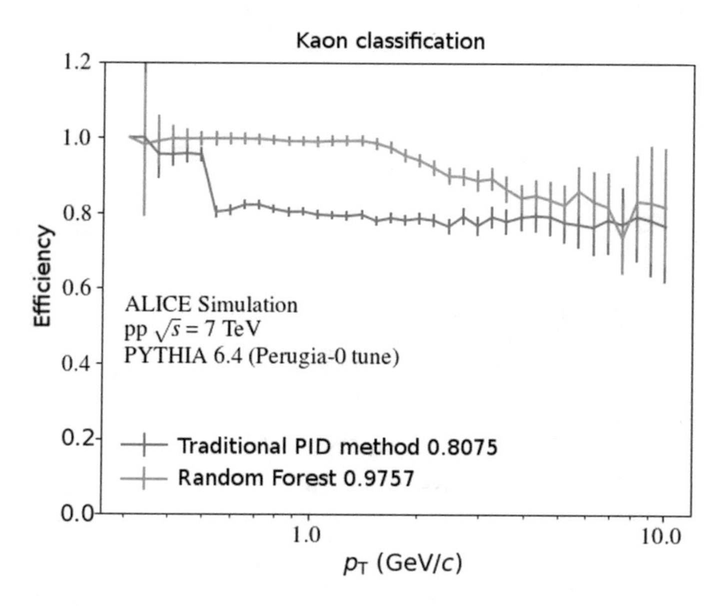

Fig. 5. Comparison of PID efficiency as a function of p_T of the kaon selection between the traditional PID method and the Random Forest classifier.

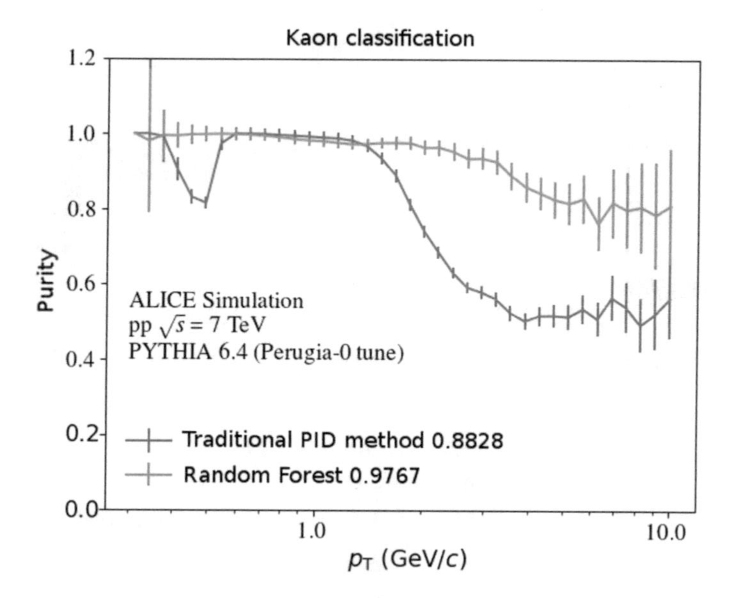

Fig. 6. Comparison of purity as a function of p_T of the kaon selection between the traditional PID method and the Random Forest classifier.

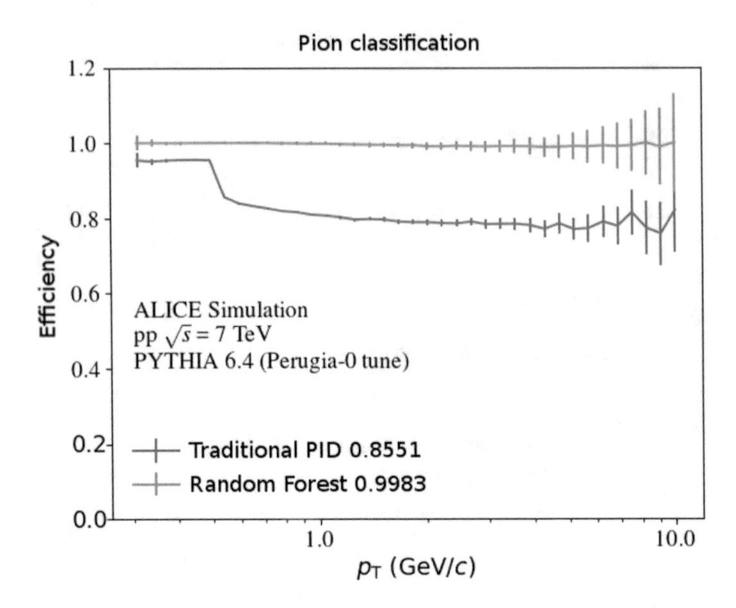

Fig. 7. Comparison of PID efficiency as a function of p_T of the pion selection between the traditional PID method and the Random Forest classifier.

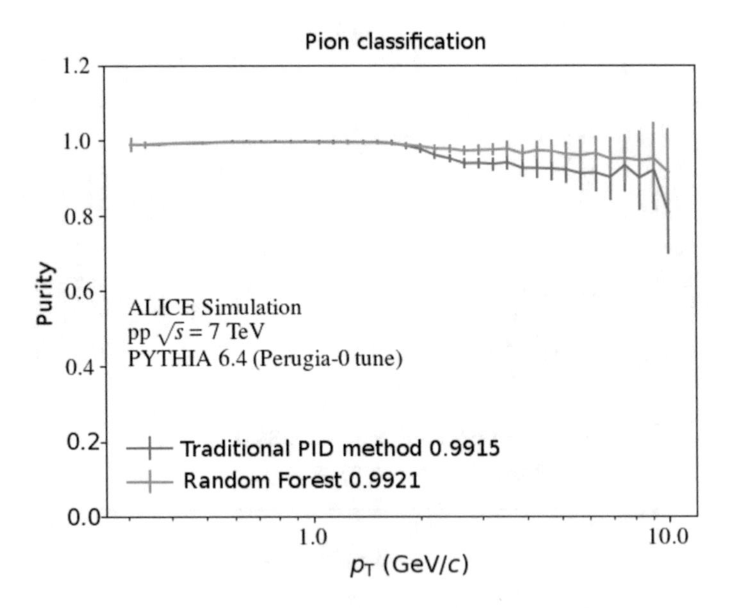

Fig. 8. Comparison of purity as a function of p_T of the pion selection between the traditional PID method and the Random Forest classifier.

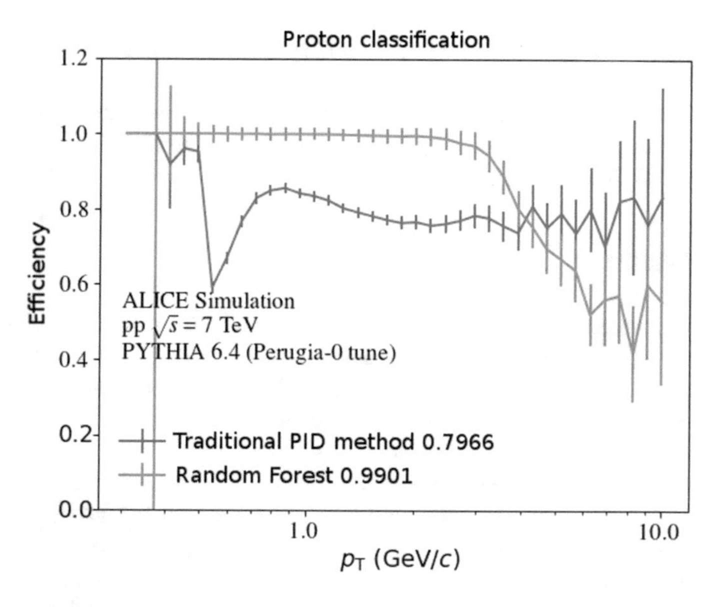

Fig. 9. Comparison of PID efficiency as a function of p_T of the proton selection between the traditional PID method and the Random Forest classifier.

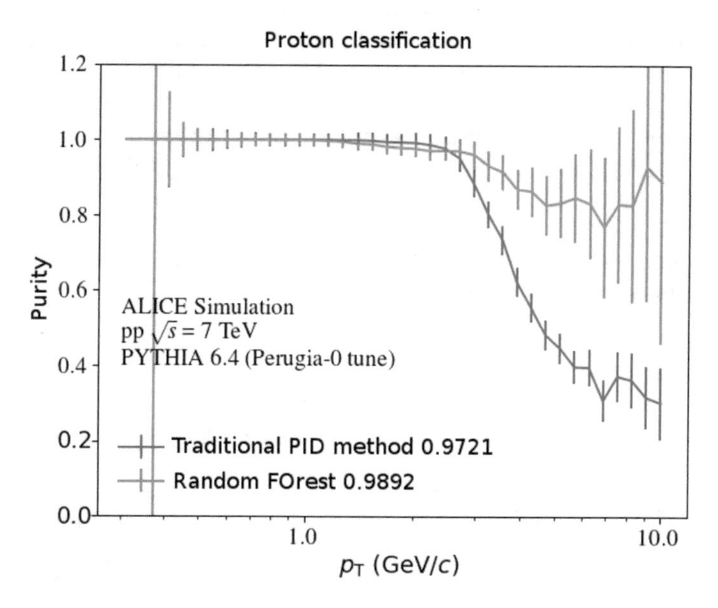

Fig. 10. Comparison of purity as a function of p_T of the proton selection between the traditional PID method and the Random Forest classifier.

While maintaining similar or higher purity levels, the Random Forest classifier achieves much higher overall efficiency than the competing traditional PID method. Overall, the proposed Random Forest method provides statistically relevant improvement across both metrics and transverse momentum levels over the traditional PID method.

6 Conclusions

In this paper, we compared the traditional PID procedure with a novel approach that is based on a machine learning classifier. Our evaluation results show that the approach based on a Random Forest classifier achieves higher purity and efficiency levels in the context of classification of particles. Not only did our approach yield better results, but it also significantly decreases the amount of human labor that is necessary to achieve this score. With the ability to scale horizontally with distributed systems, our approach seems to offer a promising alternative to the currently used and computationally expensive methods. This is especially true for analyses which require massive datasets that cannot fit in a memory of a single machine. Thanks to the proposed machine learning-based method, it possible to exploit larger datasets and obtain higher quality classifiers for the particle identification task. One important limitation of our method is that it depends greatly on how well the Monte Carlo simulation describes the real data experiment. Although there is still place for improvement, we believe that the results presented in this paper can serve as a proof of the potential offered by machine learning algorithms and the Random Forest classifier in particular. Given its performance boost, we are certain that this kind of machine learning approaches can be successfully incorporated in the ALICE Experiment, increasing the quality of high-energy physics results.

Acknowledgements. The authors acknowledge the support from the Polish National Science Centre grant no. UMO-2016/21/D/ST6/01946.

References

1. Aamodt, K., et al.: The ALICE experiment at the CERN LHC. JINST **3**, S08002 (2008)
2. Evans, L., Bryant, P.: LHC machine. JINST **3**, S08001 (2008)
3. Shuryak, E.V.: Quark-Gluon plasma and hadronic production of leptons, photons and psions. Phys. Lett. **78B**, 150 (1978). [Yad. Fiz.28,796(1978)]
4. Adams, J., et al.: Experimental and theoretical challenges in the search for the quark gluon plasma: the STAR collaboration's critical assessment of the evidence from RHIC collisions. Nucl. Phys. A **757**, 102–183 (2005)
5. Breiman, L.: Random forests. Mach. Learn, **45**, 5–32 (2001)
6. Trzciński, T., Graczykowski, Ł., Glinka, M.: Using random forest classifier for particle identification in the alice experiment. In: Kulczycki, P., Kowalski, P.A., Łukasik, S. (eds.) Contemporary Computational Science, p. 46 (2018)

7. Dellacasa, G., et al.: ALICE: Technical Design Report of the Time Projection Chamber. CERN-OPEN-2000-183, CERN-LHCC-2000-001 (2000)
8. Dellacasa, G., et al.: ALICE technical design report of the time-of-flight system (TOF) (2000)
9. Blum, W., Rolandi, L., Riegler, W.: Particle detection with drift chambers. Particle Acceleration and Detection (2008)
10. Adam, J., et al.: Determination of the event collision time with the ALICE detector at the LHC. Eur. Phys. J. Plus **132**(2), 99 (2017)
11. Adam, J., et al.: Particle identification in ALICE: a Bayesian approach. Eur. Phys. J. Plus **131**(5), 168 (2016)
12. Ho, T.K.: The random subspace method for constructing decision forests. IEEE Trans. Pattern Anal. Mach. Intell. **20**, 832–844 (1998)
13. CERN.: Worldwide LHC computing grid. http://wlcg-public.web.cern.ch/
14. Lin, W.-Z., Fang, J.-A., Xiao, X., Chou, K.-C.: iDNA-Prot: identification of DNA binding proteins using random forest with grey model. PLOS ONE **6**, 1–7 (2011)
15. Casanova, R., Saldana, S., Chew, E.Y., Danis, R.P., Greven, C.M., Ambrosius, W.T.: Application of random forests methods to diabetic retinopathy classification analyses. PLOS ONE **9**, 1–8 (2014)
16. Pedregosa, F., Varoquaux, G., Gramfort, A., Michel, V., Thirion, B., Grisel, O., Blondel, M., Prettenhofer, P., Weiss, R., Dubourg, V., Vanderplas, J., Passos, A., Cournapeau, D., Brucher, M., Perrot, M., Duchesnay, E.: Scikit-learn: machine learning in python. J. Mach. Learn. Res. **12**, 2825–2830 (2011)
17. McKinney, W.: Pandas: a python data analysis library. http://pandas.pydata.org/
18. Sjostrand, T., Mrenna, S., Skands, P.Z.: PYTHIA 6.4 physics and manual. JHEP **05**, 026 (2006)
19. Skands, P.Z.: Tuning Monte Carlo generators: the Perugia tunes. Phys. Rev. D **82**, 074018 (2010)
20. Brun, R., Bruyant, F., Carminati, F., Giani, S., Maire, M., McPherson, A., Patrick, G., Urban, L.: GEANT Detector Description and Simulation Tool (1994)
21. Kursa, M.B., Rudnicki, W.R.: Feature selection with the Boruta package. J. Stat. Softw. **36**(11), 1–13 (2010)

Fault Propagation Models Generation in Mobile Telecommunications Networks Based on Bayesian Networks with Principal Component Analysis Filtering

Artur Maździarz(✉)

Systems Research Institute, Polish Academy of Science, Warsaw, Poland
artur.mazdziarz@gmail.com

Abstract. The mobile telecommunication area has been experiencing huge changes recently. Introduction of new technologies and services (2G, 3G, 4G(LTE)) as well as multivendor environment distributed across the same geographical area bring a lot of challenges in network operation. This explains why effective yet simple tools and methods delivering essential information about network problems to network operators are strongly needed. The paper presents the methodology of generating the so-called fault propagation model which discovers relations between alarm events in mobile telecommunication networks based on Bayesian Networks with Primary Component Analysis pre-filtering. Bayesian Network (BN) is a very popular FPM which also enables graphical interpretation of the analysis. Due to performance issues related to BN generation algorithms, it is advised to use pre-processing phase in this process. Thanks to high processing efficiency for big data sets, the PCA can play the filtering role for generating FPMs based on the BN.

Keywords: Fault Propagation Model (FPM) · Primary Component Analysis (PCA) · Root Cause Analysis (RCA) · Mobile telecommunication network · Bayesian Networks

1 Introduction

The history of mobile telecommunication started in the late 1970s when we had analogue standards being introduced to cover basic voice calls. The entire family of these analog systems is called 1G. In the 1990s, we entered the digital age of mobile communication with the introduction of 2G technology. Technology development driven by the need for mobile data transfer with higher and higher speed resulted in the introduction of 2,5G (GPRS), 3G and 4G/LTE (Long Term Evolution) standards. Currently, the telecommunication community is working on the development and introduction of 5G standard which is supposed to be ready for use by the year 2020 [13].

P. Kulczycki et al. (Eds.): ITSRCP 2018, AISC 945, pp. 18–33, 2020.
https://doi.org/10.1007/978-3-030-18058-4_2

Even medium-sized mobile telecommunication network consists of hundreds or thousands of interconnected components like BSC (Base Station Controller), BCF (Base Station Control Function), BTS (Base Transceiver Station), RNC (Radio Network Controller), just to name a few. Each of these components can be a source of many alarms.

Assume, for example, that the link between a BCF and BSC fails, as presented in Fig. 1. As a consequence, many of the BTS devices would not be able to communicate with the BSC and it results in a huge number of redundant messages being usually received by network operators - often dozens messages per second.

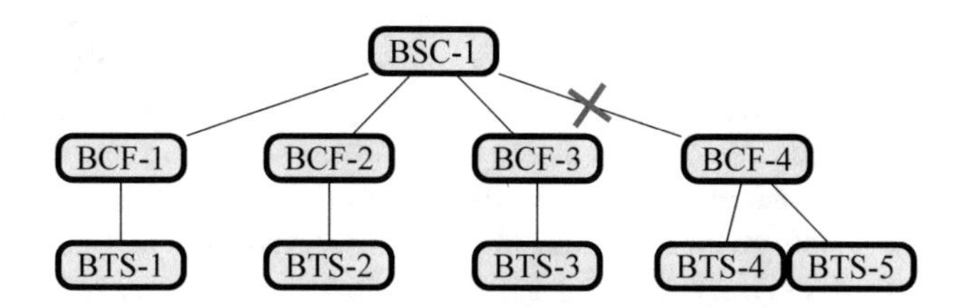

Fig. 1. Network fault example

Identifying *what happened* in these circumstances is a very difficult, if not an impossible task.

To be more precise let us start with a few terms needed in the following analysis [24]:

- **Event** is an exceptional condition occurring in the operation of hardware or software in a managed network; an *instantaneous occurrence* at a time.
- **Root causes** are events that can cause other events but are not caused by other events; they are associated with an *abnormal state* of network infrastructure.
- **Error** is a discrepancy between observed or computed value or condition and a true, assumed as correct value or condition.
- **Failure** or **Fault** is considered to be a kind of error.
- **Symptoms** are external manifestations of failures, observed as alarms.
- **Event correlation** is the process of establishing relationships between network events.
- **Alarm correlation** is the process of grouping alarms which refer to the same problem in order to highlight those which indicate possible rout cause.

The Root Cause Analysis (RCA) is the process of identifying root causes of faults. The analysis contain several steps of correlating events which occurred over certain period of time together with technical knowledge about the analyzed system [1].

There are numerous benefits of automating RCA (Root Cause Analysis)/ event correlation routines. Short troubleshooting time brings benefits, such as satisfying Customers' SLAs (Service Level Agreement). In addition, less skilled personnel can be involved in network operation routines, thus reducing network maintenance costs [22]. Major challenge during troubleshooting of faults in such a complex system like telecommunication network are amount of data and analysis time. The troubleshooting data volume during faults propagation for big networks can easily exceed several dozens of events per second. For the faults that impact network usability for massive amount of end users, the resolution time is crucial and has a big financial impact for the service provider. In order to holistically cope with the above problem the data correlation methodology should be characterized by fast processing as well as easy interpretation and reliable quantification of the results. In this paper we focus on Fault Management events (alarms) correlation. Each alarm event possesses five major attributes: alarm number, alarm description, alarm type, alarm severity, name of the alarming object (the network element). The alarm number is a unique number which identifies fault. Usually the alarm numbers are divided into ranges representing specific subsystem, network element type and alarm type. The alarm description inside the alarm frame is a very short, compact description of the fault that usually contains a few words. Alarm type can be specified as communication, or for example equipment type. Alarm severity specifies the importance of the fault and describes the alarm class. It can take one of the following logical values: critical, major, medium, minor or warning. The name of the object is object identification label which clearly identifies the network element which sent the alarm event. A medium-sized network has several thousand objects in Radio and Core subsystems which can potentially send an alarm event. In specific network element outage circumstances, the alarm flow can reach several dozens of alarms per second. The difficulty of troubleshooting such a complex system like mobile telecommunication network comes from the number of network elements as well as their geographical distribution. Practice shows that events which represent logical sequence of incidents are grouped in clusters within a limited time interval. The primary criterion for alarm events correlation is time. Troubleshooting is focused in the first stage on the alarm events which occurred close to each other considering the time. Discovered events clusters constitute events correlation hypothesis which should be further analyzed by domain experts. It can happen that multiple incidents are accumulated in the same time interval. In such a case it is always the expert's role to evaluate the events and validate proposed event correlation hypothesis. The nature of alarm events flow reflects on certain physicality of the incident in the network. The alarms are either collected at the same time or are generated with certain delay. In the light of the above circumstances it is essential to find fast methodology for discovering alarm events correlations. The picture below illustrates two alarm events clusters. First cluster includes three events occurred at the same time, the second cluster consists of three events which occurred sequentially with one second delay.

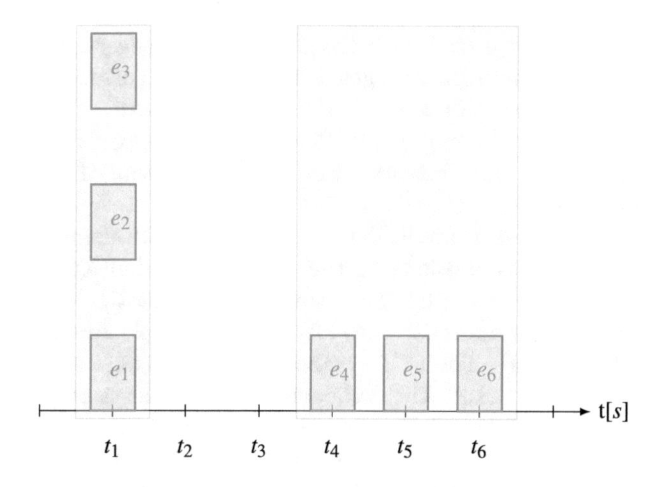

Fig. 2. Alarm events correlation visualization

The goal of the paper is to propose a fast methodology for generating Bayesian Network Fault Propagation Model for big data sets.

The preliminary version of this paper was presented at the 3rd Conference on Information Technology, Systems Research and Computational Physics, 2–5 July 2018, Cracow, Poland [15].

2 Principal Component Analysis

The Principal Component Analysis is the statistical data analysis technique based on variance analysis which is used for dimension reduction. The PCA is the so-called unsupervised method as there is no pre-classification or data labeling for the input data. The method operates directly on the input data set. The pioneer in using and developing PCA was Pearson who used it in connection to his work related to data compression and streaming at the beginning of the 20th century [19]. The method has been widely applied in signals analysis, data compression or images transformations [3]. The PCA method is also used in psychology, sociology and econometrics to discover the so-called latent variables [5,11,14,18,20,21,26,27]. Regardless of the field where the method is used the principles of PCA are the same. The general goal of PCA is to find a reduced set of variables across original data describing the data set with less redundancy from statistical point of view. There are several methods which are very closely related to PCA and are used for similar purpose of reducing data set dimension like: Karhunen - Loeve transform, Hotelling transform, Independent Component Analysis (ICA) [4,9]. PCA replaces original correlated input data set by the smallest set of uncorrelated factors the so-called Principal Components, which describe the majority of data variance. The PCA components are linear combination of original input data set variables. First primary component covers the majority of the input data variance, the second component is selected in such a way that it is not correlated with the first one and covers maximum of the rest variance. As the final result we get as many factors (principal components) as input variables but the data set variance is described by reduced number of variables, the principal components. The PCA is non parametric method so we do not need to take into consideration any parameters of data set density distribution [4,6,9,12].

Let us assume we have a sample data set vector $\hat{x}$ of the elements $\{x_1, ..., x_n\}$. We do not make any assumptions on the probabilistic density distribution for the sample, we assume that first and second order statistics exist and can be calculated. We also assume that the data in the sample are centered $m_x = E\{x\} = 0$; $x \rightarrow x - E\{x\}$. The PCA method transforms data set x linearly into an orthogonal system of axes. The orthogonal space consists of orthogonal axes which represent data set projections on axes, what maximizes variance on given direction. Projections on orthogonal axes are not correlated between each other. We can express the projections of vector $\hat{x}$ elements on the given axes' direction as linear combination of the vector elements with specific coefficients:

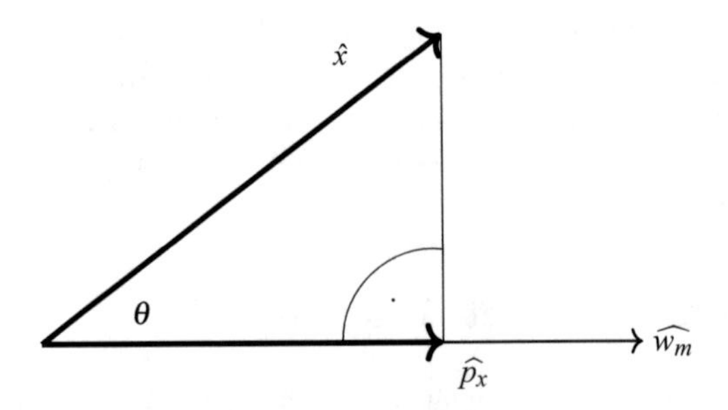

Fig. 3. Projection of $\hat{x}$ on $\widehat{w_m}$

$$\|\widehat{p_x}\| = \|\hat{x}\||cos\theta| = \frac{|\widehat{w_m}^T \hat{x}|}{\|\widehat{w_m}\|} \tag{1}$$

According to PCA, we want to find the factor (direction w_m) on which the variance of source vector projection is maximally large. The variance of vector $\hat{x}$ projection p_x is defined as follows (for $m_x = 0$):

$$Var(\|\widehat{p_x}\|) = Var\left(\frac{|\widehat{w_m}^T \hat{x}|}{\|\widehat{w_m}\|}\right) = E\left\{\frac{(\widehat{w_m}^T \hat{x})^2}{(\|\widehat{w_m}\|)^2}\right\} \tag{2}$$

The variance of the projection p_x depends on the orientation of vector $\widehat{w_m}$ and the norm of the vector $\|\widehat{w_m}\|$. Additional constraint related to the norm of vector $\widehat{w_m}$ is introduced:$\|\widehat{w_m}\| = 1$ in PCA analysis. The above constraints implies the following dependency taking Euclidean norm definition:

$$\|\widehat{w_m}\| = (\widehat{w_m}^T \widehat{w_m})^{1/2} = [\sum_{k=1}^{n} w_{km}^2]^{1/2} = 1 \tag{3}$$

$$E\left\{\frac{(\widehat{w_m}^T \hat{x})^2}{(\|\widehat{w_m}\|)^2}\right\} = E\{(\widehat{w_m}^T \hat{x})^2\} = \widehat{w_m}^T E\{\hat{x}\hat{x}^T\}\widehat{w_m} \tag{4}$$

$$C_x = E\{\hat{x}\hat{x}^T\} \tag{5}$$

The matrix C_x is the covariance matrix of vector x which is the same as correlation matrix for $m_x = 0$.

$$Var(\|\widehat{p_x}\|) = \widehat{w_m}^T C_x \widehat{w_m} \tag{6}$$

From the linear algebra the solution of maximizing the variance (2,6) is given in terms of unit-length eigenvectors $\widehat{e_1}, ..., \widehat{e_n}$ of matrix C_x which are basis of vectors $\widehat{w_m} : \widehat{e_m} = \widehat{w_m}$. Eigenvectors are given by the so-called spectral decomposition of the covariance matrix. The maximizing solution is given in the direction: $\widehat{e_m} = \widehat{w_m}$. Corresponding eigenvalues are denoted as follows: $\lambda_1, \lambda_2, ..., \lambda_m$. The direction $\widehat{w_m}$ is given by the eigenvector $\widehat{e_m} = \widehat{w_m}$ corresponding to appropriate eigenvalue λ_m of the covariance matrix C_x taking into consideration the constraint (3) [4,9].

$$\max_{\{w_m : \|\widehat{w_m}\|=1\}} \widehat{w_m}^T C_x \widehat{w_m} = \lambda_1 \geq \lambda_2 \geq ... \geq \lambda_m = \min_{\{w_m : \|\widehat{w_m}\|=1\}} \widehat{w_m}^T C_x \widehat{w_m} \tag{7}$$

$$Var(\|\widehat{p_x}\|) = Var(\widehat{w_m}^T \hat{x}) = \lambda_m \tag{8}$$

The variance $\widehat{w_m}^T \hat{x}$ is maximized under the constraint that it is not correlated with previously found principal component $\widehat{w_k}^T \hat{x}$, $k < m$.(orthogonality)

$$E\{(\widehat{w_m}^T \hat{x})(\widehat{w_k}^T \hat{x})\} = \widehat{w_m}^T C_x \widehat{w_k} = 0 \tag{9}$$

As the output of PCA method we get two major group of parameters: loadings and scores. The loadings represent the wight of the contribution of original variables to given principal component. The scores are the coordinates of original variables in new coordinate system built by the principal components.

The projection of data vector $\hat{x}$ on the direction $\widehat{w_m}$ specified in (12) is represented in the space spanned by the eigenvectors ($\widehat{e_m}$) of data covariance matrix decomposition by the following relation:

$$\widehat{w_m}^T \hat{x} = \sqrt{\lambda_m} \widehat{e_m} \tag{10}$$

Above dependency comes from a general Euclidean distance definition [4,6,9]. Euclidean distance with metric A (positive definite matrix) between 2 points in R^p (x, y) is defined as follows:

$$E_{dA} = \{x, y \in R^p | (\hat{x} - \hat{y})^T A (\hat{x} - \hat{y}) = d^2\}. \tag{11}$$

$$d^2(x, y) = (\hat{x} - \hat{y})^T A (\hat{x} - \hat{y}). \tag{12}$$

If the metric A is the identity matrix (with ones on the matrix diagonal) we get the following E_d definition:

$$d^2(x,y) = (\hat{x} - \hat{y})^T(\hat{x} - \hat{y}). \tag{13}$$

$$d^2(x,y) = \sum_{n=1}^{p}(x_i - y_i)^2. \tag{14}$$

For input vector x with center x_0 the Euclidean distance is defined as follows:

$$E_{dA} = \{x \in R^p | (\hat{x} - x_0)^T A(\hat{x} - x_0) = d^2\}. \tag{15}$$

Graphical representation of the above inquiry are spheres with radius d and center x_0 the so-called iso-distance spheres.

In the case the metric A is the identity matrix:

$$E_{dI_p} = \{x \in R^p | (\hat{x} - x_0)^T(\hat{x} - x_0) = d^2\}. \tag{16}$$

General distance definition for data transformed to orthogonal PCA space, assuming positive definite covariance matrix $C_x > 0$ and centered input data leads to the following definition:

$$E_{dC_x} = \{w_m \in R^p | \widehat{w_m}^T C_x \widehat{w_k}) = d^2\}. \tag{17}$$

The iso-distance ellipsoid in the PCA orthogonal space is presented in Fig. 4.

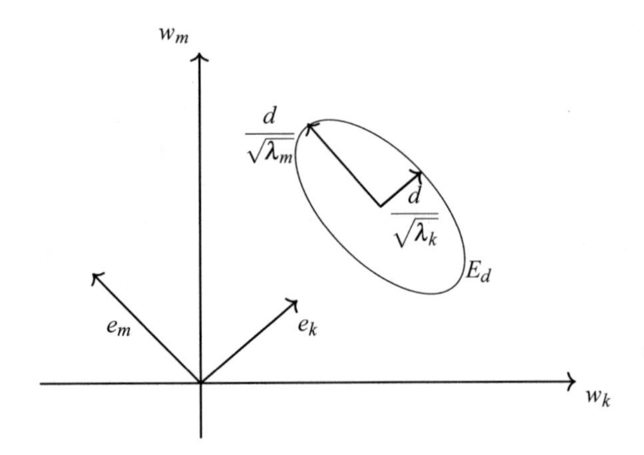

Fig. 4. Iso-distance ellipsoid

3 Bayesian Networks Introduction

The Bayesian Network (BN) is a probabilistic structure of Directed Acyclic Graph (DAG) which codes probabilistic relationships between nodes. The nodes represent random variables(alarm events, in our case). Conditional probability associated with each node represents the strength of causal dependency between the parent $pa(x_i)$ and child nodes $\{x_i\}$. For this purpose each node is characterized by the Conditional Probability Table (CPT) which denotes local conditional probability distribution. Local conditional probability attributes (CPTs) of the nodes constitute global joint probability distribution $P(\{x_i\})$ constrained by the connections between the nodes [7,10,17].

$$P(x_1, x_2, ..., x_n) = \prod_{i=1}^{n} P(x_i|pa(x_i)) \ .$$

The nodes dependency is strictly determined by a graph structure and at the same time makes the in-dependency of not connected nodes - one of the base for Bayesian Networks (Markov property, d-separation). This feature leads to the factorization of joint distribution over the nodes and shows practical usefulness of the BN analysis [7,10,17]. In principle, we can have continuous or discrete BN depending on the type of conditional probability representation.

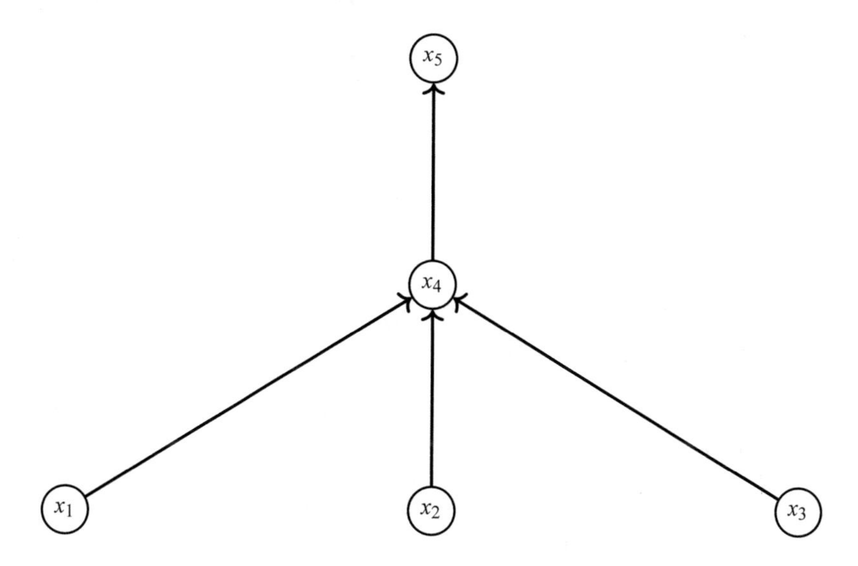

Fig. 5. An example of a Bayesian network

A simple example of BN is presented in Fig. 5 and described by the following joint probability distribution:

$$P(x_1, x_2, x_3, x_4, x_5) = P(x_1) \cdot P(x_2) \cdot P(x_3) \cdot P(x_4|x_1, x_2, x_3) \cdot P(x_5|x_4) \ .$$

Building the BN is done in two phases. In the first step we learn the structure of the BN and in the second step we discover the parameters of the BN [7,17]. The structure learning is the process of discovering connections (arcs) between nodes (random variables). In the structure learning we use two main classes of algorithms: the constraint-based algorithms and search and score-based algorithms [17]. In constraint-based algorithms we use conditional independence tests for data in order to discover relations (arcs) between variables (nodes in the graph) and the v-structures concept which is used to verify and evaluate the independence of a given two nonadjacent variables (nodes) for a given third one. Two independence tests used in the constraints-based algorithms are: Mutual Information (distance measure) and χ^2 tests. In the majority of algorithms we use Markov blanket of each node as the operation space in order to simplify the identification of neighbor nodes. In this family of the algorithms we can list the following algorithms: Inductive Causation (Verma and Pearl, 1991), Grow - Shrink (GS), Incremental Association (IAMB), Fast Incremental Association (Fast-IAMB), Interleaved Incremental Association (Inter-IAMB) [17]. In score based algorithms the structure of the graph (BN) is learned by adding or removing the arcs from the full (or empty) graph in order to maximize the so-called network score function. Examples of this type of algorithms are: Greedy Search, Hill - Climbing, Simulated Annealing and many Genetic algorithms. The scoring is done with respect to available data. Most commonly used functions in network scores metrics in the case of discrete data set are Bayesian Dirichlet equivalent (BDe) and Bayesian Information Criterion (BIC) or Akaike Information Criterion (AIC) [17]. In the BN parameters discovering phase we calculate the parameters of the network, which means calculating local distributions and global distribution based on the network structure and the data. There are two main approaches in the parameters learning used to estimate local/global conditional distributions, i.e. it is Maximum Likelihood Estimation (MLE) and Bayesian Estimation [16,17].

4 Bayesian Network Fault Propagation Model Generation with PCA Pre-filtering

Proposed alarm events correlation methodology is presented in Fig. 6. The input data are formatted as R environment data frame object. Each variable in the data frame represents the time occurrence of the alarms in the network. Before the analysis starts all the variables representing alarm occurrence for a given network element are scaled (standardized). The scaling process is applied in the data pre-processing phase and normalizes the data to achieve unit variance. It is achieved by dividing centered values of alarm occurrence time by their standard deviations. We analyze PCA loadings coefficients of the variables representing occurrence time of alarms in the network. The analysis is done for several principal components representing the majority of the data set variance. The PCA loadings coefficients represent contribution of input variable to a given principal component. Events with close values of loadings coefficients represent the same level of contribution to the principal component thus are considered to be correlated. For PCA analysis we use *prcomp* function from package R. It was observed that the PCA loadings coefficients are linked to the conditional probability of the alarm events relation discovered by the BN generation algorithms. PCA pre-filtered data with close values of loadings are then applied for

Bayesian Network generation algorithm (constraint-based or search and score-based). The reduced data set achieved though PCA pre-processing allows BN algorithms to work only on potentially correlated alarms thus reducing the time of Fault Propagation Model generation.

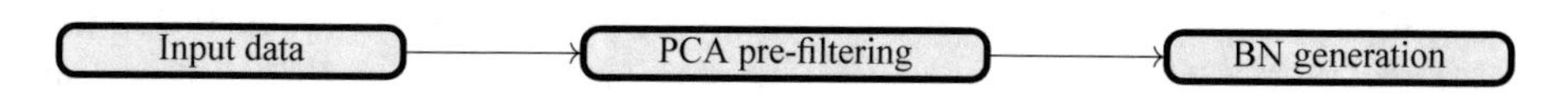

Fig. 6. BN FPM generation with PCA filtering

We can summarize our Fault Propagation Models Generation Methodology steps as follows:

1. Gather and standardize data (pre-processing phase).
2. Perform PCA analysis of the input data.
3. Construct the projection matrix containing loadings corresponding to input data for two first primary components.
4. Filter input data with close values of loadings in two first primary components subspace.
5. Derive strength and the relation direction among pre-filtered data by applying selected Bayesian Network generation algorithm.
6. Generate Fault Propagation Model and conclude on alarms correlation.

5 Results and Examples

We present examples of the correlation carried out on real alarms data samples from the live network of one of the mobile operators. The data set used for the simulations had 1440813 alarm events divided into several sample files. The data was gathered from heterogeneous mobile live network containing 2G, 3G and 4G network elements from the period between July 2014 and May 2015. There are three attributes stored for each alarm, the time of alarm occurrence, an ID which depends on the source of problem and the identification of the network element which generated the alarm. All numerical experiments were carried out on a PC with Intel(R) Core(TM) i7-4600U 2.1 GHz processor, 16 GB main memory and 64-bits MS Windows operating system. We used R package environment version 3.3.1. In the first stage the performance of PCA pre-filtering was tested as this is the main contributor to overall speed attributes for the methodology. We used four independent samples for performance testing: sample1, sample2, sample3, sample4 that we gathered at different period of time. In addition, each sample has been divided into smaller subsets of alarms in order to better evaluate the performance of the methodology. The PCA processing time has been measured for different amount of alarm events records in each sample. Following subsets sizes have been used: 1000,5000,10000,20000,30000,40000,50000,60000. The results of the tests are presented in Table 2 and in Fig. 9. The tests showed that the PCA pre-filtering works very efficiently. In all of tested the samples, the processing time for a number of alarm events on the level of 20,000 took less than 10 s. The processing time for 60000 alarms events in the worst case scenario (sample2) took 52 s.

The PCA as a method which is primarily used for data dimension reduction can be also used as pre-processing for other analysis such as Bayesian Network generation, just like in our case. In Fig. 7 we present raw data visualization for sample1. In the sample we have 7571 unique alarm events spread across the sample with the time duration of 86,392 s.

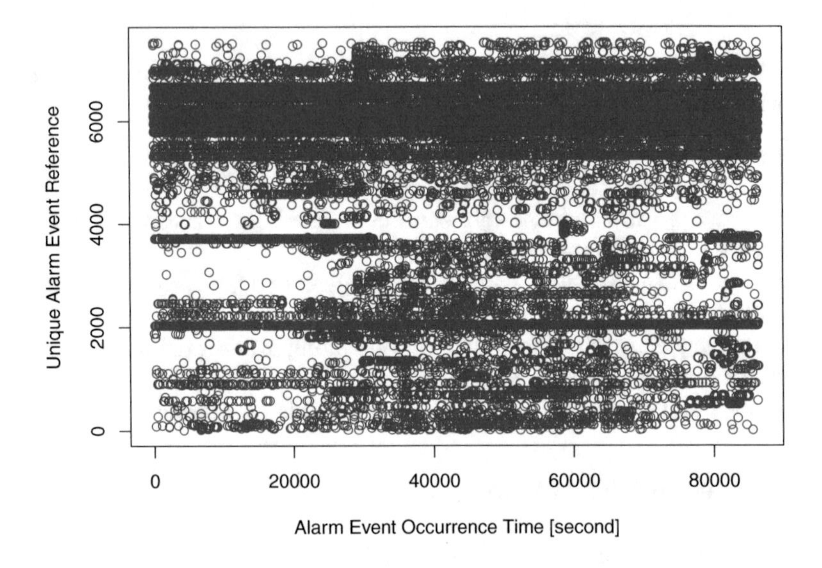

Fig. 7. Input data visualization, sample1

In the Fig. 8 the bipolt is presented which illustrates the input data in the first two primary components subspace, the picture shows expected data reduction effect.

The level of contribution of the first three primary components in the input data reduction for analysed samples is presented in Table 1. The table shows us

Table 1. Proportion of variance for first three primary components in sample1, sample2, sample3, sample4

	PC1	PC2	PC3
	Proportion of variance	Proportion of variance	Proportion of variance
Sample1	0.703	0.141	0.059
Sample2	0.731	0.154	0.051
Sample3	0.702	0.123	0.055
Sample4	0.717	0.135	0.059

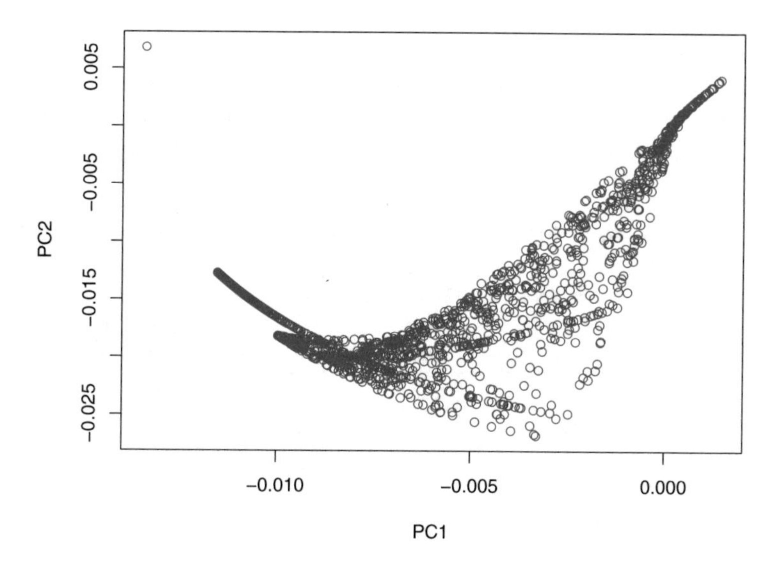

Fig. 8. Input data biplot for PC1,PC2 subspace

how much information (variance) can be attributed to the first three principal components. As visible in the table more than 80% variance is carried by only two first primary components. This confirms significant correlation of input data and justify usage of only two first primary components in our analysis.

Finally Table 3 and Fig. 10 show the result of applying score-based hill-climbing BN generation algorithm for PCA pre-filtered data. In the example we see three alarm events which occurred multiple times in the data sample. In Table 3 we present three alarm events with their occurrence time in the data sample and the loading values for three primary components. Correlation of these alarms without data analytic approach will be very challenging due to amount of data spread across big data sample. In this case 3G technology CELLs operation is affected by the problem with the license on the parent WBTS network element. Thanks to the PCA method we can isolate potentially correlated alarm events by analyzing the loading values and allow the BN generation algorithm to generate the FPM in a very efficient time. In Table 4 we present the strength of the relation in terms of conditional probability between correlated alarm events discovered BN algorithm based on maximum likelihood parameter estimation method. In addition, Fig. 10 depicts the relations directions visualization. We present there the generated Bayesian Network for the events from example 2. Presented example shows the relation between the PCA loadings and conditional probabilities derived through the BN algorithms. It shows that the PCA pre-filtering combined with BN generation algorithms is a very fast methodology with high accuracy and reliability what allows quickly generating even very complex Fault Propagation Models. In our case the BN is generated using *bnlearn* package from R environment. In addition we calculated conditional probability between the alarm events using the *bn.fit* function from the bnlearn package. It has been observed that the conditional probability of the relation between PCA pre-filtered alarm events considered as correlated by PCA pre-filtering is very high (>0.8) what proves the usefulness of the methodology.

Table 2. PCA correlation methodology performance

Alarm events number	Sample1 Correl. time [s]	Sample2 Correl. time [s]	Sample3 Correl. time [s]	Sample4 Correl. time [s]
1000	<1	<1	<1	<1
5000	<1	1	<1	1
10000	1	3	1	2
20000	5	9	3	7
30000	10	14	5	10
40000	17	27	7	21
50000	23	46	12	38
60000	NA	52	17	51

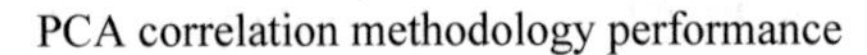

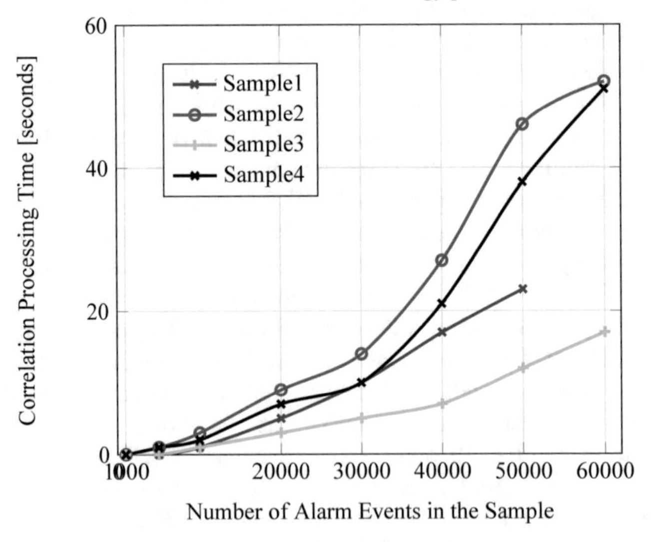

Fig. 9. Performance of PCA-based alarm events correlation methodology

Table 3. Correlation example

Alarm event	Event time	PC1	PC2	PC3
CELL OPERATION DEGRADED.RNC-864/ WBTS-20135/WCEL-23	321 662 785 837 966 1267 1521 1679 1926 2032 2199 2350 2436 2525 2673 2797 2851 2963 3075 3289 3415 3638 4414 4570 5410 5909 6066 7814 8667 8748 16403 17539 18105 19182 19342 19490 19592 19828 19972 20030 20073 20296 20460 20638 20816 20945 21293 21769 21857 22069 22612 22794 22866 23012 23203 23289 23378 23442 23518 23772 23842 23974 24042 24307 24419 24575 24729 25959 85823 86231 86348	0.000146748	0.000225049	-0.000118459
CELL OPERATION DEGRADED.RNC-864/ WBTS-20135/WCEL-13	323 453 785 837 967 1268 1522 1680 1927 2033 2200 2350 2437 2526 2674 2798 2852 2964 3076 3290 3416 3639 4415 4571 5411 5910 6067 7815 8668 8749 16404 17540 18107 19183 19342 19491 19593 19830 19973 20073 20297 20460 20640 20817 20945 21293 21770 21859 22071 22347 22612 22794 22867 23012 23203 23289 23379 23443 23519 23772 23842 23975 24042 24308 24577 24731 25960 85824 86349	0.000147244	0.000224397	-0.000246041
BTS reset needed to activate a license.RNC-864/WBTS-20135	323 455 787 839 968 1270 1523 1681 1927 2034 2201 2352 2437 2527 2675 2799 2853 2965 3077 3291 3417 3640 4416 4572 5412 5911 6068 7816 8669 8750 16405 17541 18107 19185 19343 19493 19594 19830 19975 20074 20298 20461 20640 20817 20946 21294 21772 21859 22071 22348 22614 22796 22867 23014 23205 3290 23382 23445 23520 23774 23844 23975 24044 24308 24578 24731 25961 85825 86350	0.000147198	0.000223802	-0.000246653

Table 4. Conditional probabilities distribution

BTS reset needed to activate a license.RNC-864/WBTS-20135	
CELL OPERATION DEGRADED.RNC-864/WBTS-20135/WCEL-13	CELL OPERATION DEGRADED.RNC-864/WBTS-20135/WCEL-23
0.823460154	0.999966717

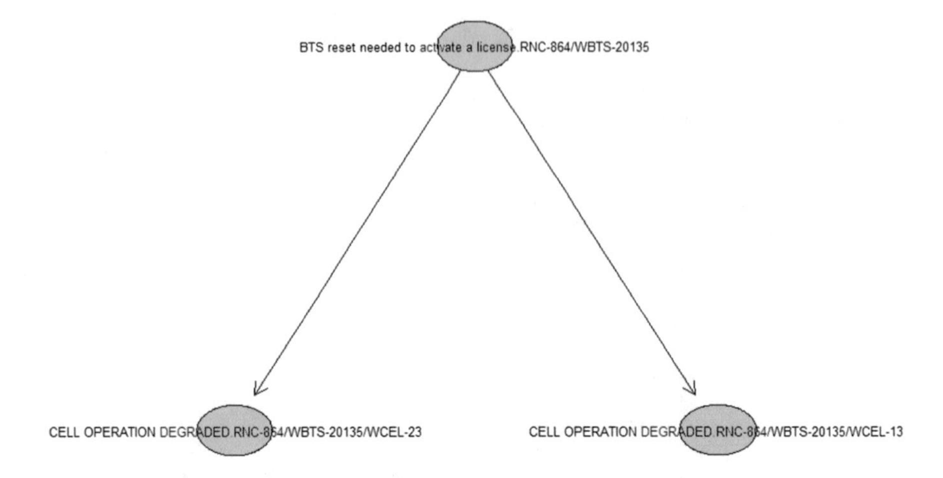

Fig. 10. A Bayesian network generated from PCA pre-filtered data.

6 Summary and Conclusions

The Primary Component Analysis method is a very efficient, fast and reliable methodology for discovering correlation among input variables representing alarms time occurrence in the network. We used time as the correlation domain in our experiments. It has been proven that the loadings values which are close represent a level of correlation with the input variables. The level of correlation is measured by the similarity of the loadings coefficients. We've used values of loadings coefficients for the first two Primary Components of PCA analysis of input data for detecting correlation. All input variables representing given alarm occurrence with similar loadings coefficients for the first two Primary Components are classified as correlated. The PCA is very efficient in selecting potentially correlated alarm events in big data sets. It has been proven by experiments that PCA works very well as data dimension reduction method and can be used as a pre-filtering mechanism for Bayesian Network Fault Propagation Models generation. The experiments shown that there is a relation between PCA loadings coefficients and conditional probability in Bayesian Network environment. Higher value of the PCA loading results in higher value of the conditional probability in Bayesian Network Fault Propagation Model.

References

1. Bhaumik, S.K.: Root cause analysis in engineering failures. Trans. Indian Inst. Met. **63**(2–3), 297–299 (2010)
2. Datta, R., Niharika, N.: Comparative study between the generations of mobile communication 2G, 3G & 4G. Int. J. Recent Innov. Trends Comput. Commun. **1**(4) (2013). ISSN 2321-8169
3. Górecki, T.: Podstawy statystyki z przykładami w R. BTC, Legionowo (2011). ISBN 978-83-60233-69-6
4. Hardle, W., Simar, L.: Applied Multivariate Statistical Analysis. Springer (2007)

5. Harman, H.: Modern Factor Analysis. University of Chicago Press, Chicago (1975)
6. Hastie, T., Tibshirani, R., Friedman, J.: The Elements of Statistical Learning, Data Mining, Inference, and Prediction. Springer (2001). https://doi.org/10.1007/b94608
7. Holmes, D.E., Jain, L.C.: Innovations in Bayesian Networks Theory and Applications (2008). ISBN 978-3-540-85065-6
8. Hong, P., Sen, P.: Incorporating non-deterministic reasoning in managing heterogeneous network faults. In: Krishnan, I., Zimmer, W. (eds.) Integrated Network Management II, pp. 481–492, PNorth-Holland, Amsterdam (1991)
9. Hyvarinen, A., Karhunen, J., Oja, E.: Independent Component Analysis. Wiley (2001)
10. Jensen, F.V., Nielsen, T.D.: Bayesian Networks and Decision Graphs (2007). ISBN-10: 0-387-68281-3
11. Kim, J.O., Mueller, C.W.: Factor Analysis. Statistical Methods and Practical Issues. Sage Publications, Beverly Hills (1978)
12. Koronacki, J., Ćwik, J.: Statystyczne systemy uczące się, 2nd edn. EXIT, Warszawa (2008)
13. Lopa, M., Vora, J.: Evolution of Mobile Generation Technology: 1G to 5G and Review of Upcoming Wireless Technology 5G. Int. J. Modern Trends Eng. Res. **121**(6) (2015). ISSN 2393-8161
14. Manly, B.F.J.: Multivariate Statistical Methods. A Primer. Chapman and Hall, London (1986)
15. Maździarz, A.: Fault Propagation Models Generation in Mobile Telecommunications Networks based on Bayesian Networks with Principal Component Analysis Filtering. In: Kulczycki, P., Kowalski, P.A., Lukasik, S. (eds.) Contemporary Computational Science, p. 47. AGH-UST Press, Cracow (2018)
16. Myung, I.J.: Tutorial on maximum likelihood estimation. J. Math. Psychol. **47**, 90–100 (2003)
17. Nagarajan, R., Scutari, M., Lebre, S.: Bayesian Networks in R with Applications in Systems Biology (2013). ISBN 978-1-4614-6445-7
18. Okoń, J.: Analiza czynnikowa w psychologii. PWN, Warszawa (1960)
19. Pearson, K.: On lines and planes of closest fit to systems of points in space. Phil. Mag. **2**, 559–572 (1901)
20. Pluta, W.: Wielowymiarowa analiza porównawcza w modelowaniu ekonometrycznym. PWN, Warszawa (1986)
21. Rothman, J.: Some considerations affecting the use of factor analysis in market research. J. Mark. Res. Soci. **38**(4), 371–381 (1996)
22. Samba, A.: A Network Management Framework for Emerging Telecommunications Networks. Department of Computer Science Kent State University, Kent OH 44242, USA ,Chapter 8 of Modeling and Simulation Tools for Emerging Telecommunication Networks Needs, Trends, Challenges and Solutions (2006). ISBN-10: 0-387-32921-8
23. Singh, K., Thakur, S., Singh, S.: Comparison of 3G and LTE with other Generation. Int. J. Comput. Appl. (0975 - 8887) **121**(6) (2015). ISSN 0975-8887
24. Steinder, M., Sethi, A.S.: A survey of fault localization techniques in computer networks. Sci. Comput. Program. **53**, 165–194 (2004)
25. Sutter, M.T., Zeldin, P.E.: Designing expert systems for real time diagnosis of self-correcting networks. IEEE Netw. 43–51 (1998)
26. Zakrzewska, M.: Analiza czynnikowa w budowaniu i sprawdzaniu modeli psychologicznych. UAM, Poznań (1994). ISBN-10: 8323204772
27. Zeliaś, A.: Metody Statystyczne. PWN, Warszawa (2000)

Optimizing Clustering with Cuttlefish Algorithm

Piotr A. Kowalski[1,2(✉)], Szymon Łukasik[1,2], Małgorzata Charytanowicz[2,3], and Piotr Kulczycki[1,2]

[1] Faculty of Physics and Applied Computer Science, AGH University of Science and Technology, al. Mickiewicza 30, 30-059 Cracow, Poland
{pkowal,slukasik,kulczycki}@agh.edu.pl

[2] Systems Research Institute, Polish Academy of Sciences, ul. Newelska 6, 01-447 Warsaw, Poland
{pakowal,slukasik,mchat,kulczycki}@ibspan.waw.pl

[3] Electrical Engineering and Computer Science Faculty, Lublin University of Technology, Nadbystrzycka 36A, 20-618 Lublin, Poland
m.charytanowicz@pollub.pl

Abstract. The Cuttlefish Algorithm, a modern metaheuristic procedure, is a very recent solution to a broad-range of optimization tasks. The aim of the article is to outline the Cuttlefish Algorithm and to demonstrate its usability in data mining problems. In this paper, we apply this metaheuristic procedure for a clustering problem, with the Calinski-Harabasz index used as a measure of solution quality. To examine the algorithm performance, selected datasets from the UCI Machine Learning Repository were used. Furthermore, the well-known and commonly utilized k-means procedure was applied to the same data sets - to obtain a broader, independent comparison. The quality of generated results were assessed via the use of the Rand Index.

Keywords: Clustering · Cuttlefish Algorithm · Biologically inspired algorithm · Optimization · Metaheuristic

1 Introduction

Data mining and computational intelligence are currently one of the fastest developing IT fields. The dynamic growth of computing capabilities in computers enables automatic analysis of huge amounts of data and building very accurate models of investigated phenomena from various fields. Examples of applications can be seen in different situations, including, in particular, those of engineering [11], but also in control tasks [18], economics [16] or even within the environmental sciences [5]. The use of computers allows to perform tasks that often exceed human capabilities in terms of the amount of data analysed, the speed of learning and the accuracy of results. There are a huge number of tools to

P. Kulczycki et al. (Eds.): ITSRCP 2018, AISC 945, pp. 34–43, 2020.
https://doi.org/10.1007/978-3-030-18058-4_3

automatically perform tasks such as classification, regression or clustering, and the choice of a specific algorithm often depends on the data set. Currently, a lot of research is carried out in this field and a large number of scientific papers are being put forward that propose different algorithms.

Exploratory Data Analysis mainly consists of addressing the issues of clustering, classification, data and dimension reduction, as well as outliers detection. The main goal of the clusterization procedure is to break up the confederate data collection into smaller subsets called 'clusters'. This action is unsupervised, being achieved by way of information directly derived from the data set itself. This division algorithm has been successfully applied within a wide variety of situations, including, in particular, those of engineering [11], but also in control tasks [18], economics [16] or even within the environmental sciences [5].

In this work, the exploratory data analysis and computational intelligence houses have been combined. The first of these domains is represented by the task of clustering, i.e. the division of a certain set of dancers into a number of subgroups. Such grouping belongs to the tasks of combinatorial optimization and is characterized by the time complexity of NP [26]. The second of the algorithms considered here is the metaheuristic optimization algorithm inspired biologically by the life and behavior of the Cuttlefish herd. This algorithm belongs to a large family of procedures inspired by herd retention. This group includes many other procedures such as Particle Swarm Optimization, Bee Colony Optimization, Elephant Herding Optimization, Ant Colony Optimization, Bacterial Memetic Algorithm, Firefly Algorithm Krill Herd Algorithm, Flower Pollination Algorithm, Grasshopper Optimisation Algorithm, Symbiotic Organisms Search and Crow Search Algorithm.

In addition, it is worth noting that many biologically inspired optimization algorithms have been referred to in the task of clasterization, e.g. Firefly Algorithm [23], Flower Pollination Algorithm [15,19], Krill Herd Algorithm [12,15], Grasshopper Optimisation Algorithm [13].

Optimization problems are encountered when deriving solutions to many engineering issues. The optimization task may be considered as that of choosing the best possible use of limited resources (time, money, energy, resources etc.), while attempting to bring about certain goals. In achieving this, the optimization problem can be notated in a formal way. Let us introduce the 'cost function' K:

$$K : A \longmapsto \mathbb{R}, \tag{1}$$

where $A \subset \mathbb{R}^n$. The optimization task consists of finding the value $x^* \in A$, such that for every $x \in A$, the following relationship is true:

$$K(x) \geq K(x^*). \tag{2}$$

Although the optimization problem can be easily defined and described, determining its solution is already a very difficult issue. To resolve this problem, certain optimization algorithms are commonly used.

The presented research compares the quality of results achieved through applying a Nature-derived clustering algorithm, with that gained through utilizing the k-means [22] procedure. The biological inspired group of algorithm is employed so as to find the best position of the cluster centre, and then the particular element of the investigated data set are assigned to a particular group. In the presented approach, the Cuttlefish Algorithm (CFA) [6], is employed, this being a continuous optimisation procedure. For assigning the particular solution in each iteration of the optimisation task, the Celinski-Harabasz Clustering Index [4] is subsequently applied. For global comparative purposes, final results based on all considered clustering procedures are measured by way of the Rand Index [1].

This article summarizes research on the CFA metaheuristic algorithm as applied for clustering tasks. After a short introduction, in Sect. 2, the CFA procedure will be presented with a description of the biological inspired characteristics. Section 3 reveals the sets of benchmark data sets for which the research presented in this article is based. It, as well provides the results of the comparative tests. In the last part of the paper, a summary of the results obtained and plans for further research within the clustering procedure based on CFA procedure are included.

The preliminary version of this paper was presented at the 3rd Conference on Information Technology, Systems Research and Computational Physics, 2–5 July 2018, Cracow, Poland [14].

2 Optimization Metaheurisctics

The CFA is a global optimization procedure inspired by the natural behavioural activity of the Cuttlefish organisms. This procedure was introduced by Adel Sabry Eesa, Adnan Mohsin Abdulazeez Brifcani and Zeynep Orman, in the paper [6]. A characteristic feature of Cuttlefish, and the inspiration of the present algorithm [7], is the ability for the activity of individual creatures to be modelled within that of a large herd.

The algorithm is inspired by the change of color by the cuttlefish - a predatory cephalopod with eight arms and two tentacles. Cuttlefish changes color both to hide from the threat and to lure a partner during the mating season. The patterns and colors of cuttlefish are produced by light reflected from different layers of cells (chromatophores, leucophores and iridophores) put together and this combination of cells allows the cuttlefish to have a large number of patterns and colors. The algorithm was designed based on two processes (reflection and visibility), which serve as a strategy for searching for new solutions.

The formula (3) describes the way of finding a new solution in the optimization task by means of reflection and visibility:

$$X_{new} = reflection + visibility. \quad (3)$$

The cuttlefish cells are divided into four groups in which new patterns and colors are obtained - which in computational intelligence is associated with the acquisition of a new solution:

- Group I - based on the current solution

$$reflection = R * X_c, \tag{4}$$

$$visibility = V * (X_{best} - X_c) \tag{5}$$

- Group II - based on current and the best solution

$$reflection = R * X_{best}, \tag{6}$$

$$visibility = V * (X_{best} - X_c) \tag{7}$$

- Group III - based on the best solution

$$reflection = R * X_{best}, \tag{8}$$

$$visibility = V * (X_{best} - X_{avg}) \tag{9}$$

- Group IV - based on a random search of investigated space

$$reflection = \xi, \tag{10}$$

$$visibility = 0 \tag{11}$$

where: X_{new} denotes a proposition of a new solution, X_{cur} is a current solution, X_{best} indicates a best solution, ξ is a random position of investigated state space, R designates a random value generated from the uniform distribution for the interval $[r_1, r_2]$, V denotes a random value generated from the uniform distribution for the interval $[v_1, v_2]$ and finally X_{avg} is an average value of the vectors representing best solution.

The operation of the algorithm begins via the initialization of data structures, i.e. describing individuals, as well as the whole population. Initializing the data structure representing a single cuttlefish, means situating it in a certain place (at the 'solution space') by giving it a set of coordinates. For this purpose, it is recommended to employ random number generation according to a uniform distribution. Like other algorithms inspired by Nature, each individual represents one possible solution of the problem under consideration. After the initialization phase of the algorithm, it continues into a series of iterations.

Algorithm 1 presents the pseudocode of the discussed algorithm. At the beginning, the population is initialized with random values of the N-dimensional space and the evaluation of individual solutions. In the next step, the population is divided into the 4 groups which were discussed above. If the current solution (in each of the 4 groups) is better than the best solution so far, then the best solution is overwritten with the new solution. These steps are repeated until the stop criterion is fulfilled.

Algorithm 1. High-level CFA pseudocode

Define and populate the algorithm's data structures
Initialize random initial population, i.e. position of the swarm of Cuttlefishes
Evaluate fitness of each Cuttlefish individual on the basis of its position
Find the best global solution
Divide Cuttlefish population into four subgroups G_I, G_{II}, G_{III} and G_{IV}
while stop criteria is not reached **do**
  Calculate the average value of the best global solution X_{avg}
  for i=1: population size in G_I **do**
    Calculate the reflection based on eq (4)
    Calculate the visibility based on eq (5)
    Generate the new position of i-th individual based on eq (3)
    Update current position (if the achieved cost is lower)
  end for
  for i=1:population size in G_{II} **do**
    Calculate the reflection based on eq (6)
    Calculate the visibility based on eq (7)
    Generate the new position of i-th individual based on eq (3)
    Update current position (if the achieved cost is lower)
  end for
  for i=1:population size in G_{III} **do**
    Calculate the reflection based on eq (8)
    Calculate the visibility based on eq (9)
    Generate the new position of i-th individual based on eq (3)
    Update current position (if the achieved cost is lower)
  end for
  for i=1:population size in G_{IV} **do**
    Generate the new position of i-th individual based on eqs (11) and (3)
    Update current position (if the achieved cost is lower)
  end for
  Check the feasible bounds for new positions
  Evaluate fitness of each individual on the basis of its position
  Update the best global solution (if the achieved cost is lower)
end while
Return: x^* /best solution/ and $K(x^*)$

The population is divided into groups in order to increase the possibility of defining procedures that will lead to finding the best solution. Group I is aimed at exploiting the search space in order to find a global extreme, using the best solution found so far. Group IV has a similar role, here, a search of the space in a completely random way is carried out. However, a different role is played by Groups II and III. They are focused on exploring the local extreme in the hope that it is a global extreme.

The algorithm depends on the selection of four parameters: r_1, r_2, v_1 and v_2. As a result of applying the CFA, the best value of the cost function K and the argument for which it was calculated, are achieved.

3 Clusterization Procedure

Let us assume that Y is a data set matrix with dimensions D and M, respectively. Each data collection element is represented by one column of Y. The main goal of the clustering procedure is to divide the data set and assign particular elements of Y to the distinguished clusters $CL_1, \ldots, CL_C$,. In this procedure each cluster is characterised by a point called as the 'centre of cluster', which is calculated as:

$$O_c = \frac{1}{\#CL_c} \sum_{x_i \in CL_c} y_i, \tag{12}$$

where $\#CL_c$ denotes the number of elements assigned to the cth cluster. Similarly, the centre of gravity for all the investigated elements $y_1, \ldots, y_M$ is defined as:

$$O_Y = \frac{1}{M} \sum_{i=1}^{M} y_i. \tag{13}$$

In this approach, the optimisation task in the clusterization procedure is based on the metaheuristic CFA. Herein, each element of the optimisation task is encoded as a collection of cluster centroids. Therefore, the product value $D \cdot C$ expresses the dimensionality of a particular optimization task. In this case, the number of the clusters is established at the beginning of the grouping process. Moreover, the assignment of individual elements to particular cluster is made on basis of the rule of the newest centroid point. Thus, for each point of Y, the distances to all cluster- centroids is calculated. In addition, the investigated point y_i belongs to the group CL_c if the Euclidean distance $dist(y_i, O_c)$ is the smallest. In this work, the Celinski-Harabasz Index [3,4] is applied as a criterion for assessing the quality of the data set division. This clustering index is implemented within the optimisation metaheuristic algorithm as a cost function. The Celinski-Harabaszis applied as a criterion for assessing the quality of the data set division. This clustering index is implemented within the optimisation metaheuristic algorithm as a cost function. The Celinski-Harabasz criterion has its foundation within the concept of data set variance. This index is defined as:

$$I_{CH} = \frac{V_B}{V_W} \frac{M - C}{C - 1}, \tag{14}$$

where V_B and V_W denote overall between-cluster and within-cluster variance respectively. These are calculated according to the following formulas:

$$V_B = \sum_{c=1}^{C} \#CL_c \|O_c - O_Y\|^2, \tag{15}$$

and

$$V_W = \sum_{c=1}^{C} \sum_{y_i \in CL_c} \|y_i - O_c\|^2, \tag{16}$$

here, $\|\cdot\|$ is the L^2 norm (Euclidean distance) between the two vectors.

It should be underlined that high values of Celinski-Harabasz Index results point towards well-defined partitions. Because of the aforementioned properties, the following forms of cost functions are formulated:

$$K_{CH} = \frac{1}{I_{CH}} + \#CL_{\text{empty}}, \tag{17}$$

where $\#CL_{\text{empty}}$ denotes the number of clusters without any assigned element. More information about this index can be found at [4].

4 Numerical Studies

In order to obtain numerical results regarding quality of the verification tasks, twelve sets of data obtained from the UCI Machine Learning Repository and S-sets were taken into consideration [17] (Table 1). In Table 2, results in the form of Rand Index mean value and resulting standard deviation are presented. In the first part of the table, results based on the well-known kmeans clustering procedure are revealed, in subsequent parts, results based on CFA-clustering are shown. The best achieved results are bolded. All presented investigations were repeated 30 times. Based on other previous studies, the following sets of CFA parameters were used for all investigations $r_1 = -0.5$, $r_2 = 1.0$, $v_1 = -2.0$ and $v_2 = 2.0$.

Table 1. Data sets used for experimental verification

Name of data set	Abbreviation in paper	Number of			Bibliographical reference
		Elements (M)	Features (D)	Classes (C)	
Synthetic 1	S1	5000	2	15	[8]
Synthetic 2	S2	5000	2	6	[8]
Synthetic 3	S3	5000	2	3	[8]
Synthetic 4	S4	5000	2	6	[8]
Ionosphere	ION	351	34	2	[25]
Iris	Iris	150	4	3	[10]
Seeds	Seeds	210	7	3	[5]
Sonar	SON	208	60	2	[9]
Thyroid	TH	7200	21	3	[20,21]
Vehicle	VH	846	18	4	[24]
Wisconsin Breast Cancer	WBC	683	10	2	[27]
Wine	Wine	178	13	3	[2]

Table 2. Achieved results of comparison

Data set	k-means clustering		CFA-clustering	
	$\overline{R}$	σ_R	$\overline{R_{CFA}}$	$\sigma_{R_{CFA}}$
S1	0.9748	0.0093	**0.9930**	0.0018
S2	0.9760	0.0072	**0.9842**	0.0037
S3	0.9522	0.0072	**0.9587**	0.0026
S4	0.9454	0.0056	**0.9432**	0.0023
Iris	0.8458	0.0614	**0.8991**	0.0022
Ionosphere	0.5945	0.0004	**0.5949**	0.0003
Seeds	0.8573	0.0572	**0.8731**	0.0042
Sonar	0.5116	0.0016	**0.5191**	0.0001
Vehicle	0.5843	0.0359	**0.6161**	0.0006
WBC	0.5448	0.0040	**0.5491**	0.0056
Wine	0.7167	0.0135	**0.7452**	0.0003
Thyroid	**0.5844**	0.0982	0.5218	0.0032

Based on the results reported in Table 2, it can be noted that it is only in the case of the Thyroid data collection that the application of the classic k-means procedure achieved the generation of better results than did the utilization of the CFA-clustering algorithm. In other applications, in a comparison between these metaheuristic methods, the CFA-clustering won 11 times, the k-means won once.

A quite interesting observation is that regarding the stability of the obtained results. In the case of the k-means algorithm, the standard deviation of the results is several times higher than that of other methods. Especially worth emphasizing, is the pronounced negligible standard deviation of results based on CFA-clustering for most data set cases. It, thus, can be mainly concluded from the aforementioned results, that, unequivocally, the heuristic method generates better solutions to the problem of clustering than does the classic k-means method.

5 Conclusions

This paper was a presentation of a comparison in quality terms, of the two algorithms - CFA and k-means - when applied for optimisation purposes within a data clustering problem. In the numerical verification, twelve data sets, taken from the UCI repository, were used for comparison purposes. The location of the cluster centre was then investigated by way of optimisation procedures. Here, the Celinski-Harabasz Index was first used to determine the quality in the heuristic solution. For comparative purposes, the well-known rand index evaluation measure was performed. Based upon the obtained results, one can note that in almost all the data set cases, the metaheuristic algorithms demonstrated greater quality of result when compared with that gained by way of the k-means procedure. In particular, the CFA procedure revealed a much greater stability and greater quality.

References

1. Achtert, E., Goldhofer, S., Kriegel, H.P., Schubert, E., Zimek, A.: Evaluation of clusterings – metrics and visual support. In: 2012 IEEE 28th International Conference on Data Engineering, pp. 1285–1288, April 2012
2. Aeberhard, S., Coomans, D., De Vel, O.: Comparison of classifiers in high dimensional settings. Department Mathematics and Statistics, James Cook University of North Queensland, Australia, Technical report 92(02) (1992)
3. Arbelaitz, O., Gurrutxaga, I., Muguerza, J., Pérez, J.M.: An extensive comparative study of cluster validity indices. Pattern Recogn. **46**(1), 243–256 (2013)
4. Calinski, T., Harabasz, J.: A dendrite method for cluster analysis. Commun. Stat.-Theory Methods **3**(1), 1–27 (1974)
5. Charytanowicz, M., Niewczas, J., Kulczycki, P., Kowalski, P.A., Łukasik, S., Żak, S.: Complete gradient clustering algorithm for features analysis of x-ray images. In: Pietka, E., Kawa, J. (eds.) Information Technologies in Biomedicine. Advances in Intelligent and Soft Computing, vol. 69, pp. 15–24. Springer, Heidelberg (2010)
6. Eesa, A.S., Brifcani, A.M.A., Orman, Z.: Cuttlefish algorithm-a novel bio-inspired optimization algorithm. Int. J. Sci. Eng. Res. **4**(9), 1978–1986 (2013)
7. Eesa, A.S., Brifcani, A.M.A., Orman, Z.: A new tool for global optimization problems-cuttlefish algorithm. Int. J. Math. Comput. Nat. Phys. Eng. **8**(9), 1208–1211 (2014)
8. Fränti, P., Virmajoki, O.: Iterative shrinking method for clustering problems. Pattern Recogn. **39**(5), 761–775 (2006)
9. Gorman, R.P., Sejnowski, T.J.: Analysis of hidden units in a layered network trained to classify sonar targets. Neural Netw. **1**(1), 75–89 (1988)
10. Kowalski, P.A., Kulczycki, P.: Interval probabilistic neural network. Neural Comput. Appl. **28**(4), 817–834 (2017)
11. Kowalski, P.A., Łukasik, S.: Experimental study of selected parameters of the krill herd algorithm. In: Intelligent Systems'2014, pp. 473–485. Springer Science Business Media (2015)
12. Kowalski, P.A., Łukasik, S., Charytanowicz, M., Kulczycki, P.: Clustering based on the krill herd algorithm with selected validity measures. In: Ganzha, M., Maciaszek, L., Paprzycki, M. (eds.) Federated Conference on Computer Science and Information Systems 2016 (FedCSIS 2016), Annals of Computer Science and Information Systems, vol. 8, pp. 79–87, Gdansk, Poland, September 2016. IEEE (2016)
13. Kowalski, P.A., Łukasik, S., Charytanowicz, M., Kulczycki, P.: Data clustering with grasshopper optimization algorithm. In: Ganzha, M., Maciaszek, L., Paprzycki, M. (eds.) Federated Conference on Computer Science and Information Systems 2017 (FedCSIS 2017), Annals of Computer Science and Information Systems, vol. 11, pp. 71–74, Prague, Czech Republic, September 2017. IEEE (2017)
14. Kowalski, P.A., Łukasik, S., Charytanowicz, M., Kulczycki, P.: Optimizing clustering with cuttlefish algorithm. In: Kulczycki, P., Kowalski, P.A., Łukasik, S. (eds.) Contemporary Computational Science, p. 74. AGH-UST Press, Cracow (2018)

15. Kowalski, P.A., Łukasik, S., Charytanowicz, M., Kulczycki, P.: Nature inspired clustering - use cases of krill herd algorithm and flower pollination algorithm. In: Kóczy, L.T., Medina, J., Ramírez-Poussa, E. (eds.) Interactions Between Computational Intelligence and Mathematics. Studies in Computational Intelligence, pp. 83–98. Springer International Publishing, Cham (2019)
16. Kulczycki, P., Charytanowicz, M., Kowalski, P.A., Łukasik, S.: The complete gradient clustering algorithm: properties in practical applications. J. Appl. Stati. **39**(6), 1211–1224 (2012)
17. Lichman, M.: UCI Machine Learning Repository (2013)
18. Łukasik, S., Kowalski, P.A., Charytanowicz, M., Kulczycki, P.: Fuzzy models synthesis with kernel-density-based clustering algorithm. In: Fifth International Conference on Fuzzy Systems and Knowledge Discovery, 2008, FSKD 2008, vol. 3, pp. 449–453, October 2008
19. Łukasik, S., Kowalski, P.A., Charytanowicz, M., Kulczycki, P.: Clustering using flower pollination algorithm and calinski-harabasz index. In: 2016 IEEE Congress on Evolutionary Computation (CEC), pp. 2724–2728, July 2016
20. Quinlan, J.S.: Induction of decision trees. Mach. Learn. **1**(1), 81–106 (1986)
21. Quinlan, J.S., Compton, P.J., Horn, K.A., Lazarus, L.: Inductive knowledge acquisition: a case study. In: Proceedings of the Second Australian Conference on Applications of Expert Systems, pp. 137–156. Addison-Wesley Longman Publishing Co., Inc. (1987)
22. Rokach, L., Maimon, O.: Clustering methods. In: Maimon, O., Rokach, L. (eds.) Data Mining and Knowledge Discovery Handbook, pp. 321–352. Springer US (2005)
23. Senthilnath, J., Omkar, S.N., Mani, V.: Clustering using firefly algorithm: performance study. Swarm Evol. Comput. **1**(3), 164–171 (2011)
24. Setiono, R., Leow, W.K.: Vehicle recognition using rule based methods. Turing Institute Research Memorandum TIRM-87-018, 121 (1987)
25. Sigillito, V.G., Wing, S.P., Hutton, L.V., Baker, K.B.: Classification of radar returns from the ionosphere using neural networks. Johns Hopkins APL Tech. Dig. **10**(3), 262–266 (1989)
26. Welch, W.J.: Algorithmic complexity: three NP- hard problems in computational statistics. J. Stat. Comput. Simul. **15**(1), 17–25 (1982)
27. Zhang, J.: Selecting typical instances in instance-based learning. In: Proceedings of the Ninth International Conference on Machine Learning, pp. 470–479 (1992)

A Memetic Version of the Bacterial Evolutionary Algorithm for Discrete Optimization Problems

Boldizsár Tüű-Szabó[1(✉)], Péter Földesi[2], and László T. Kóczy[1,3]

[1] Department of Information Technology, Széchenyi István University, Győr, Hungary
{tuu.szabo.boldizsar, koczy}@sze.hu
[2] Department of Logistics, Széchenyi István University, Győr, Hungary
foldesi@sze.hu
[3] Department of Telecommunications and Media Informatics, Budapest University of Technology and Economics, Budapest, Hungary

Abstract. In this paper we present our test results with our memetic algorithm, the Discrete Bacterial Memetic Evolutionary Algorithm (DBMEA). The algorithm combines the Bacterial Evolutionary Algorithm with discrete local search techniques (2-opt and 3-opt).

The algorithm has been tested on four discrete NP-hard optimization problems so far, on the Traveling Salesman Problem, and on its three variants (the Traveling Salesman Problem with Time Windows, the Traveling Repairman Problem, and the Time Dependent Traveling Salesman Problem). The DBMEA proved to be efficient for all problems: it found optimal or close-optimal solutions. For the Traveling Repairman Problem the DBMEA outperformed even the state-of-the-art methods.

The preliminary version of this paper was presented at the 3rd Conference on Information Technology, Systems Research and Computational Physics, 2–5 July 2018, Cracow, Poland [1].

Keywords: Traveling Salesman Problem · Time windows · Traveling Repairman Problem · Time dependent

1 Introduction

The first version of the Bacterial Evolutionary Algorithm (BEA) was introduced by Nawa and Furuhashi in 1999 [20]. They used it for discovering the optimal parameters of a fuzzy rule based system.

Since then the algorithm has been widely used for various optimization problems. In 2002 Inoue *et al.* used the bacterial evolutionary algorithm for an interactive nurse scheduling optimization problem [15]. In 2009 Das *et al.* introduced a clustering algorithm based on the bacterial evolutionary algorithm for automatic data clustering [5]. In 2009 the Bacterial Memetic Algorithm was proposed for the modified Traveling Salesman Problem with time dependent costs [8]. In 2011 the Eugenic Bacterial Memetic Algorithm was presented for solving the Traveling Salesman Problem with

P. Kulczycki et al. (Eds.): ITSRCP 2018, AISC 945, pp. 44–55, 2020.
https://doi.org/10.1007/978-3-030-18058-4_4

fuzzy and time dependent cost values [10, 11]. In 2011 the group of the authors introduced an algorithm based on the BEA for solving the Three Dimensional Bin Packing Problem [6]. In 2012 a Bacterial Memetic Algorithm was introduced for offline path planning of mobile robots by Botzheim *et al.* [3]. In 2013 the group of the authors proposed a memetic algorithm which combines the BEA with gradient-based local search for solving a fuzzy resource allocation problem [7]. In 2014 Hsieh proposed a BEA inspired algorithm for solving the Hierarchical redundancy allocation for multi-level reliability systems [14].

2 The DBMEA Algorithm

The DBMEA combines the discrete version of the BEA with local search techniques (2-opt and 3-opt), so it can be called memetic algorithm.

Recently the memetic algorithms [19] are very popular in the field of optimization because it can be efficient in solving NP-hard optimization problems. This combination can eliminate the weaknesses of both methods. Evolutionary algorithms search in the global search which often cause slow convergence speed. Local search methods search in the neighborhood of the current tour, so they often fail to find the global optimum.

The process of the DBMEA consists of the following steps:

- Creating the Initial Population

In each iteration until the stoppage criterion is met the following are repeated:

- Bacterial mutation for each bacterium
- Combined 2-opt and 3-opt local search
- Gene transfer performed on the population

2.1 Creating the Initial Population

In DBMEA the bacteria form a population. Each bacterium represents a possible solution for the examined problem.

In DBMEA the bacteria are coded with a very simple permutation encoding. The starting point of the tour is indexed with 0 (for each tour it is the starting node, so it does not appear in the codes), and other nodes are mapped with a unique index (1..n). The tours are expressed as the sequence of these indices clearly identifying the tour. Each node needs to visit once, so the length of the tours is $n - 1$. An example of encoding can be seen in Fig. 1.

The initial population contains randomly created individuals and three deterministic individuals. The deterministic individuals are created with the following heuristics:

- Nearest neighbour (NN) heuristic: It creates a tour in which always the nearest unvisited city is visited.
- Secondary nearest neighbour (SNN) heuristic: in this tour always the second nearest unvisited city is visited.

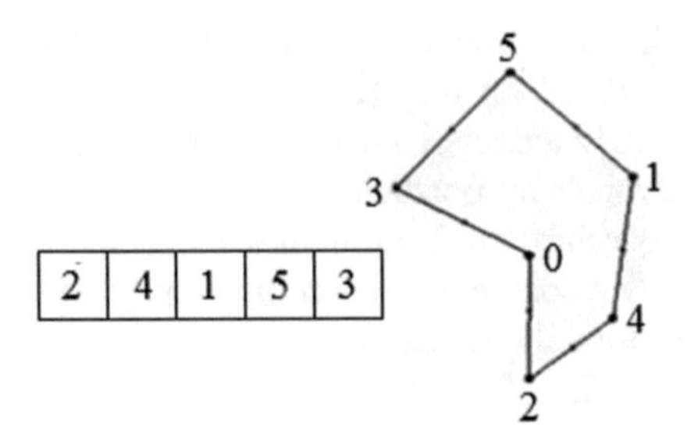

Fig. 1. The encoding of the tour

- Alternating nearest neighbour (ANN) heuristic: It is the combination of the above mentioned two methods. It represents a tour in which the nearest and second nearest unvisited cities are visited in alternating order.

2.2 Bacterial Mutation

The bacterial mutation operation -works on the bacterium individually- consists of the following steps (Fig. 2):

- Creating a pre-defined number (N_{clones}) of clones from the original bacterium
- Dividing the bacterium into fixed length (not necessary coherent) segments
  The following step are repeated until all segments are examined:
  - Selecting randomly a non-mutated segment
  - Randomly changing the node sequence of the selected segment in the clones
  - Calculating the costs of the clones and the original bacterium
  - Selecting the individual with the lowest cost among the clones and the original bacterium
  - Copying the segment content of the best individual into the other clones and original bacterium

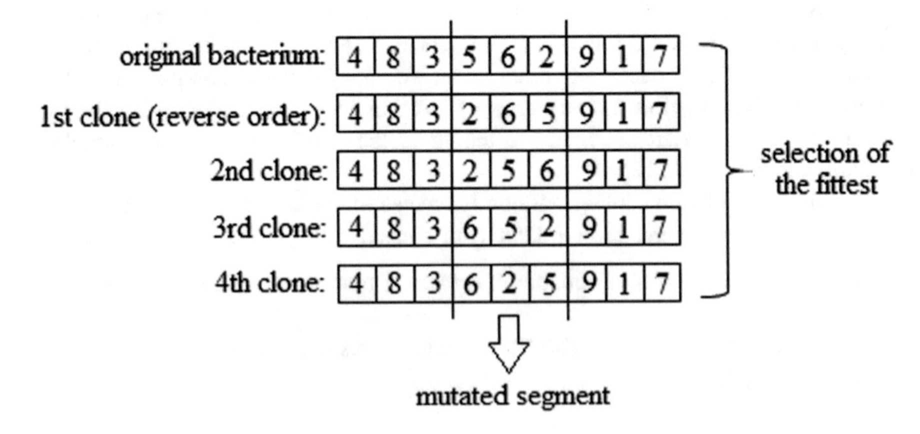

Fig. 2. Bacterial mutation

- After the mutation of all segments replacing the original bacterium with the best individual among the mutated original and clones in the population

The bacterial mutation results that each bacterium in the population becomes more or in worst case equally fit.

In our algorithm two different types of mutation, the coherent segment mutation and the loose segment mutation are used.

Coherent Segment Mutation: In this case adjacent elements form the segments. It is easy to execute: the chromosome is cut into segments with equal length (Fig. 3.).

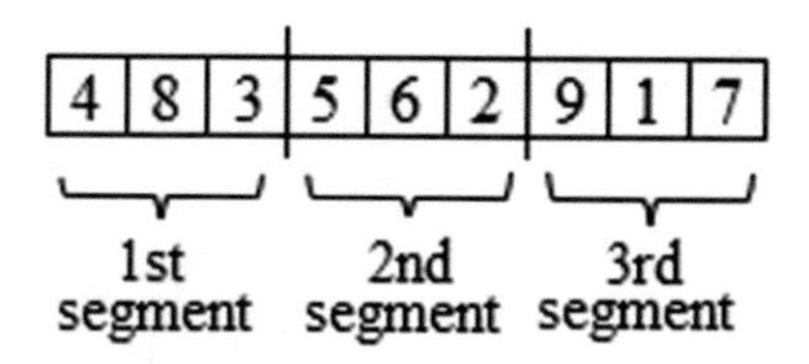

Fig. 3. Coherent segments

Loose Segment Mutation: As opposed to the coherent segment mutation, the segments of the bacterium do not need to consist of adjacent elements. The elements of the segments may come from different parts of the bacterium (Fig. 4.).

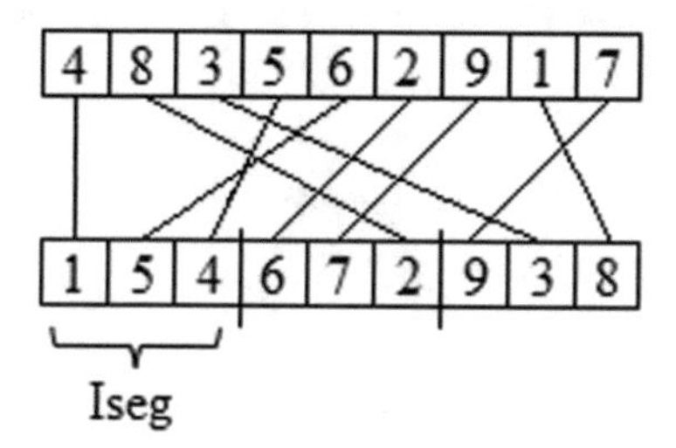

Fig. 4. Loose segments

The time complexity of the bacterial mutation is $O(N_{ind}N_{clones}n^2)$ in one generation and the space requirement is $O(N_{ind}N_{clones}n)$ [3].

2.3 Local Search

Adding local search techniques to the process of an evolutionary algorithm usually greatly increases the effectiveness of the algorithm [13].

They search for improvements in the local environment of the candidate solutions, therefore they usually find only a local optimal solution.

In DBMEA the local search combines the 2-opt and 3-opt techniques. First, the tour is improved by 2-opt steps. And when no further improvement is possible with 2-opt, then 3-opt will be applied for the tour. The local search is stopped, when the tour cannot be improved further by 3-opt steps.

2-opt Local Search

2-opt local search replaces two edge pairs in the original graph to reduce the cost of the tour (Fig. 5).

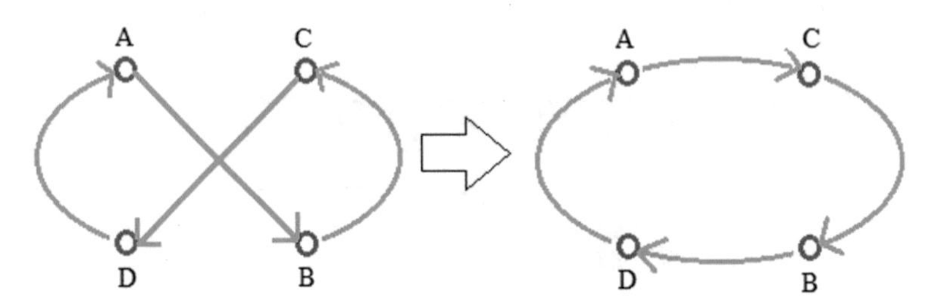

Fig. 5. 2-opt local steps

The edge exchange is continued till no further improvement is possible.

3-opt Local Search

The 3-opt local search improves the tour by replacing three edges with three other (Fig. 6). There are four ways to reconnect the tour. The output of the 3-opt step is always the less costly tour.

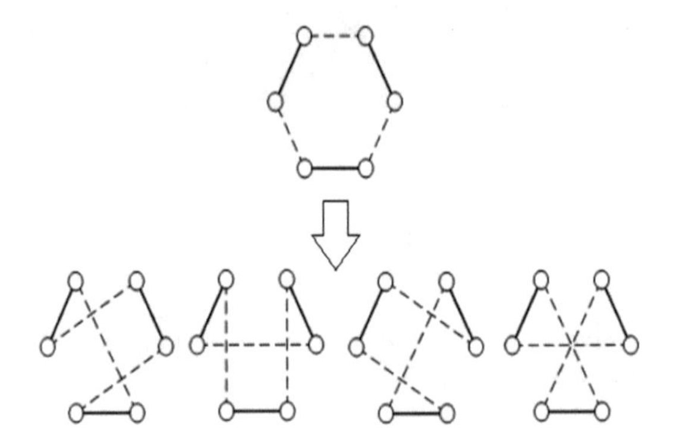

Fig. 6. 3-opt local steps

2.4 Gene Transfer

The gene transfer operation provides the information transfer within the population promoting the development of the bacteria.

The gene transfer consists of the following steps:

- Sorting the population in descending order according to their fitness values
- Dividing the sorted population into two halves (superior and inferior half) N_{inf} times the following is repeated:
  - Choosing randomly a bacterium from both parts (source and destination bacterium)
  - Copying a part of the source bacterium with pre-defined ($I_{transfer}$) length into the destination bacterium
  - Eliminating the double occurrence of nodes to keep the same length (Fig. 7)

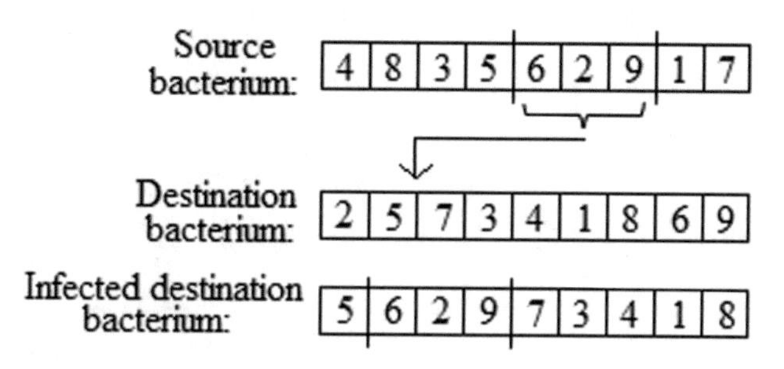

Fig. 7. Gene transfer

The time complexity of the gene transfer consists of the following components:

- The calculation time of fitness values $O(N_{ind}n)$
- The time complexity of sorting the population in a descending order, based on the fitness values $O(N_{ind}logN_{ind})$
- The calculation time of the new fitness value of the modified bacterium, and its reinsertion into the population $O\left(N_{inf}(n+N_{ind})\right)$

The total time complexity of the gene transfer operation in each generation is $C_{GT} = O\left(N_{ind}(n+logN_{ind})+N_{inf}(n+N_{ind})\right)$ [9].

3 Computational Results

The DBMEA algorithm was tested on 4 discrete optimization problems, on the Traveling Salesman Problem and on three TSP variants.

3.1 The Traveling Salesman Problem

In 2016 we presented the Discrete Bacterial Memetic Evolutionary Algorithm (DBMEA) for the Traveling Salesman Problem (TSP). The algorithm was tested on TSP benchmark instances up to 1400 nodes [16].

The efficiency of the DBMEA has improved significantly by speed-up the local search. This improved version of DBMEA was able to solve instances up to 5000 nodes close-optimally within reasonable time [17]. Our results were compared with the state-of-the-art heuristic and exact solver, the Helsgaun's Lin-Kernighan [11] and the Concorde algorithm [2] running on the same hardware as the DBMEA. Our algorithm in all cases found near-optimal solutions (average solution is within 0.16% to the optimum) and the runtime was well predictable. The Helsgaun's Lin-Kernighan and the Concorde behaved worse in terms of runtime predictability, in addition the Concorde failed to solve the four examined biggest instances within 3 CPU days (Fig. 8) [17].

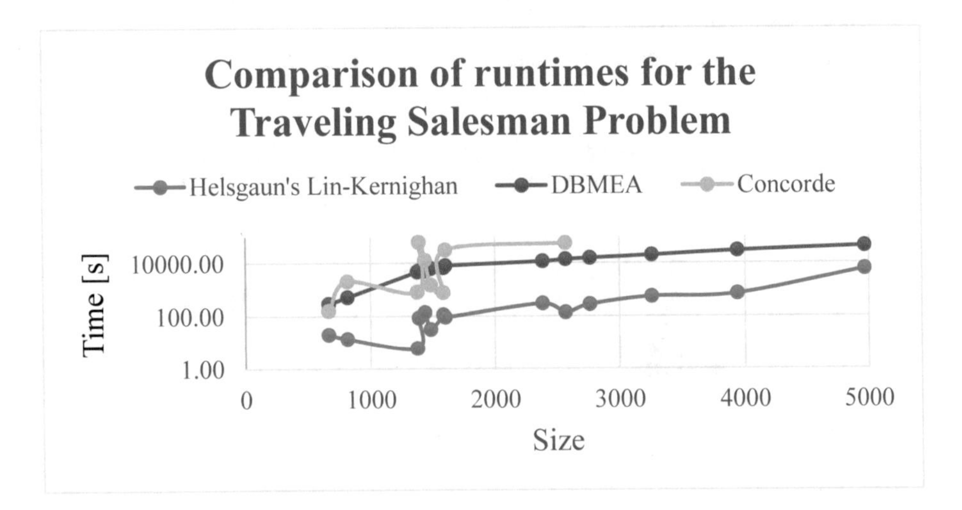

Fig. 8. Comparison of runtime for the Traveling Salesman Problem [16]

3.2 The Traveling Salesman Problem with Time Windows

After some minor modifications the DBMEA was tested on the TSP with Time Windows which is a TSP variant problem [17]. A time window is assigned to each node and the salesman has to reach and leave the nodes within its time windows. The aim is to find the feasible solution with the lowest total cost.

The DBMEA produced similarly good results as the most efficient methods (general VNS [3], compressed annealing [21]) in the literature: it found in all test cases the best-known (24 out of 28) or the near-best-known solutions even for larger time windows [16].

3.3 The Traveling Repairman Problem

We also examined the Traveling Repairman Problem (its other names are Minimum Latency Problem, Delivery Man Problem). In the case of the TRP the cost function is the sum of arrival times at each node. The aim is to find the tour which has the minimal cost.

The DBMEA was tested on TSPLIB benchmark instances, and the results were compared with the most efficient methods (GILS-RVND [24], heuristic of Salehipour *et al.* [22]) in the literature. Except two cases (lin318, pr439) the DBMEA found the

Table 1. Comparison of results for the Traveling Repairman Problem [24]

Instance	Best known	DBMEA		GILS-RVND		Salehipour *et al.*	Gap [%] GILS-RVND		Gap [%] Salehipour *et al.*
		Best value	Avg. value	Best value	Avg. value	Best value	Best value	Avg. value	Best value
st70	19215	19215	19215	19215	19215	19553	0.00	0.00	**−1.73**
rat99	54984	54984	54984	54984	54984	56994	0.00	0.00	**−3.53**
kroD100	949594	949594	949594	949594	949594	976830	0.00	0.00	**−2.79**
lin105	585823	585823	585823	585823	585823	585823	0.00	0.00	0.00
pr107	1980767	1980767	1980767	1980767	1980767	1983475	0.00	0.00	**−0.14**
rat195	210191	210191	**210284.3**	210191	210335.9	213371	0.00	**−0.02**	**−1.49**
pr226	7100308	7100308	7100308	7100308	7100308	7226554	0.00	0.00	**−1.75**
lin318	5560679	5562148	**5566344.8**	5560679	5569820	5876537	0.03	**−0.06**	**−5.35**
pr439	17688561	17693137	**17710528**	17688561	17734922	18567170	0.03	**−0.14**	**−4.70**
att532	5581240	**5579113**	**5584915.3**	5581240	5597867	18448435[1]	**−0.04**	**−0.23**	**−69.76**

[1]Calculate Euclidean distances instead of ATT pseudo-Euclidean distances

best-known solutions, and for att532 it found new best solution (Table 1) [25]. With the DBEA the average values (averaging 10 runs) was lower than with the GILS-RVND for the bigger instances which means that with few runs the DBMEA is more likely to produce "good" solutions than the GILS-RVND. The three methods were run on different, but similar hardware configurations. With the growing of the problem size the DBMEA became faster compared to the other methods (Fig. 9.).

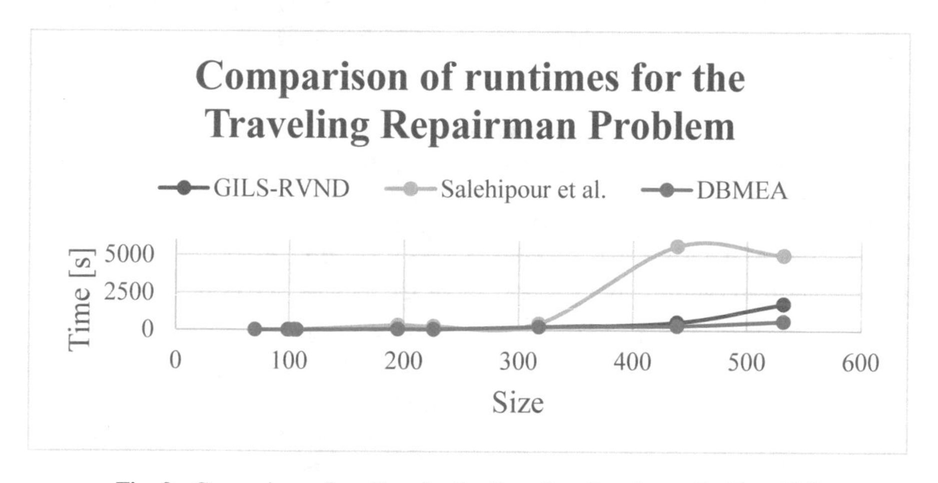

Fig. 9. Comparison of runtime for the Traveling Repairman Problem [24]

3.4 The Time Dependent Traveling Salesman Problem

Recently the Time Dependent Traveling Salesman Problem (TDTSP) was examined by Schneider [23] and Li *et al.* [18]. In 2002 Schneider presented a simulated annealing algorithm, Li *et al.* developed a record-to-record travel algorithm (RTR algorithm) for solving the TDTSP. Both methods were tested on the bier127 problem which consists of 127 beer gardens in the area of Augsburg.

The TDTSP problem was formulated by Schneider. A traffic jam region was placed in the centre (coordinate of the left corner point (7080, 7200), width is 6920, and the height is 9490). The salesman works between 9 am and 3 pm. The rush hours are between 12 pm and 3 pm. In this period the edge lengths between two nodes in the traffic region are multiplied with a jam factor. The travel speed is computed by dividing the total distance of the optimal TSP tour (118293.524) by the working hours of the salesman. The travel speed is held constant for all values of the jam factor. The aim is to find the tour with the lowest cost.

The DBMEA was tested on the bier127 and two self-generated instance with 250 nodes. The coordinates of the nodes have been generated from a uniform distribution (for s250_1 instance 70%, for s250_2 instance 50% of the nodes were placed within the jam region). The working hours and the calculation of travel speed is the same as in the case of bier127.

For the Time Dependent TSP the following speed-up techniques were used to accelerate the local search:

- Candidate list [13]: It contains the indices of the closest vertices in ascending order and is created for all vertices. During the local search only the pre-defined number of closest vertices (candidate lists contain them) are examined for each vertex, because a shorter edge is more likely to be part of a good solution.
- "Don't look back bits" [13]: Each vertex in assigned to a "don't look back bit". If no improving was found for a given vertex *v*, then until an incident edge changes, do not consider *v* (changes its "don't look back bit" to 1).

The algorithm was tested with the following parameters:

- the number of bacteria in the population ($N_{ind} = 100$)
- the number of clones in the bacterial mutation ($N_{clones} = n_{cities}/10$)
- the number of infections in the gene transfer ($N_{inf} = 40$)
- the length of the chromosomes ($I_{seg} = n_{cities}/20$)
- the length of the transferred segment ($I_{trans} = n_{cities}/10$)
- length of the candidate lists (square root of the number of cities)

Our tests were carried out on an Intel Core i7-7500U 2.7 GHz, 8 GB of RAM memory workstation under Linux Mint 18.2. The DBMEA was coded in C++. Our results were calculated by averaging 10 test runs.

For bier127 instance like the simulated annealing the DBMEA found the best-known solutions with each jam factor while RTR algorithm in most cases failed to find them (Table 2). The runtimes of the DBMEA are also given, although they are not comparable with the other two methods: the runtimes of the simulated annealing was not given by Schneider, and the RTR algorithm was tested on a much slower hardware than the DBMEA.

Table 3 contains our result for our two self-generated instances with 250 nodes. Based on the results we can state that the DBMEA also worked well on these instances: it found high quality solutions even with bigger jam factors (the cost of the tours are close to the cost of the tour with jam factor 1.00).

Table 2. Comparison of results on the bier127 problem with different jam factors

Jam factor	DBMEA			Simulated annealing	RTR algorithm
	Best value	Average value	Average time [s]	Best value	Best value
1.00	118293.524	118293.524	26.878	118293.524	118293.524
1.03	118749.356	118749.356	21.993	118749.356	118796.154
1.04	118901.300	118901.300	20.237	118901.300	119971.191
1.05	119053.244	119053.244	18.802	119053.244	119503.279
1.06	119153.582	119189.167	49.288	119153.582	119857.323
1.10	119313.720	119313.720	36.190	119313.720	119957.387
1.20	119714.065	119714.065	40.395	119714.065	119714.065
1.30	120114.410	120130.594	38.964	120114.410	120637.093
1.38	120434.687	120438.749	48.904	120434.687	120434.687
1.39	120453.554	120453.554	34.028	120453.554	120453.554
1.50	120571.743	120571.743	37.150	120571.743	120617.178
1.60	120679.186	120821.535	52.399	120679.186	121108.329
1.70	120786.630	120866.307	47.350	120786.630	120898.269
1.80	120894.074	120966.040	43.558	120894.074	121195.816
1.90	121001.518	121102.499	52.044	121001.518	121148.519
2.00	121125.195	121125.195	24.668	121125.195	121298.538
3.00	121125.195	121125.195	15.839	121125.195	122222.204
10.00	121125.195	121125.195	26.750	121125.195	121167.051
100.00	121125.195	121125.195	45.636	121125.195	122280.886
2000.00	121125.195	121125.195	42.277	121125.195	121417.575

Table 3. Results on the instances with 250 nodes with different jam factors

Jam factor	DBMEA					
	s250_1			s250_2		
	Best value	Average value	Average time [s]	Best value	Average value	Average time [s]
1.00	5593.721	5598.908	385.005	5883.419	5888.389	364.514
1.01	5602.693	5606.621	333.424	5886.761	5889.916	308.640
1.02	5607.850	5612.466	332.826	5887.832	5894.831	298.073
1.03	5614.172	5619.788	346.240	5888.904	5894.328	310.768
1.04	5618.168	5621.647	374.396	5889.975	5895.813	343.722
1.05	5618.168	5621.516	320.237	5891.047	5896.819	310.488
1.1	5618.168	5623.638	277.473	5896.389	5907.867	311.524
1.2	5618.168	5626.335	309.946	5904.205	5921.092	310.449
1.3	5618.168	5627.508	327.514	5912.021	5936.651	328.916
1.5	5618.168	5631.149	303.666	5922.831	5938.454	319.962
2.00	5618.168	5630.845	317.846	5943.707	5948.585	324.703

4 Conclusion

In this paper our results were summarized with DBMEA testing it on four discrete NP-hard optimization problems. Our former results were briefly evaluated on the TSP, the TSP with Time Windows, and the TRP problem. Our newest results on the Time Dependent TSP were also presented.

The DBMEA can be called an efficient algorithm because it produced optimal or near-optimal solutions for every examined optimization problem. For TRP the DBMEA outperformed even the state-of-the-art methods in the literature.

Acknowledgement. The authors would like to thank to EFOP-3.6.1-16-2016-00017 1 'Internationalisation, initiatives to establish a new source of researchers and graduates, and development of knowledge and technological transfer as instruments of intelligent specialisations at Széchenyi István University' for the support of the research. This work was supported by National Research, Development and Innovation Office (NKFIH) K124055. Supported by the ÚNKP-18-3 New National Excellence Program of the Ministry of Human Capacities.

References

1. Tüű-Szabó, B., Földesi, P., Kóczy, L.T.: A Memetic version of the Bacterial Evolutionary Algorithm for discrete optimization problem. In: Kulczycki, P., Kowalski, P.A., Łukasik, S. (eds.) Contemporary Computational Science, p. 75. AGH-UST Press, Cracow (2018)
2. Applegate, D.L., Bixby, R.E., Chvátal, V., Cook, W.J., Espinoza, D., Goycoolea, M., Helsgaun, K.: Certification of an optimal tour through 85,900 cities. Oper. Res. Lett. **37**(1), 11–15 (2009)
3. Botzheim, J., Toda, Y., Kubota, N.: Bacterial memetic algorithm for offline path planning of mobile robots. Memetic Comput. **4**, 73–86 (2012)
4. da Silva, R.F., Urrutia, S.: A General VNS heuristic for the traveling salesman problem with time windows. Discrete Optim. **7**(4), 203–211 (2010)
5. Das, S., Chowdhury, A., Abraham, A.: A bacterial evolutionary algorithm for automatic data clustering. In: IEEE Congress on Evolutionary Computation, CEC 2009, pp. 2403–2410 (2009)
6. Dányádi, Zs., Földesi, P., Kóczy, L.T.: A bacterial evolutionary solution for three dimensional bin packing problems using fuzzy fitness evaluation. Aus. J. Intell. Inf. Process. Syst. **13**(1), 7 (2011). Paper 1181
7. Dányádi, Zs., Földesi, P., Kóczy, L.T.: Solution of a fuzzy resource allocation problem by various evolutionary approaches. In: IFSA World Congress and NAFIPS Annual Meeting (IFSA/NAFIPS) 2013 Joint, pp. 807–812 (2013)
8. Farkas, M., Földesi, P., Botzheim, J., Kóczy, L.T.: Approximation of a modified traveling salesman problem using bacterial memetic algorithms. In: Towards Intelligent Engineering and Information Technology (Studies in Computational Intelligence, vol. 243), pp. 607–625. Springer, Berlin (2009)
9. Földesi, P., Botzheim, J.: Modeling of loss aversion in solving fuzzy road transport traveling salesman problem using eugenic bacterial memetic algorithm. Memetic Comput. **2**(4), 259–271 (2010)

10. Földesi, P., Kóczy, L.T., Botzheim, J., Farkas, M.: Eugenic bacterial memetic algorithm for fuzzy road transport traveling salesman problem. In: The 6th International Symposium on Management Engineering (ISME 2009), August 5–7, Dalian, China (2009)
11. Földesi, P., Botzheim, J., Kóczy, L.T.: Eugenic bacterial memetic algorithm for fuzzy road transport traveling salesman problem. Int. J. Innovative Comput. **7**(5(B)), 2775–2798 (2011)
12. Helsgaun, K.: An effective implementation of the Lin-Kernighan traveling salesman heuristic. Eur. J. Oper. Res. **126**, 106–130 (2000)
13. Hoos, H.H., Stutzle, T.: Stochastic Local Search: Foundations and Applications. Morgan Kaufmann, San Francisco (2005)
14. Hsieh, T.-J.: Hierarchical redundancy allocation for multi-level reliability systems employing a bacterial-inspired evolutionary algorithm. Inf. Sci. **288**, 174–193 (2014)
15. Inoue, T., Furuhashi, T., Maeda, H., Takaba, M.: A study on interactive nurse scheduling support system using bacterial evolutionary algorithm engine. Trans. Inst. Electron. Inf. Syst. **122-C**, 1803–1811 (2002)
16. Kóczy, L.T., Földesi, P., Tüű-Szabó, B.: An effective discrete bacterial memetic evolutionary algorithm for the traveling salesman problem. Int. J. Intell. Syst. **32**(8), 862–876 (2017)
17. Kóczy, L.T., Földesi, P., Tüű-Szabó, B.: Enhanced discrete bacterial memetic evolutionary algorithm-an efficacious metaheuristic for the traveling salesman optimization. Inf. Sci. **460**, 389–400 (2017)
18. Li, F., Golden, B., Wasil, E.: Solving the time dependent traveling salesman problem. In: Golden, B., Raghavan, S., Wasil, E. (eds.) The Next Wave in Computing, Optimization, and Decision Technologies. Operations Research/Computer Science Interfaces Series, vol. 29. Springer, Boston (2005)
19. Moscato, P.: On Evolution, Search, Optimization, Genetic Algorithms and Martial Arts - Towards Memetic Algorithms, Technical Report Caltech Concurrent Computation Program, Report. 826, California Institute of Technology, Pasadena, USA (1989)
20. Nawa, N.E., Furuhashi, T.: Fuzzy system parameters discovery by bacterial evolutionary algorithm. IEEE Trans. Fuzzy Syst. **7**, 608–616 (1999)
21. Ohlmann, J.W., Thomas, B.W.: A compressed-annealing heuristic for the traveling salesman problem with time windows. INFORMS J. Comput. **19**(1), 80–90 (2007)
22. Salehipour, A., Sörensen, K., Goos, P., Bräysy, O.: Efficient GRASP+VND and GRASP+VNS metaheuristics for the traveling repairman problem. 4OR Q. J. Oper. Res. **9**(2), 189–209 (2011)
23. Schneider, J.: The time-dependent traveling salesman problem. Phys. A **314**, 151–155 (2002)
24. Silva, M.M., Subramanian, A., Vidal, T., Ochi, L.S.: A simple and effective metaheuristic for the Minimum Latency Problem. Eur. J. Oper. Res. **221**(3), 513–520 (2012)
25. Tüű-Szabó, B., Földesi, P., Kóczy, L.T.: A population based metaheuristic for the Minimum Latency Problem. In: ESCIM 2017, Faro, Portugal (2017)

On Wavelet Based Enhancing Possibilities of Fuzzy Classification Methods

Ferenc Lilik[1(✉)], Levente Solecki[1], Brigita Sziová[1], László T. Kóczy[1,2], and Szilvia Nagy[1]

[1] Széchenyi István University, Egyetem tér 1, Győr 9026, Hungary
lilikf@sze.hu

[2] Budapest University of Technology and Economics, Magyar tudósok krt. 2., Budapest 1117, Hungary

Abstract. If the antecedents of a fuzzy classification method are derived from pictures or measured data, it might have too many dimensions to handle. A classification scheme based on such data has to apply a careful selection or processing of the measured results: either a sampling, re-sampling is necessary or the usage of functions, transformations that reduce the long, high dimensional observed data vector or matrix into a single point or to a low number of points. Wavelet analysis can be useful in such cases in two ways.

As the number of resulting points of the wavelet analysis is approximately half at each filters, a consecutive application of wavelet transform can compress the measurement data, thus reducing the dimensionality of the signal, i.e., the antecedent. An SHDSL telecommunication line evaluation is used to demonstrate this type of applicability, wavelets help in this case to overcome the problem of a one dimensional signal sampling.

In the case of using statistical functions, like mean, variance, gradient, edge density, Shannon or Rényi entropies for the extraction of the information from a picture or a measured data set, and they don not produce enough information for performing the classification well enough, one or two consecutive steps of wavelet analysis and applying the same functions for the thus resulting data can extend the number of antecedents, and can distill such parameters that were invisible for these functions in the original data set. We give two examples, two fuzzy classification schemes to show the improvement caused by wavelet analysis: a measured surface of a combustion engine cylinder and a colonoscopy picture. In the case of the first example the wear degree is to be determine, in the case of the second one, the roundish polyp content of the picture. In the first case the applied statistical functions are Rényi entropy differences, the structural entropies, in the second case mean, standard deviation, Canny filtered edge density, gradients and the entropies.

In all the examples stabilized KH rule interpolation was used to treat sparse rulebases.

The preliminary version of this paper was presented at the 3rd Conference on Information Technology, Systems Research and Computational Physics, 2–5 July 2018, Cracow, Poland [1].

P. Kulczycki et al. (Eds.): ITSRCP 2018, AISC 945, pp. 56–73, 2020.
https://doi.org/10.1007/978-3-030-18058-4_5

Keywords: Fuzzy classification · Wavelet analysis · Fuzzy rule interpolation · Structural entropy

1 Introduction

Real-life control or classification problems are often solved by fuzzy methods, as fuzzy inference is usually practically more efficient, flexible and close to the human way of thinking than classical, crisp decision schemes. As the digital measuring and picture taking devices become more and more widespread, the measured data becomes larger, contains more information bout the measured or photographed objects, but most of the such acquired information disturbs an automatic control or classification scheme. For training a neural network or other, nature based learning method, however, very large number of measurement with time and resource consuming pre-processing is often needed, thus in some cases it is not possible to use them.

Digital measuring devices sample either in time, like in the case of an oscilloscope or temperature monitoring system; in frequency, like in the case of spectrum analysers; or in space, like in the case pictures or 3D scanners. The results of such measurements often consist of several hundreds or thousands or even millions of points, but such large data sets are not suitable for serving as the antecedent set for a fuzzy decision or classification scheme.

The measured data has to be made processable by a fuzzy inference system of reasonable size and complexity, mostly by decreasing the amount of data with the condition of keeping as much information as possible. The simplest step to achieve this may be re-sampling: selecting only a few from the measured values as representatives of the whole data set. A more sophisticated method would be averaging. However, both lead to loss of information. From image and data compression it is well known that wavelets are suitable for distilling a lot of the available information and achieving a large compression ratio, thus it seems to be reasonable to try wavelet transform for achieving suitable compression rate as well as sufficiently small information loss. In case of telecommunication line insertion loss over frequency functions, this method proved to be effective.

Instead of wavelet-based compression of the measurement results, its useful information content can be extracted by calculating statistical parameters, like its mean value, standard deviation, average gradient or gradient direction, some kind of shape-related quantity or its entropy, or entropies, if Rényi's generalisation of the definition of entropy is used. In the case of pictures edge densities or colour elated parameters might be also necessary. Rényi's generalised entropies can be combined into such quantities that characterise the shape or topology of the measured distribution, too, this step can map the measured data into a couple of points. This scheme is useful especially in the case of two- or three-dimensional measured data, as the number of measured points is usually too high, and the number of the remaining points is still too high after wavelet-based compression. In many cases, using only entropies, or other, similar functions leads to critically high information loss, thus a method for regaining

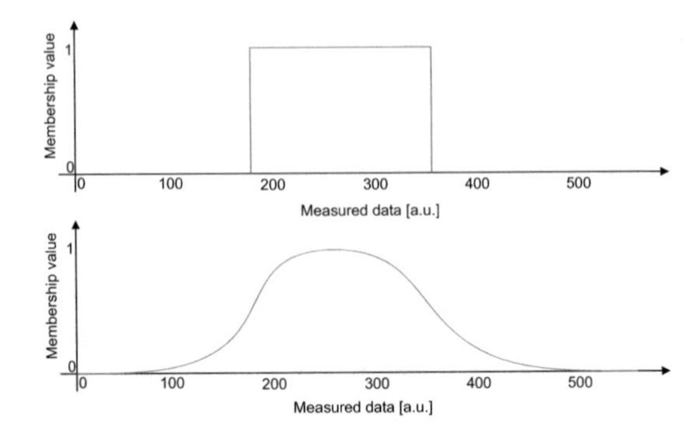

Fig. 1. Crisp and fuzzy membership function of a measured variable.

some of the information should be introduced. As the high-pass filter outputs of the wavelet transform distill the fine details from the original data and the low-pass outputs behave as a kind of averaging, applying the same functions – the previously mentioned statistical parameters and entropies – on the wavelet transformed versions of the images leads to other information. This step can enhance the performance of an inference system without increasing the number of antecedents too much. This scheme proved to be efficient in significantly increasing the effectiveness of a classification scheme in surface roughness and colorectal polyp content characterisation.

These two approaches are studied in the following considerations. As a first step, in Sect. 2 a summary about fuzzy classifications is given completed with a section of fuzzy rule interpolation. Next the introduction of wavelet transforms is given in Sect. 3. Section 4 gives the generalization of the Shannon entropy and its use in the characterisation of shapes of functions or distributions. The applicability of the first approach is demonstrated in Sect. 5, and the usage of wavelet transform for increasing the information content is given in Sects. 6 and 7. A conclusions to be found in the last section.

2 The Fuzzy Component of the Approach

2.1 Fuzzy Sets

In set theory L.A. Zadeh came forth with a new concept in 1965. According to his idea [2], an element can not only be fully member of a set, or fully not member of it, but there can be infinite many possibilities inbetween. This concept is closely related to the human way of thinking, as there is a smooth transition between a tea being hot or cold, or a dog breed being small, medium-sized or large. Zadeh's fuzzy sets do not only allow membership values of exclusively 0 or 1 (like with the traditional, crisp sets), but any value in the $[0, 1]$ unit interval. A measured value can have a membership degree in a fuzzy set, thus it also becomes a fuzzy quantity, as it can be seen in Fig. 1 Based on such fuzzy sets,

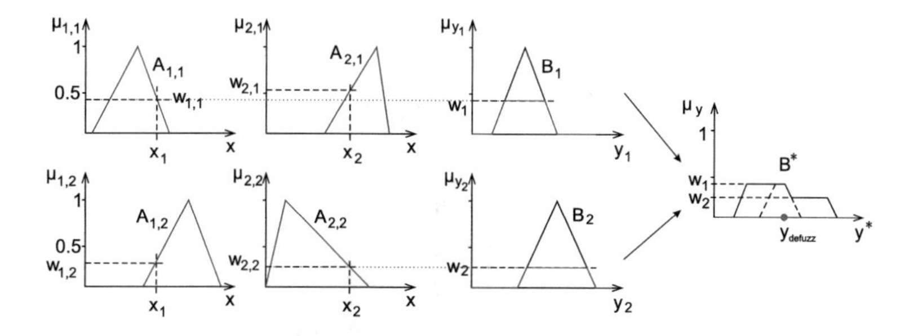

Fig. 2. Mamdani's fuzzy consequent from a two-dimensional antecedent and two consequent sets. The rules are denoted by the membership values $A_{i,j}$ of the variable x_i, the consequent sets by B_{y_j}, and the final consequent set by B^*. The final consequent is usually defuzzified, thus one value y_{defuzz} becomes the result of the inference. Arbitrary units.

decisions can be made, like "IF the water level is low", "THEN fill the water tank with a small amount of water", or "IF the temperature is high", "THEN classify it to the highest class". In the case of multiple conditions, like "IF the gasoline concentration is high AND the pressure is high" it is necessary to redefine he operators "AND" and "OR". Zadeh defined "AND" as the lowest of the membership values (called nowadays rather *t-norm*), and "OR" as the highest (called *s-norm* or *t-conorm*).

2.2 Fuzzy Inference

Using the fuzzy membership functions a rather flexible control systems can be built. Mamdani proposed [3] the first concept for carrying out fuzzy control (and decisions), which was a computationally more efficient implementation of the Compositional Rule of Inference method also proposed by Zadeh [4]. His concept used multiple input variables, i.e., antecedents. For each of the outputs, i.e., consequents, he defined a set of rules, consisting of membership functions for all the antecedents. The consequent fuzzy set arises as the s-norm (e.g., maximum) of the results for consequents, and the results for a consequent is the t-norm (e.g., minimum) of the rules belonging to that consequent, as it can be seen in Fig. 2.

The rules are generated from measured data, either using statistics or some intelligent learning algorithm. This means that there must be some measurements, where not only the antecedents, but also the consequents are known, moreover, it is useful to have such data for testing the inference system. In our examples very simple rules are used: the membership function of each antecedent for each consequent is a triangle, with the minimum and maximum of the measured data forming the support of the membership function and the mean forming the core, the peak of the triangle, as seen in Fig. 3.

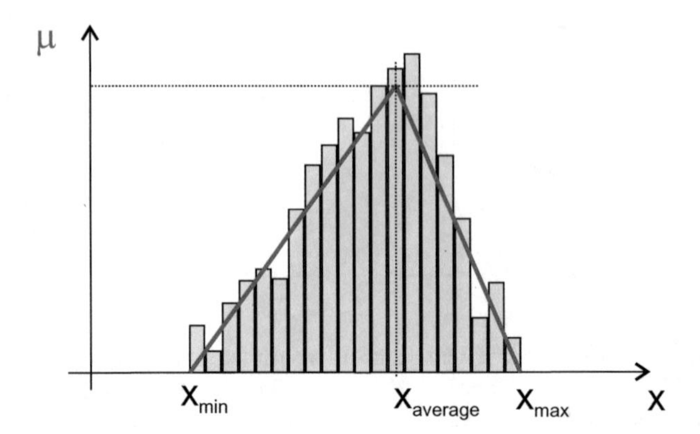

Fig. 3. A simple rule generation method from the measured values of the training set. The statistics of the measured data is represented by the histogram of the plot, while the resulting fuzzy rule by the triangle-shaped membership function. Arbitrary units.

2.3 Fuzzy Rule Interpolation

In measurements it often happens that the rulebase does not completely cover the space of the possible measured data, i.e., sparse rulebases are generated. In this case, making the other measured data evaluable can be carried out by rule interpolation. Stabilized KH interpolation [5–7]

$$\inf\{B^*_\alpha\} = \frac{\sum_{i=1}^{2n}\left(\frac{1}{d_{\alpha L}(A^*,A_i)}\right)^k \inf\{B_{i\alpha}\}}{\sum_{i=1}^{2n}\left(\frac{1}{d_{\alpha L}(A^*,A_i)}\right)^k} \tag{1}$$

$$\sup\{B^*_\alpha\} = \frac{\sum_{i=1}^{2n}\left(\frac{1}{d_{\alpha U}(A^*,A_i)}\right)^k \sup\{B_{i\alpha}\}}{\sum_{i=1}^{2n}\left(\frac{1}{d_{\alpha U}(A^*,A_i)}\right)^k}, \tag{2}$$

give very good results in our examples, too.

3 The Wavelet Component of Our Approach

Wavelet analysis [8] developed from several branches of signal processing and numerical mathematics in the late 1970s-early 1980s. Up till now its main use is image compression, from fingerprint databases through the Mars rover [10] to the JPEG2000 image coding standard [9]. It is also possible to suppress noise, enhance edges, or retrieve special types of patterns from images by wavelet transform, moreover, similarly to Fourier transform [12], wavelets are used to simplifying differential equations, too [11].

3.1 Multiresolution Analysis

The discrete wavelet transform is mathematically defined by a so called multiresolution analysis of the function space, mostly the square integrable functions' Hilbert space. It consists of subspaces embedded into each other, each subspace belonging to a resolution level, hence the name. The finest resolution level subspace is dense in the original function space (i.e., to any function of the original function space to any limit there can bee found a function in the infinitely fine resolution subspace that is closer than the limit). The lowest (infinitely low) resolution level consists of only constant functions.

The most interesting part of this approach of the function space is that each subspace is expanded by a set of basis functions, which have the same shape, just shifted over a regular grid. The shape of the basis functions change from subspace to subspace, i.e., from resolution level to resolution level only by shrinking or dilation: the fine resolution level subspaces have higher and narrower basis function distributed over a grid of smaller grid distance, while the rougher resolution levels have lower, wider basis functions over grids with larger grid distance, as it can be seen from the following definitions of the basis functions at resolution level j and shift position k

$$\phi_{jk}(x) = 2^{j/2}\phi(2^j x - k). \tag{3}$$

These basis functions ϕ_{jk} of the embedded subspaces are called scaling functions.

Wavelets are also similar basis functions: they provide the way between two resolution levels. The spaces completing a rougher resolution level subspace to the next, finer resolution subspace are the detail spaces, and their basis functions are the wavelets, defined as

$$\psi_{jk}(x) = 2^{j/2}\psi(2^j x - k). \tag{4}$$

This subspace setup means, that any function of any resolution level j can be expressed either as a linear combination of its resolution level subspace, or using any rougher resolution level scaling function subspace as a basis, and adding wavelets to it as a refinement, i.e., as

$$f^{[j]}(x) = \sum_{-\infty}^{\infty} c_{jk}\phi_{jk}(x), \tag{5}$$

or by decreasing the basic resolution level by 1 as

$$f^{[j]}(x) = \sum_{-\infty}^{\infty} c_{j-1\,k}\phi_{j-1\,k}(x) + \sum_{-\infty}^{\infty} d_{j-1\,k}\psi_{j-1\,k}(x), \tag{6}$$

or by decreasing the rougher resolution level to zero, as

$$f^{[j]}(x) = \sum_{-\infty}^{\infty} c_{0k}\phi_{0k}(x) + \sum_{i=0}^{j-1}\sum_{-\infty}^{\infty} d_{ik}\phi_{ik}(x), \tag{7}$$

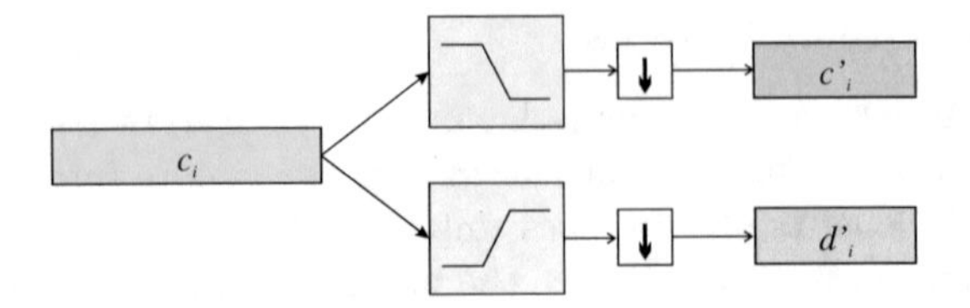

Fig. 4. One step of the wavelet transform as two branches of convolutional filter signal processing and downsampling steps. The low-pass filter results in the scaling function expansion coefficients c'_i, while the high-pass filters give the fine details, i.e., the wavelet expansion coefficients d'_i.

Practically, measurement results can be treated as a very fine resolution level coefficient set of the sampled function, their wavelet transform results in the rougher resolution level scaling function and wavelet coefficients by using the so called refinement equation

$$\phi(x) = 2^{1/2} \sum_{i=0}^{N_s} h_i \phi(2x - i), \tag{8}$$

and its wavelet counterpart

$$\psi(x) = 2^{1/2} \sum_{i=-N_s+1}^{1} (-1)^i h_{-i+1} \phi(2x - i). \tag{9}$$

For the simplest wavelet family, the so called Haar wavelets [13], the coefficients $h_0 = h_1$, these mother basis functions have a support length 1 unit; the other wavelets have more coefficients, thus longer support.

3.2 Wavelet Analysis in Signal Processing

This mathematical definition can be translated to signal processing devices: practically the usage of (8) and (9) can be translated directly to digital signal processors as convolutional filtering and downsampling, as it can be seen in Fig. 4. The coefficients in the convolutional filters are proportional to the coefficients at the refinement equations.

In multiple dimension data sets either multiple dimension wavelets, or more often separate wavelet analysis steps in the separate dimensions has to be carried out.

4 Entropies as Compact Descriptions of Two-Dimensional Datasets

The entropy in information theory was introduced by Shannon [14] as the expectation value of the information for a complete set of events, i.e., for such sets, where all the events have probabilities between 0 and 1, and the sum of all the probabilities is 1. If the probabilities are $\{p_1, p_2, \ldots, p_N\}$, then the entropy can be written as

$$S = -\sum_{i=1}^{N} p_i \log_2 p_i. \tag{10}$$

4.1 Rényi Entropies

Shannon's entropy definition was generalised for many purposes, Rényi's [15] series of entropies, i.e.,

$$S_\alpha = \frac{1}{1-\alpha} \log \sum_{i=1}^{N} p_i^\alpha, \tag{11}$$

gives the Shannon entropy as a limit at $\alpha = 1$.

For $\alpha = 0$ this entropy is the entropy of the uniform distribution. This is sometimes called Hartley entropy, as Hartley has introduced the concept of information and its expectation value using such set, where the probabilities were equal. This approximation is still used if nothing is known about the probability distributions, only the number of the possible outcomes.

For $\alpha = 2$ the formula turns into

$$S_2 = -\log \sum_{i=1}^{N} p_i^2. \tag{12}$$

4.2 Structural Entropies of Probability Distributions

In the beginning of the 1990s Pipek and Varga [16,17] found out, that the difference of Rényi entropies can characterize the structure of the probability distribution $\{p_1, p_2, \ldots, p_N\}$ in a very peculiar way. They introduced structural entropy as the difference of two Rényi entropies

$$S_{str} = S_1 - S_2, \tag{13}$$

and similarly, the they proved that the so called filling factor q, which was used in solid state physics and quantum mechanics, is also related to a Rényi entropy difference the following way

$$\log(q) = S_0 - S_2. \tag{14}$$

Later these quantities were applied in characterisation of the localisation of pixel intensities in scanning microscopy images [18,19]. Bonyár developed a localization factor for describing the roughness of gold electrodes based on these entropy differences [20,21].

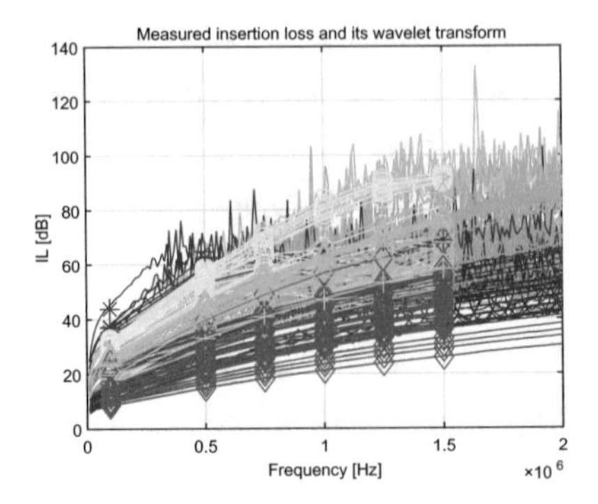

Fig. 5. Measured insertion loss as a function of frequency for the lines of the 5 performance groups. The darker shades of colours mean the measured lines properties, while the brighter counterparts denote the wavelet transforms.

5 Telecommunication Lines

5.1 Measurements

In telecommunication line performance prediction the goal is to develop a method that approximates the real-life performance of the line sufficiently well, without actually building the connection, as it is costly and time consuming. For SHDSL lines, which are mainly for business use, errors in the prediction lead to financial loss to the telecommunications provider. Lilik measured over 170 lines [22], and developed a fuzzy method for performance prediction [23,24].

During the measurements the SHDSL links were built, and their performances were measured. Also many physical parameters were determined using general measuring devices available at telecommunications service providers. Sorting out the noise, the return loss and many other parameters, it was proved that based on solely the insertion loss values of an area's telecommunications lines their performance can be evaluated with rather high reliability. The measured insertion loss values over the 0 to 2 MHz frequency ban can be seen in Fig. 5.

5.2 Characterisation Scheme

In order to be able to build a classification method, the measured data was separated to a training set and a test set. In the original characterisation scheme simple, triangular rules were built from the measured data of the training set according to Fig. 3. As the insertion loss values have rather large fluctuations around a quite smooth trend, five characteristic frequencies were chosen, and the insertion loss values at those frequencies served as antecedents. Selecting fewer frequency points makes the calculations unstable, more points make it unnecessarily complicated. Still, there were a lot of lines that were not evaluable, and there were very few cases, when the fluctuations were so bad, that the evaluation was not successful, i.e., instead of sorting the line into its real performance group, or one group below it (which is still acceptable for the provider), it sorted it to better performing group or predicted much worse data rate than the real one.

For overcoming these problems, as a first step, we introduced fuzzy rule interpolation to out classification scheme, thus making practically all the lines evaluable.

Instead of selecting five representative samples from the complete insertion loss-frequency function, we also applied wavelet analysis to filter out the large-scale trends from the function. The first wavelet transform we use can also be seen in Fig. 5. It can be seen, that the distribution of the transformed points is not equidistant: in the lower frequency domain we included one step finer resolution level results than in the higher frequencies, because the communication's spectral power density is much larger at lower frequencies.

65 lines were used for testing, in the case of the characteristic frequencies, 12 lines were put to one class lower than their real group, while in the case of the wavelet transformed antecedent selection 8 lines went to the acceptable group and the remaining 57 ones to their true classes. In both cases all the lines were classified well, moreover, if the wavelet transform was carried out until 2 or 4 points remained, the classification was still as correct as the 5-point version [24].

With these results we demonstrated that wavelets can be used for stabilising calculations, if the measured data fluctuates and lowering the antecedent dimension as well.

6 Wear

If the number of the antecedent dimensions is already too low, like in the case of classification based on the S_{str} and $\ln q$ values calculated for an image, wavelets can provide 4 more pictures to be analysed, as it can be seen in Figs. 6 and 7. In 2D data sets the wavelet analysis is carried out in both dimensions. This results in 4 output pictures of approximately quarter of the size of the original data set (half in each direction), one output for the case of using low-pass filters in both dimensions, which is a kind of average of the original image, two outputs where one of the directions have low-pass, the other high-pass filter, and one output where both filters are high-pass. In the followings the latest picture will be called diagonal, the first averaging, and the two between will be mentioned as vertical and horizontal results, depending on which direction has the fine details, i.e., which direction used high-pass filter.

6.1 Measurements

In [25] Solecki and Dreyer measured a combustion engine using silicone replica and surface scanners. Silicone replicas are often used in geometrical measurements as the shape of many instruments does not allow to access certain interesting points of an object. In the case of a combustion engine the inner surface

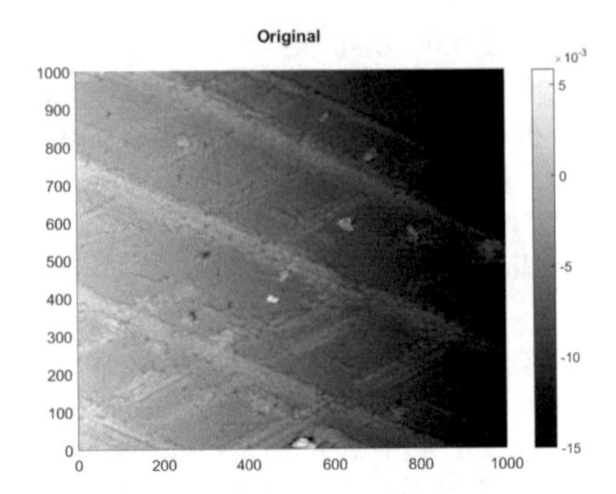

Fig. 6. A measured surface segment before wavelet analysis.

of the cylinders can be accessed only by specially designed devices that are not available in general laboratories, but they are developed exquisitely for one type of automated measurement in the industry. These highly specialized tools are expensive thus if hardly accessible surfaces are to be measured, either the object has to be cut, or replicas are to be taken from the surface. If the object under test is needed for further tests, clearly only the second option is possible.

The measurements of the 4 cylinders of the engine under test were carried out using Struers RepliSet F5 which is able to reproduce patterns of size down to 0.1 microns. Replicas were taken of the new engine before building it and after 500 hours of polycyclic endurance test (later the engine was cut so that the worn surface could be studied directly, as well). The resulting samples were measured by a TalysurfCLI2000 white-light surface scanner at 5 points for each of the cylinders. These points were selected so that one point would be between the topmost and the second ring's turning point, another point would be between the next two piston rings's turning points, and three along the path where all 3 piston rings had worn the surface. An example of the 1 mm by 1 mm surface parts of a new and worn engine can be seen in Fig. 8. Slight vertical scratches can be seen on the worn surface. Such a vertical scratch can be seen in the previous worn image of Fig. 6, too, and slightly visible in the vertical transformed picture of Fig. 7.

6.2 Characterisation Scheme

The classification scheme was very similar to the one in the previous section. The antecedents consisted of solely the structural entropy and the logarithm of the filling factor, i.e., the two Rényi entropy differences. The $S_{str}(\ln q)$ plot of 128 measured surface sub-domains are plotted in Fig. 9. It can be seen, that though the points corresponding to worn and new surfaces occupy overlapping domains, there are clearly such parts of the plot which belong to only one type of surface.

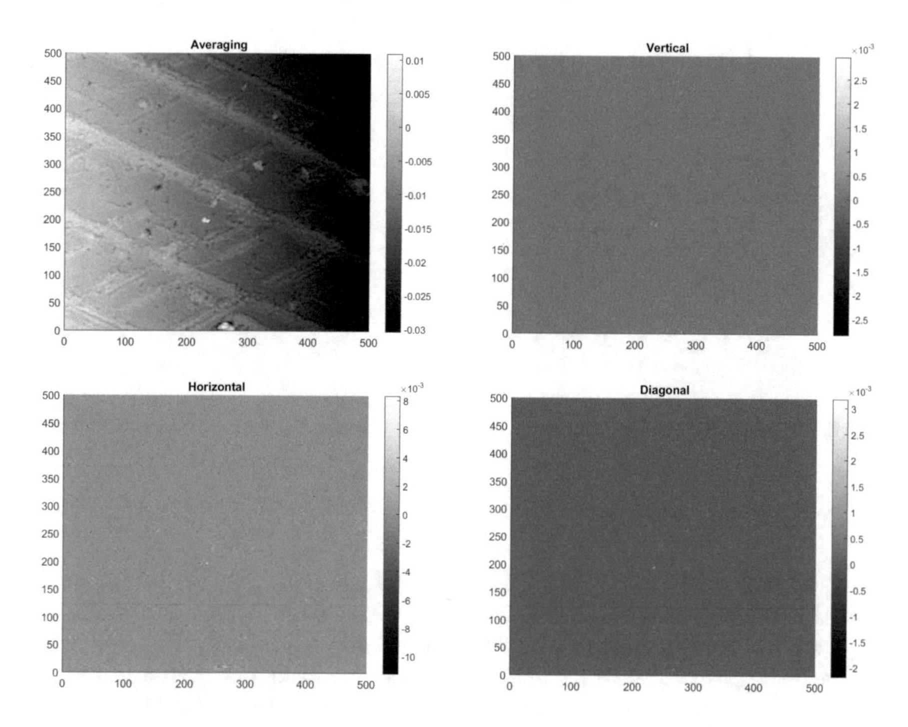

Fig. 7. A measured surface segment after wavelet analysis.

However, these two characterising quantities are not sufficient for building fuzzy classification scheme: from the 128 surface subdomains 64 were used for building the rulebase, and of the 64 test data, only 33 could be classified correctly, which is worse than a random guess.

In the case of two-dimensional data, such as the above surface scans, wavelet transformation has to be carried out in both dimensions, thus resulting in 4 output data matrices: one for the transformation, where both directions had low-pass filters, one for the high-pass-high-pass case, and two mixed filter pairs. The structural entropy and the filling factor can be calculated for all 4 of the resulting matrices, thus the antecedent dimension can be increased from 2 to up to 10. We tested [26] the method with all 4 wavelet transformed surface types as well as with only the low-pass–low-pass and high-pass–high-pass matrices, and the results were not different from each other. The number of incorrectly classified surface elements went down to 13, which indicates, that the structural entropies are not suitable for being antecedents without other quantities. The second wavelet transform usually does not improve the results in this example.

However we could demonstrate that wavelet analysis is able to introduce independent information to the overly simplified antecedents.

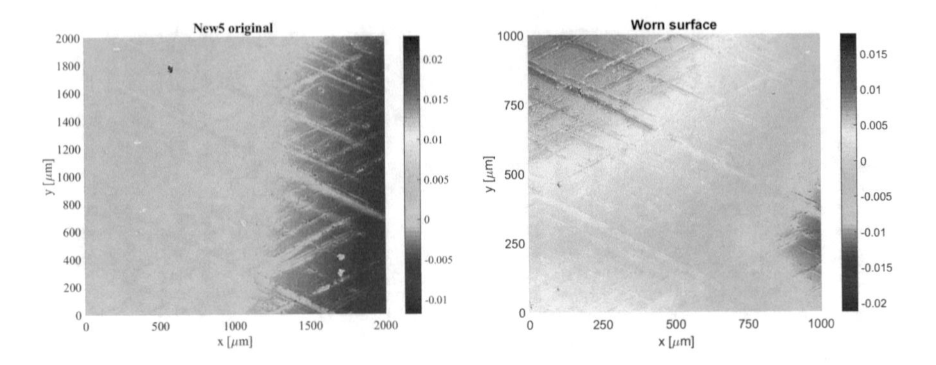

Fig. 8. A measured surface segment before and after the polycyclic endurance run.

7 Colorectal Polyps

Colorectal polyps are wart-like objects inside the last parts of the bowel system. Some of these polyps can develop into colorectal cancer, which is a really dangerous type of cancer, as it can be detected usually quite lately. If these polyps, that have the possibility to develop into cancer could be detected and removed early, then they would not develop into malign objects, thus detecting and classifying colorectal polyps is a really important task. Having a visual aid for the medical experts based on automatic image processing can help the diagnosis. There are several groups trying to find polyps on colonoscopy images, some of them even have their own database built. In the following considerations, we apply our method founded in [27] on the pictures of [28]. An example can be seen in Fig. 10, while the wavelet transform of the picture is given in Fig. 11.

The classification scheme consists of the following steps. First, the images are cut into tiles of size $N \times N$, where N is generally between tenth and fifth of the original image size, in our case 200 compared to the image size of magnitude 1000. Next, using the masks provided by the database, for each tile the polyp content, i.e., the percentage of the area with masked pixels is calculated: based on this value the tiles are classified as "with polyp" and "without polyp". For each of the tiles, for all 3 colour channels the antecedents were calculated. The antecedents are the mean, standard deviation, edge density, structural entropy and $\ln q$ and the gradients. The edge density is calculated the following way: the tile is transformed to a black and white edge image by Canny filtering, then the rate of the edges (white points) compared to the tile-size.

Next, every second image is used for determining the fuzzy rules according to Fig. 3. As out previous experiences show, that for different types of images the classification success rates are different, we sorted the pictures into groups of the same patient of the same take, and generated rules from each of the groups. The thus arising rules are applied for classification using the same method as in the case of the cylinder surfaces, only the antecedent dimension increased to 21, or 99 in the case of using wavelet analysed pictures, too.

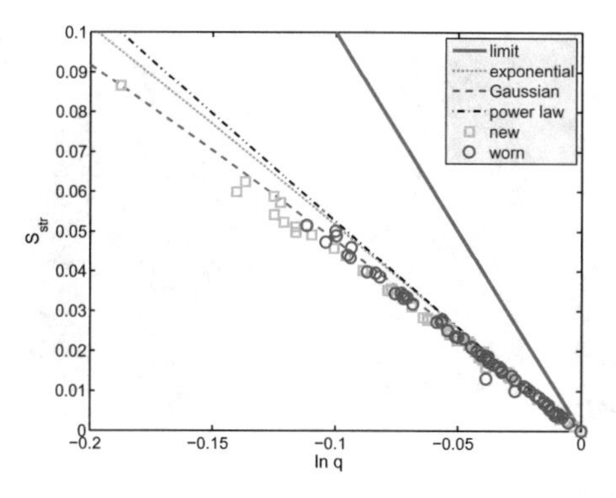

Fig. 9. Structural entropy map of 64 new and 64 worn surface segments. This is a typical way of plotting structural entropy, and determining the localization type of the studied distribution. The teal thick solid line denotes the theoretical limiting curve, above which no points should appear. The dashed, dotted and dash-dotted lines shows three typical distributions, i.e., if a point is near the line corresponding to the 2nd order power law distribution (dash-dotted), then the average localization type in the distribution is similar to the second order power law function. The green squares give the points generated from the new surfaces's scans, while the red circles the worn ones.

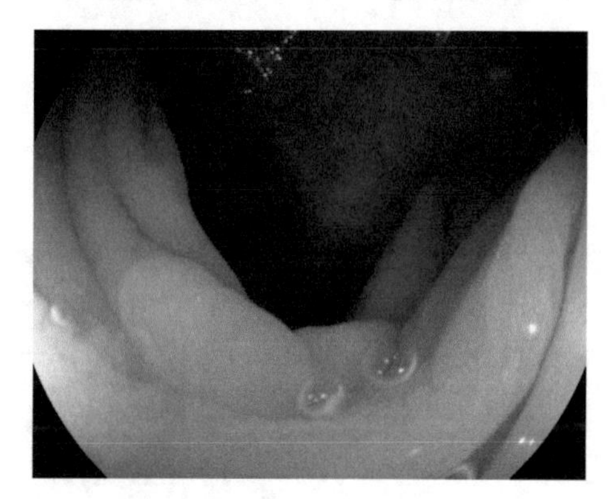

Fig. 10. A colonoscopy picture of [28] turned into grayscale image, before wavelet analysis.

The results for both cases can be seen as ROC plots in Fig. 12. The Classical ROC plots are not that much visible due to the large number of points, however, if a 3rd axis, i.e., the image group number is also given, we can conclude the followings. The false positive rate is rather low in all cases, especially in the case of using wavelet analysed images, too. The true positive rate is for some pictures extremely low, so this method without wavelet analysis is not usable, however, wavelets improve the results up to a more acceptable level in all cases.

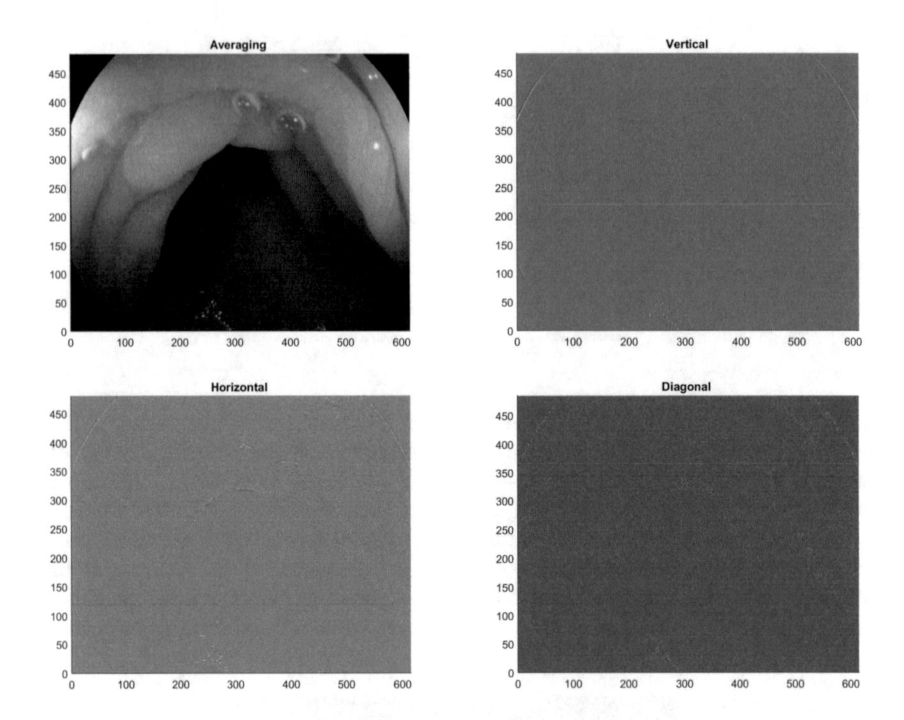

Fig. 11. Colonoscopy picture after wavelet analysis. Note that the upper left corner of the picture in Fig. 10, i.e., the pixel of index (1, 1) moved to the lower left part of the coordinate system (the picture is upside down).

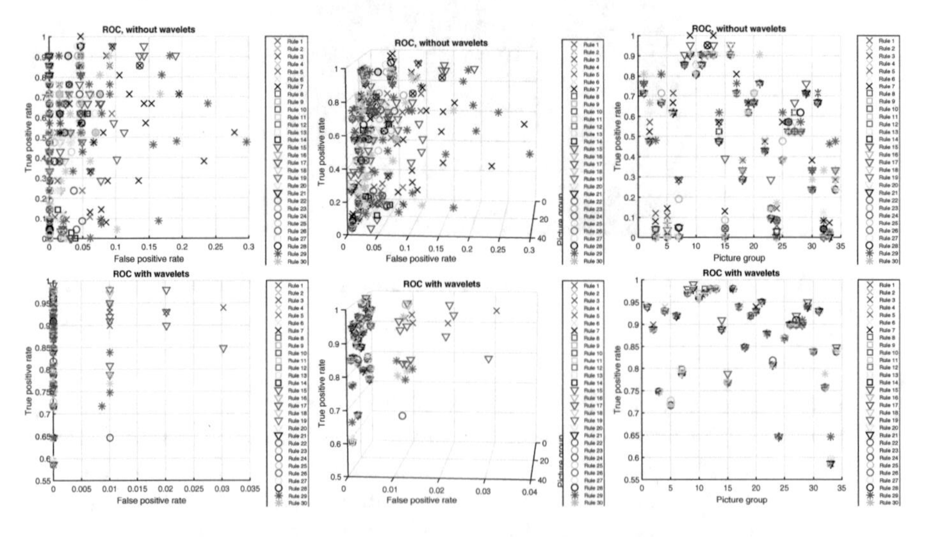

Fig. 12. True positive vs false positive rate for the various picture groups for the various rulebases. First plot is the ROC, the second is a 3D view, and 3rd plot focuses n true positive rate. First row: without wavelet analysis, 21 antecedents, second row: with wavelet analysis, 99 antecedents.

8 Conclusion

In this articles the usage of wavelet transform in fuzzy antecedent selection was studied. Two completely different strategies were mentioned. First, the simplification of the decision and decrease of the number of antecedents by using wavelet transform instead of samples form a measured data vector, which scheme's effectiveness was demonstrated on insertion loss-based performance prediction of telecommunication lines. Second, the introduction of new, independent information by using wavelet transformed data beside the original one for classification schemes with overly simplifying antecedent selection such as selecting structural entropies. Combustion engine cylinder surface scan classification was used as a demonstrating example, where the performance of the classification could be improved significantly by introducing two of the wavelet transforms of the surface matrix. The other example was colonoscopy picture segment classification, where the improvement due to wavelet analysis was more visible, and to almost all the image types the classification error rate became acceptable.

Acknowledgment. The authors would like to thank the financial support of the projects GINOP-2.3.4-15-2016-00003 and the ÚNKP-18-4 New National Excellence Programme of the Ministry of Human Capacities of Hungary. This work was supported by National Research, Development and Innovation Office (NKFIH) K124055. The authors would like to thank to EFOP-3.6.1-16-2016-00017 1 "˜Internationalisation, initiatives to establish a new source of researchers and graduates, and development of knowledge and technological transfer as instruments of intelligent specialisations at Széchenyi István University" for the support of the research.

References

1. Lilik, F., Solecki, L., Sziová, B., Kóczy, L.T., Nagy, Sz.: On wavelet based enhancing possibilities of fuzzy classification of measurement results. In: Kulczycki, P., Kowalski, P.A., Łukasik, S. (eds.) Contemporary Computational Science, AGH-UST Press, Cracow, p. 138 (2018)
2. Zadeh, L.A.: Fuzzy sets. Inf. Control **8**, 338–353 (1965). https://doi.org/10.1016/S0019-9958(65)90241-X
3. Mamdani, E.H., Assilian, S.: An experiment in linguistic synthesis with a fuzzy logic controller. Int. J. Man-Mach. Stud. **7**, 1–13 (1975). https://doi.org/10.1016/S0020-7373(75)80002-2
4. Zadeh, L.A.: Outline of a new approach to the analysis of complex systems and decision processes. IEEE Trans. Syst. Man and Cybern. **SMC–3**, 28–44 (1973). https://doi.org/10.1109/TSMC.1973.5408575
5. Kóczy, L.T., Hirota, K.: Approximate reasoning by linear rule interpolation and general approximation. Int. J. Approx. Reason. **9**, 197–225 (1993). https://doi.org/10.1016/0888-613X(93)90010-B
6. Kóczy, L.T., Hirota, K.: Interpolative reasoning with insufficient evidence in sparse fuzzy rule bases. Inf. Sci. **71**, 169–201 (1993). https://doi.org/10.1016/0020-0255(93)90070-3

7. Tikk, D., Joó, I., Kóczy, L.T., Várlaki, P., Moser, B., Gedeon, T.D.: Stability of interpolative fuzzy KH-controllers. Fuzzy Sets Syst. **125**, 105–119 (2002). https://doi.org/10.1016/S0165-0114(00)00104-4
8. Daubechies, I.: Ten Lectures on Wavelets, CBMS-NSF Regional Conference Series in Applied Mathematics 61. SIAM, Philadelphia (1992)
9. Christopoulos, C., Skodras, A., Ebrahimi, T.: The JPEG2000 still image coding system: an overview. IEEE Trans. Consum. Electron. **46**, 1103–1127 (2000)
10. Kiely, A., Klimesh, M.: The ICER Progressive Wavelet Image Compressor, IPN Progress Report 42-155, 15 November 2003. http://ipnpr.jpl.nasa.gov/tmo/progressreport/42-155/155J.pdf
11. Nagy, S., Pipek, J.: An economic prediction of the finer resolution level wavelet coefficients in electronic structure calculations. Phys. Chem. Chem. Phys. **17**, 31558–31565 (2015)
12. Fourier, J-B.J.: Theorie Analitique de la Chaleur, Firmin Didot, Paris (1822)
13. Haar, A.: Zur Theorie der orthogonalen Funktionensysteme (On the theory of orthogonal function systems, in German). Math. Ann. **69**, 331–371 (1910)
14. Shannon, C.E.: A mathematical theory of communication. Bell Syst. Techn. J. **27**, 379–423 (1948). https://doi.org/10.1002/j.1538-7305.1948.tb01338.x
15. Rényi, A.: On measures of information and entropy. In: Proceedings of the fourth berkeley symposium on mathematics, statistics and probability 1960, pp. 547–561 (1961)
16. Pipek, J., Varga, I.: Universal classification scheme for the spatial localization properties of one-particle states in finite d-dimensional systems. Phys. Rev. A **46**, 3148–3164 (1992). APS, Ridge NY-Washington DC
17. Varga, I., Pipek, J.: Rényi entropies characterizing the shape and the extension of the phase space representation of quantum wave functions in disordered systems. Phys. Rev. E **68**, 026202 (2003). APS, Ridge NY-Washington DC
18. Mojzes, I., Dominkonics, Cs., Harsányi, G., Nagy, Sz., Pipek, J., Dobos, L.: Heat treatment parameters effecting the fractal dimensions of AuGe metallization on GaAs. Appl. Phys. Lett. **91**(7) (2007). Article No. 073107
19. Molnár, L.M., Nagy, S., Mojzes, I.: Structural Entropy in Detecting Background Patterns of AFM Images, Vacuum, vol. 84, pp. 179–183. Elsevier, Amsterdam (2010)
20. Bonyár, A., Molnár, L.M., Harsányi, G.: Localization factor: a new parameter for the quantitative characterization of surface structure with atomic force microscopy (AFM). MICRON **43**, 305–310 (2012). Elsevier, Amsterdam
21. Bonyár, A.: AFM characterization of the shape of surface structures with localization factor. Micron **87**, 1–9 (2016)
22. Lilik, F., Botzheim, J.: Fuzzy based prequalification methods for EoSHDSL technology. Acta Technica Jaurinensis **4**(1), 135–144 (2011)
23. Lilik, F., Nagy, Sz., Kóczy, L.T.: Wavelet based fuzzy rule bases in pre-qualification of access networks' wire pairs. In: IEEE Africon 2015, Addis Ababa, Ethiopia, 14–17 September 2015, Paper P-52 (2015)
24. Lilik, F., Nagy, S., Kóczy, L.T.: Improved method for predicting the performance of the physical links in telecommunications access networks. Complexity **2018**, 1–14 (2018). ID 3685927
25. Dreyer, M.R., Solecki, L.: Verschleissuntersuchungen an Zylinderlaufbahnen von Verbrennungsmotoren. In: 3. Symposium Produktionstechnik – Innovativ und Interdisziplinär, Zwickau, 6–7 April 2011, pp. 69–74 (2011)

26. Nagy, Sz., Solecki, L.: Wavelet analysis and structural entropy based intelligent classification method for combustion engine cylinder surfaces. In: Proceedings of the 8th European Symposium on Computational Intelligence and Mathematics, ESCIM, Sofia, 5–8 October 2016, pp. 115–120 (2016)
27. Nagy, Sz., Lilik, F., Kóczy, L.T.: Entropy based fuzzy classification and detection aid for colorectal polyps. In: IEEE Africon 2017, Cape Town, South Africa, 15–17 September 2017 (2017)
28. Silva, J.S., Histace, A., Romain, O., Dray, X., Granado, B.: Towards embedded detection of polyps in WCE images for early diagnosis of colorectal cancer. Int. J. Comput. Assist. Radiol. Surg. **9**, 283–293 (2014)
29. Georgieva, V.M., Draganov, I.: Multistage approach for simple kidney cysts segmantation in CT images. In: Kountchev, R., Nakamatsu, K. (eds.) Intelligent Systems Reference Library, New Approaches in Intelligent Image Analysis, vol. 108, pp. 223–252. Springer Nature (2016)
30. Georgieva, V.M., Vassilev, S.G.: Kidney segmentation in ultrasound images via active contours. In: 11th International Conference on Communications, Electromagnetics and Medical Applications, Athens, Greece, 13–15 October 2016, pp. 48–53 (2016)

On the Convergence of Fuzzy Grey Cognitive Maps

István Á. Harmati[1](✉) and László T. Kóczy[2,3]

[1] Department of Mathematics and Computational Sciences, Széchenyi István University, Egyetem tér 1, Győr 9026, Hungary
harmati@sze.hu

[2] Department of Information Technology, Széchenyi István University, Egyetem tér 1, Győr 9026, Hungary
koczy@sze.hu

[3] Department of Telecommunication and Media Informatics, Budapest University of Technology and Economics, Magyar tudósok körútja 2, Budapest 1117, Hungary

Abstract. Fuzzy grey cognitive maps (FGCMs) are extensions of fuzzy cognitive maps (FCMs), applying uncertain weights between the concepts. This uncertainty is expressed by so-called grey numbers. Similarly to FCMs, the inference is determined by an iteration process, which may converge to an equilibrium point, but limit cycles or chaotic behaviour may also turn up.

In this paper, based on the grey weighted connections between the concepts and the parameter of the sigmoid threshold function, we give sufficient conditions for the existence and uniqueness of fixed points for sigmoid FGCMs.

Keywords: Fuzzy cognitive map · Grey system theory · Fuzzy grey cognitive map · Fixed point

1 Introduction

Many decision-making problems are too complex to be solved by classical methods, especially when several uncertain or imprecise factors present [2]. Numerous successful techniques are based on cognitive or fuzzy models [3], which are extremely useful when a high number of interrelated factors should be considered by the decision maker and these factors form a complex system [4]. Fuzzy Cognitive Map (FCM) is a soft computing technique, which can effectively represent causal expert knowledge and uncertain information of complex systems [5] by using direct causal representation, moreover, the quick simulation of complex models [6] is also possible.

Fuzzy cognitive maps use directed graphs in which constant weights are assigned to the edges from the interval $[-1, 1]$ to express the strength and direction of causal connections. The nodes represent specific factors of the modelled

P. Kulczycki et al. (Eds.): ITSRCP 2018, AISC 945, pp. 74–84, 2020.
https://doi.org/10.1007/978-3-030-18058-4_6

system and are usually called 'concepts' in FCM theory. The current states of the concepts are also characterized by numbers in the $[0, 1]$ interval (in some applications the interval $[-1, 1]$ is also applicable [7]). These are the so-called 'activation values'.

The system can be formally defined by a 4-tuple (C, W, A, f) where $C = C_1, C_2, \ldots, C_n$ is the set of n concepts, $W : (C_i, C_j) \rightarrow w_{ij} \in [-1; +1]$ is a function which associates a causal value (weight) w_{ij} to each edge connecting the nodes (C_i, C_j), describing how strongly influenced is concept C_i by concept C_j. The sign of w_{ij} indicates whether the relationship between C_j and C_i is direct or inverse. So the connection or weight matrix $W_{n\times n}$ gathers the system causality. The function $A : (C_i) \rightarrow A_i$ assigns an activation value $A_i \in \mathbb{R}$ to each node C_i at each time step t $(t = 1, 2, \ldots, T)$ during the simulation. A transformation or threshold function $f : \mathbb{R} \rightarrow [0, 1]$ calculates the activation value of concepts and keeps them in the allowed range (sometimes a function $f : \mathbb{R} \rightarrow [-1, 1]$ is applied). The iteration which calculates the values of the concept may or may not include self-feedback. In general form it can be written as

$$A_i(k) = f\left(\sum_{j=1, j\neq i}^{n} w_{ij}A_j(k-1) + d_iA_i(k-1)\right) \tag{1}$$

where $A_i(k)$ is the value of concept C_i at discrete time k, w_{ij} is the weight of the connection from concept C_j to concept C_i and d_i expresses the possible self-feedback. If we include the self-feedback in the weight matrix W, the equation can be rewritten in a simpler form:

$$A(k) = f\left(WA(k-1)\right) \tag{2}$$

Continuous FCM may behave chaotically, can produce limit cycles or reach a fixed point attractor [7]. Chaotic behaviour means that the activation vector never stabilizes. If a limit cycle occurs, a specific number of consecutive state vectors turn up repeatedly. In case of a fixed point attractor, the state vector stabilizes after a certain number of iterations [8,9]

The behaviour of the iteration depends on the threshold function applied and its parameter(s), on the elements (weights) of the extended weight matrix and on the topology of the map.

2 Mathematical Background

The existence and uniqueness of fixed point of sigmoid fuzzy cognitive maps was firstly discussed by Boutalis, Kottas and Christodoulou in [10] for the case when the parameter of the log-sigmoid threshold function is $\lambda = 1$, (so the function was $f(x) = 1/(1 + e^{-x})$. The possible number of fixed points was analysed by Knight, Lloyd and Penn in [11]. The results of [10] were generalized in [12].

2.1 The Contraction Mapping Theorem

The mathematical investigation presented in Sect. 3 is based on the so-called contraction mapping theorem. First we recall the notion of contraction:

Definition 1. *Let (X, d) be a metric space. A mapping $f: X \to X$ is a* contraction mapping *or* contraction *if there exists a constant c (independent from x and y), with $0 \leq c < 1$, such that*

$$d(f(x), f(y)) \leq cd(x, y) \tag{3}$$

Let $f: X \to X$, then a point $x^* \in X$ such that $f(x^*) = x^*$ is a fixed point of f. The following theorem provides sufficient condition for the existence and uniqueness of a fixed point.

Theorem 1 (Banach's fixed point theorem). *If $f: X \to X$ is a contraction mapping on a nonempty complete metric space (X, d), then f has only one fixed point x^*. Moreover, x^* can be found as follows: start with an arbitrary $x_0 \in X$ and define the sequence $x_n = f(x_{n-1})$, then $\lim_{n\to\infty} x_n = x^*$.*

The following statement and its corollary play a crucial role in the proofs: *The derivative of the sigmoid function $f: \mathbb{R} \to \mathbb{R}$, $f(x) = 1/(1 + e^{-\lambda x})$, $(\lambda > 0)$ is bounded by $\lambda/4$.*

Lagrange's mean value theorem states that if a function f is continuous on the closed interval $[a, b]$, and differentiable on the open interval (a, b), then there exists a point $c \in (a, b)$ such that $f(b) - f(a) = f'(c)(b - a)$. It implies that for a log-sigmoid function $f: |f(x) - f(y)| \leq \lambda/4 \cdot |x - y|$.

In our previous work the following theorem was proved regarding the existence and uniqueness of fixed points of FCMs (see [12]). We will show in Sect. 3, that for fuzzy grey cognitive maps a very similar result can be derived. Moreover, the theorem below is a special case of the result presented in Sect. 3.

Theorem 2. *Let W be the extended (including possible feedback) weight matrix of a FCM, let $\lambda > 0$ be the parameter of the log-sigmoid function. If the inequality*

$$\|W\|_F < \frac{4}{\lambda} \tag{4}$$

holds, then the FCM has one and only one fixed point. Here $\|\cdot\|_F$ stands for the Frobenius norm of the matrix, $\|W\|_F = \left(\sum_i \sum_j w_{ij}^2\right)^{1/2}$.

We should note here that this is a sufficient, but not necessary condition. The fact that $\|W\|_F < 4/\lambda$ implies that there is one and only one fixed point, while if $\|W\|_F \geq 4/\lambda$, we do not know whether there are more than one fixed points or limit cycles.

2.2 Fuzzy Grey Cognitive Maps

Fuzzy grey cognitive maps (FGCMs) are very effective problem-solving techniques within environments with high uncertainty and imprecision, combining the findings of Grey System Theory (GST) and fuzzy cognitive maps [13,14]. It has been designed to analyze small data samples with poor information [15,16]. The main difference between fuzzy and grey systems concepts arises in the intension and extension of the modelled or analyzed object. While grey system theory focuses on objects with clear extension and unclear intension, fuzzy theory in most of the cases deals with objects with clear intension and unclear extension.

FGCM based models unavoidably consist of numerical operations on so-called grey numbers. A grey number (usually denoted by $\otimes G$) is a number whose accurate value is unknown, but it is known the range within the value is included. A grey number with both a lower limit ($\underline{G}$) and an upper limit ($\overline{G}$) is called an interval grey number [17], so $\otimes G \in [\underline{G}, \overline{G}]$. In applications, a grey number is usually an interval. Basic operations on grey numbers:

1. $\otimes G_1 + \otimes G_2 \in [\underline{G_1} + \underline{G_2}, \overline{G_1} + \overline{G_2}]$
2. $- \otimes G \in [-\overline{G}, -\underline{G}]$
3. $\otimes G_1 - \otimes G_2 \in [\underline{G_1} - \overline{G_2}, \overline{G_1} - \underline{G_2}]$
4. $\otimes G_1 \times \otimes G_2 \in [\min(S), \max(S)]$
   where $S = \{\underline{G_1} \cdot \underline{G_2}, \underline{G_1} \cdot \overline{G_2}, \overline{G_1} \cdot \underline{G_2}, \overline{G_1} \cdot \overline{G_2}\}$
5. If $\lambda > 0$, $\lambda \in \mathbb{R}$, then $\lambda \cdot \otimes G \in [\lambda \underline{G}, \lambda \overline{G}]$

The following statement is straightforward and plays an important role in our further investigations: if $f\colon \mathbb{R} \to \mathbb{R}$ is strictly monotone increasing function, then $f(\otimes G) \in [f(\underline{G}), f(\overline{G})]$.

An FGCM models unstructured knowledge by causalities through grey relationships (grey numbers) between them based on fuzzy cognitive maps. FGCMs are a generalization of FCMs, since an FGCM with all the relations' intensities represented by exact numbers (in the grey system theory they called white numbers) would be a usual FCM. In general, FGCM represents the human intelligence better than FCM, because it expresses unclear relations between factors and models incomplete information better than FCM.

The dynamics of an FGCM begins with the initial grey vector state $A(0)$, which represents initial uncertainty. The elements of this vector are grey numbers, i.e. $A_i(0) \in [\underline{A_i(0)}, \overline{A_i(0)}]$ for every i. The updated nodes' states are computed by an iterative process with an activation function, resulting grey numbers as concept values:

$$A_i(k) \in \left[f(\underline{w_i A(k-1)}), f(\overline{w_i A(k-1)})\right] \tag{5}$$

After a certain number of iterations, an FGCM with continuous threshold function arrives at one of the following cases:

1. It settles down to a so-called grey fixed-point attractor.
2. The state could keep cycling between several states, known as a limit grey cycle.

3. The FGCM continues to produce different grey vector states for each iteration, this is the grey chaotic attractor.

3 Convergence of Fuzzy Grey Cognitive Maps

Easy to see that a fuzzy cognitive map with grey weights (FGCM) has grey concept values. We assume that the human expert or the training process assigns the proper signs to the weights, so a weight is either positive or negative (more exactly nonnegative or nonpositive). This means that the type of relationship (direct or inverse) between the concepts is properly described by the fuzzy cognitive map, so in most of the cases, it seems to be a right assumption.

Let $\otimes w_{ij}$ be a weight describing the connection between concepts C_j and C_i. Due to our assumptions, $\otimes w_{ij}$ is a grey number and it is element of a subset of the interval $[-1, 0]$ or the interval $[0, 1]$, so

$$\otimes w_{ij} \in [\underline{w_{ij}}, \overline{w_{ij}}] \subset [-1, 0] \quad \text{or} \quad \otimes w_{ij} \in [\underline{w_{ij}}, \overline{w_{ij}}] \subset [0, 1] \tag{6}$$

Let us introduce the following notation:

$$w_{ij}^* = \begin{cases} |\underline{w_{ij}}| \text{ if } \otimes w_{ij} \leq 0 \\ \overline{w_{ij}} \text{ if } \otimes w_{ij} \geq 0 \end{cases} \tag{7}$$

I.e. w_{ij}^* is the absolute value of the most extreme value of $\otimes w_{ij}$ that is possible.

The following theorem provides sufficient condition for the existence and uniqueness of fixed point of FGCMs. Here fixed point $\otimes A^*$ is

$$\otimes A^* = [\otimes A_1^*, \ldots, \otimes A_n^*]^T \in \left[[\underline{A_1^*}, \overline{A_1^*}], \ldots, [\underline{A_n^*}, \overline{A_n^*}]\right]^T$$

The grey fixed point is unique in the sense that the endpoints of the intervals containing grey concept values are unique, i.e. the values $\underline{A_i^*}$ and $\overline{A_i^*}$ are unique for every i.

Theorem 3. *Let $\otimes W$ be the extended (including possible feedback) weight matrix of a fuzzy grey cognitive map (FGCM), where the weights $\otimes w_{ij}$ are nonnegative or nonpositive grey numbers and let w_{ij}^* be defined as in Eq. 7. Moreover, let $\lambda > 0$ be the parameter of the sigmoid function $f(x) = 1/(1 + e^{-\lambda x})$ applied for the iteration. If the inequality*

$$\left(\sum_{i=1}^{n} \sum_{j=1}^{n} {w_{ij}^*}^2 \right)^{1/2} < \frac{4}{\lambda} \tag{8}$$

holds, then the FGCM has one and only one grey fixed point, regardless of the initial concept values.

Actually, the left handside term is the Frobenius norm of a matrix which entries are the w_{ij}^* values.

Proof. Let $\otimes A$ be the vector of concept values: $\otimes A = [\otimes A_1, \ldots, \otimes A_n]^T$, and let G be a function for the iteration process:

$$\otimes G(A) = G(\otimes A) = [G(\otimes A)_1, \ldots, G(\otimes A)_n]^T = [\otimes G(A)_1, \ldots, \otimes G(A)_n]^T \quad (9)$$

Here the elements of the vector are grey numbers, i.e.

$$\otimes G(A)_i \in \left[f(\underline{w_i A}), f(\overline{w_i A})\right] \quad (10)$$

where w_i is the ith row of matrix $\otimes W$. We are going to show that for a suitable distance metric d and under certain conditions the inequality

$$d(\otimes G(A), \otimes G(A')) \leq c \cdot d(\otimes A, \otimes A') \quad (11)$$

holds with $c < 1$, so mapping G is a contraction, so it has one and only one fixed point. For metric d we choose the following one:

$$d(\otimes G(A), \otimes G(A')) = \left[\sum_{i=1}^{n} d^2(\otimes G(A)_i, \otimes G(A')_i)\right]^{1/2} \quad (12)$$

where

$$d^2(\otimes G(A)_i, \otimes G(A')_i) = \frac{\left(\underline{G(A)_i} - \underline{G(A')_i}\right)^2 + \left(\overline{G(A)_i} - \overline{G(A')_i}\right)^2}{2} \quad (13)$$

Note that if $\underline{G(A)_i} = \overline{G(A)_i}$ for A and A' and for all of the coordinates, then it becomes the ordinary Euclidean metric.

Moreover, the following inequalities hold:

$$\left(\underline{G(A)_i} - \underline{G(A')_i}\right)^2 = \left(f(\underline{w_i A}) - f(\underline{w_i A'})\right)^2 \leq \left(\frac{\lambda}{4}\right)^2 \left(\underline{w_i A} - \underline{w_i A'}\right)^2 \quad (14)$$

$$\left(\overline{G(A)_i} - \overline{G(A')_i}\right)^2 = \left(f(\overline{w_i A}) - f(\overline{w_i A'})\right)^2 \leq \left(\frac{\lambda}{4}\right)^2 \left(\overline{w_i A} - \overline{w_i A'}\right)^2 \quad (15)$$

The main problem is that the equations

$$\underline{w_i A} = \sum_{j=1}^{n} \underline{w_{ij}} \cdot \underline{A_j} \quad (16)$$

$$\overline{w_i A} = \sum_{j=1}^{n} \overline{w_{ij}} \cdot \overline{A_j} \quad (17)$$

hold only in the case when all of weights are nonnegative. In general

$$\otimes w_{ij} A_j \in \begin{cases} \left[\underline{w_{ij}} \cdot \underline{A_j},\ \overline{w_{ij}} \cdot \overline{A_j}\right] & \text{if } \otimes w_{ij} \geq 0 \\ \left[\underline{w_{ij}} \cdot \overline{A_j},\ \overline{w_{ij}} \cdot \underline{A_j}\right] & \text{if } \otimes w_{ij} \leq 0 \end{cases} \quad (18)$$

In the case when all of the weights are nonnegative we get the following upper estimation:

$$\left(f(\underline{w_i A}) - f(\underline{w_i A'})\right)^2 \leq \left(\frac{\lambda}{4}\right)^2 \left(\underline{w_i A} - \underline{w_i A'}\right)^2 \tag{19}$$

$$= \left(\frac{\lambda}{4}\right)^2 \left(\sum_{j=1}^{n} \underline{w_{ij}} \cdot \left(\underline{A_j} - \underline{A'_j}\right)\right)^2 \tag{20}$$

$$\leq \left(\frac{\lambda}{4}\right)^2 \left(\sum_{j=1}^{n} \underline{w_{ij}}^2\right) \cdot \left(\sum_{j=1}^{n} \left(\underline{A_j} - \underline{A'_j}\right)^2\right) \tag{21}$$

where the last row comes from applying the well-known Cauchy-Schwarz inequality. Similar inequality is true for the upper endpoints:

$$\left(f(\overline{w_i A}) - f(\overline{w_i A'})\right)^2 \leq \left(\frac{\lambda}{4}\right)^2 \left(\sum_{j=1}^{n} \overline{w_{ij}}^2\right) \cdot \left(\sum_{j=1}^{n} \left(\overline{A_j} - \overline{A'_j}\right)^2\right) \tag{22}$$

Applying the definition of w_{ij}^*, further upper estimations can be given:

$$\left(f(\underline{w_i A}) - f(\underline{w_i A'})\right)^2 \leq \left(\frac{\lambda}{4}\right)^2 \left(\sum_{j=1}^{n} {w_{ij}^*}^2\right) \cdot \left(\sum_{j=1}^{n} \left(\underline{A_j} - \underline{A'_j}\right)^2\right) \tag{23}$$

$$\left(f(\overline{w_i A}) - f(\overline{w_i A'})\right)^2 \leq \left(\frac{\lambda}{4}\right)^2 \left(\sum_{j=1}^{n} {w_{ij}^*}^2\right) \cdot \left(\sum_{j=1}^{n} \left(\overline{A_j} - \overline{A'_j}\right)^2\right) \tag{24}$$

Now we are ready to give an upper estimation for the distance of $\otimes G(A)$ and $\otimes G(A')$:

$$d^2(\otimes G(A), \otimes G(A')) = \sum_{i=1}^{n} d^2(\otimes G(A)_i, \otimes G(A')_i) \tag{25}$$

$$= \sum_{i=1}^{n} \frac{\left(\underline{G(A)_i} - \underline{G(A')_i}\right)^2 + \left(\overline{G(A)_i} - \overline{G(A')_i}\right)^2}{2} \tag{26}$$

$$\leq \sum_{i=1}^{n} \frac{1}{2} \left(\frac{\lambda}{4}\right)^2 \left(\sum_{j=1}^{n} {w_{ij}^*}^2\right) \left[\left(\sum_{j=1}^{n} \left(\underline{A_j} - \underline{A'_j}\right)^2\right) + \left(\sum_{j=1}^{n} \left(\overline{A_j} - \overline{A'_j}\right)^2\right)\right] \tag{27}$$

$$= \left(\frac{\lambda}{4}\right)^2 \left(\sum_{i=1}^{n}\sum_{j=1}^{n} {w_{ij}^*}^2\right) \cdot \frac{1}{2} \left[\left(\sum_{j=1}^{n} \left(\underline{A_j} - \underline{A'_j}\right)^2\right) + \left(\sum_{j=1}^{n} \left(\overline{A_j} - \overline{A'_j}\right)^2\right)\right] \tag{28}$$

$$= \left(\frac{\lambda}{4}\right)^2 \left(\sum_{i=1}^{n}\sum_{j=1}^{n} {w_{ij}^*}^2\right) \cdot d^2(\otimes A, \otimes A') \tag{29}$$

By taking the squareroot of each side we get that

$$d(\otimes G(A), \otimes G(A')) \leq \frac{\lambda}{4} \left[\sum_{i=1}^{n} \sum_{j=1}^{n} {w_{ij}^{*}}^{2} \right]^{1/2} \cdot d(\otimes A, \otimes A') \tag{30}$$

By Banach's fixed point theorem, if $\frac{\lambda}{4} \left(\sum_{i=1}^{n} \sum_{j=1}^{n} {w_{ij}^{*}}^{2} \right)^{1/2} < 1$, then mapping G is a contraction, so it has one and only one fixed point, which was the statement in Theorem 3.

Let's turn to the case when all of the weights are nonpositive. The argument is similar to the previous one, with a small modification:

$$\left(f(\underline{w_i A}) - f(\underline{w_i A'}) \right)^2 \leq \left(\frac{\lambda}{4} \right)^2 \left(\underline{w_i A} - \underline{w_i A'} \right)^2 \tag{31}$$

$$= \left(\frac{\lambda}{4} \right)^2 \left(\sum_{j=1}^{n} \underline{w_{ij}} \cdot \left(\overline{A_j} - \overline{A_j'} \right) \right)^2 \tag{32}$$

$$\leq \left(\frac{\lambda}{4} \right)^2 \left(\sum_{j=1}^{n} \underline{w_{ij}}^2 \right) \cdot \left(\sum_{j=1}^{n} \left(\overline{A_j} - \overline{A_j'} \right)^2 \right) \tag{33}$$

Similarly:

$$\left(f(\overline{w_i A}) - f(\overline{w_i A'}) \right)^2 \leq \left(\frac{\lambda}{4} \right)^2 \left(\sum_{j=1}^{n} \overline{w_{ij}}^2 \right) \cdot \left(\sum_{j=1}^{n} \left(\underline{A_j} - \underline{A_j'} \right)^2 \right) \tag{34}$$

Moreover, applying again the definition of w_{ij}^{*} we can state that:

$$\left(f(\underline{w_i A}) - f(\underline{w_i A'}) \right)^2 \leq \left(\frac{\lambda}{4} \right)^2 \left(\sum_{j=1}^{n} {w_{ij}^{*}}^{2} \right) \cdot \left(\sum_{j=1}^{n} \left(\overline{A_j} - \overline{A_j'} \right)^2 \right) \tag{35}$$

$$\left(f(\overline{w_i A}) - f(\overline{w_i A'}) \right)^2 \leq \left(\frac{\lambda}{4} \right)^2 \left(\sum_{j=1}^{n} {w_{ij}^{*}}^{2} \right) \cdot \left(\sum_{j=1}^{n} \left(\underline{A_j} - \underline{A_j'} \right)^2 \right) \tag{36}$$

From this point the proof goes on the same way as in the previous case and we get the same inequality:

$$d(\otimes G(A), \otimes G(A')) \leq \frac{\lambda}{4} \cdot \left[\sum_{i=1}^{n} \sum_{j=1}^{n} {w_{ij}^{*}}^{2} \right]^{1/2} \cdot d(\otimes A, \otimes A') \tag{37}$$

In the general case, when positive and negative weights also occur, in the upper bound of $\left(f(\underline{w_i A}) - f(\underline{w_i A'}) \right)^2$ may occur terms like $\underline{w_{ij}} \cdot \left(\underline{A_j} - \underline{A_j'} \right)$ and terms like $\underline{w_{ij}} \cdot \left(\overline{A_j} - \overline{A'_j}\right)$. The following facts come trivially from the two cases we have discussed already:

– If $\underline{w_{ij}} \cdot \left(\underline{A_j} - \underline{A'_j}\right)$ occurs in the upper bound of $\left(f(\underline{w_i A}) - f(\underline{w_i A'})\right)^2$, then $\overline{w_{ij}} \cdot \left(\overline{A_j} - \overline{A'_j}\right)$ occurs in the upper bound of $\left(f(\overline{w_i A} - f(\overline{w_i A'})\right)^2$.
– If $\underline{w_{ij}} \cdot \left(\overline{A_j} - \overline{A'_j}\right)$ occurs in the upper bound of $\left(f(\underline{w_i A}) - f(\underline{w_i A'})\right)^2$, then $\overline{w_{ij}} \cdot \left(\underline{A_j} - \underline{A'_j}\right)$ occurs in the upper bound of $\left(f(\overline{w_i A}) - f(\overline{w_i A'})\right)^2$.

The following inequalities come from the definition of w^*_{ij}:

$$\left|\underline{w_{ij}} \cdot \left(\underline{A_j} - \underline{A'_j}\right)\right| \leq w^*_{ij} \cdot \left|\underline{A_j} - \underline{A'_j}\right| \tag{38}$$

$$\left|\overline{w_{ij}} \cdot \left(\overline{A_j} - \overline{A'_j}\right)\right| \leq w^*_{ij} \cdot \left|\overline{A_j} - \overline{A'_j}\right| \tag{39}$$

$$\left|\underline{w_{ij}} \cdot \left(\overline{A_j} - \overline{A'_j}\right)\right| \leq w^*_{ij} \cdot \left|\overline{A_j} - \overline{A'_j}\right| \tag{40}$$

$$\left|\overline{w_{ij}} \cdot \left(\underline{A_j} - \underline{A'_j}\right)\right| \leq w^*_{ij} \cdot \left|\underline{A_j} - \underline{A'_j}\right| \tag{41}$$

Applying these upper estimations and rearranging the terms in the the upper estimation of $d^2(\otimes G(A), \otimes G(A'))$ by $\left(\underline{A_j} - \underline{A'_j}\right)^2$ and $\left(\overline{A_j} - \overline{A'_j}\right)^2$ we get that

$$d^2(\otimes G(A), \otimes G(A')) \leq \left(\frac{\lambda}{4}\right)^2 \sum_{i=1}^{n} \sum_{j=1}^{n} {w^*_{ij}}^2 \cdot \sum_{j=1}^{n} \frac{\left(\underline{A_j} - \underline{A'_j}\right)^2 + \left(\overline{A_j} - \overline{A'_j}\right)^2}{2} \tag{42}$$

$$= \left(\frac{\lambda}{4}\right)^2 \left(\sum_{i=1}^{n} \sum_{j=1}^{n} {w^*_{ij}}^2\right) \cdot d^2(\otimes A, \otimes A') \tag{43}$$

Taking the square root of both sides we get the same inequality as in the previous cases:

$$d(\otimes G(A), \otimes G(A')) \leq \frac{\lambda}{4} \left[\sum_{i=1}^{n} \sum_{j=1}^{n} {w^*_{ij}}^2\right]^{1/2} \cdot d(\otimes A, \otimes A') \tag{44}$$

If $\frac{\lambda}{4} \left[\sum_{i=1}^{n} \sum_{j=1}^{n} {w^*_{ij}}^2\right]^{1/2} < 1$, then mapping G is a contraction, so according to Banach's fixed point theorem, it has one and only one fixed point, which completes the proof for the general case. □

Note that if $\underline{w_{ij}} = \overline{w_{ij}}$ for all the weights (in grey system theory it means that w_{ij} is not grey, but white number, a number without uncertainty), then this condition is the same as in Theorem 2.

4 Summary

Fuzzy grey cognitive maps are extensions of fuzzy cognitive maps that are able to handle uncertainties in the weight matrix. These uncertainties can express the imprecision or incomplete information of human experts. Similarly to standard FCMs, FGCMs can reach an equilibrium point (fixed point), produce limit cycles or chaotic behaviour. In this paper, we proved a theorem that under conditions expressed by the elements of the weight matrix and the parameter of the applied sigmoid threshold function, the FGCM always reaches an equilibrium point, regardless of the initial concept values.

Acknowledgment. The primary version of this paper was presented at the 3rd Conference on Information Technology, Systems Research and Computational Physics, 2–5 July 2018, Cracow, Poland [1].

This research was supported by National Research, Development and Innovation Office (NKFIH) K124055.

References

1. Harmati, I.Á., Kóczy, L.T: On the convergence of fuzzy grey cognitive maps. In: Kulczycki, P., Kowalski, P.A., Łukasik, S. (eds.) Contemporary Computational Science, p. 139. AGH-UST Press, Cracow (2018)
2. Carlsson, C., Fullér, R.: Possibility for decision: a possibilistic approach to real life decisions. In: Studies in Fuzziness and Soft Computing Series, vol. 270/2011. Springer (2011)
3. Papageorgiou, E.I., Salmeron, J.L.: Methods and algorithms for fuzzy cognitive map-based decision support. In: Papageorgiou, E.I. (ed.) Fuzzy Cognitive Maps for Applied Sciences and Engineering (2013)
4. Busemeyer, J.R.: Dynamic decision making. In: International Encyclopedia of the Social & Behavioral Sciencesm, pp. 3903–3908 (2001)
5. Felix, G., Nápoles, G., Falcon, R., Froelich, W., Vanhoof, K., Bello, R.: A review on methods and software for fuzzy cognitive maps. Artif. Intell. Rev., 1–31 (2017)
6. Stylios, C.D., Groumpos, P.P.: Modeling complex systems using fuzzy cognitive maps. IEEE Trans. Syst. Man Cybern. Part A Syst. Hum. **34**(1), 155–162 (2004)
7. Tsadiras, A.K.: Comparing the inference capabilities of binary, trivalent and sigmoid fuzzy cognitive maps. Inf. Sci. **178**(20), 3880–3894 (2008)
8. Nápoles, G., Papageorgiou, E., Bello, R., Vanhoof, K.: Learning and convergence of fuzzy cognitive maps used in pattern recognition. Neural Process. Lett. **45**(2), 431–444 (2017)
9. Nápoles, G., Papageorgiou, E., Bello, R., Vanhoof, K.: On the convergence of sigmoid fuzzy cognitive maps. Inf. Sci. **349–350**, 154–171 (2016)
10. Boutalis, Y., Kottas, T.L., Christodoulou, M.: Adaptive estimation of fuzzy cognitive maps with proven stability and parameter convergence. IEEE Trans. Fuzzy Syst. **17**(4), 874–889 (2009)
11. Knight, C.J., Lloyd, D.J., Penn, A.S.: Linear and sigmoidal fuzzy cognitive maps: an analysis of fixed points. Appl. Soft Comput. **15**, 193–202 (2014)

12. Harmati, I.A., Hatwágner, F.M., Kóczy, L.T.: On the existence and uniqueness of fixed points of fuzzy cognitive maps. In: Medinam, J., et al. (eds.) Information Processing and Management of Uncertainty in Knowledge-Based Systems. Theory and Foundations, IPMU 2018. Communications in Computer and Information Science, vol. 853, pp. 490–500. Springer, Cham (2018)
13. Salmeron, J.L.: Modelling grey uncertainty with fuzzy grey cognitive maps. Expert. Syst. Appl. **37**(12), 7581–7588 (2010)
14. Papageorgiou, E.I., Salmeron, J.L.: Learning fuzzy grey cognitive maps using non-linear hebbian-based approach. Int. J. Approx. Reason. **53**(1), 54–65 (2012)
15. Salmeron, J.L., Papageorgiou, E.I.: A fuzzy grey cognitive maps-based decision support system for radiotherapy treatment planning. Knowl. Based Syst. **30**, 151–160 (2012)
16. Salmeron, J.L., Gutierrez, E.: Fuzzy grey cognitive maps in reliability engineering. Appl. Soft Comput. **12**(12), 3818–3824 (2012)
17. Liu, S., Lin, Y.: Grey Information. Springer, London (2006)

Hierarchical Fuzzy Decision Support Methodology for Packaging System Design

Kata Vörösk̋oi(✉), Gergő Fogarasi, Adrienn Buruzs, Péter Földesi, and László T. Kóczy

Széchenyi István University, Egyetem 1, Győr 9028, Hungary
voroskoi.kata@sze.hu

Abstract. In the field of logistics packaging (industrial-, or even customer packaging), companies have to take decisions on determining the optimal packaging solutions and expenses. The decisions often involve a choice between one-way (disposable) and reusable (returnable) packaging solutions. Even nowadays, in most cases the decisions are made based on traditions and mainly consider the material and investment costs. Although cost is an important factor, it might not be sufficient for finding the optimal solution. Traditional (two-valued) logic is not suitable for modelling this problem, so here the application of a fuzzy approach, because of the metrical aspects, a fuzzy signature approach is considered. In this paper a fuzzy signature modelling the packaging decision is suggested, based on logistics expert opinions, in order to support the decision making process of choosing the right packaging system. Two real life examples are also given, one in the field of customer packaging and one in industrial packaging.

Keywords: Fuzzy signature · One-way packaging · Returnable packaging

1 Introduction

Packaging is a significant element in any logistics system [12]. Without proper packaging handling and transportation would be difficult and expensive along the supply chain (SC). Although cost is an important factor, it is not enough to consider only the costs of material and investment while choosing the right packaging system, many other aspects should be considered.

It has been found that paying limited attention to packaging can cause higher costs in the physical distribution. Furthermore, researchers argue that packaging should not only be considered from the point of view of cost, but focus should be put on its role as a value-added function in the SC [5].

The best packaging solutions are those that, beside the optimal cost levels maximize the use of packaging space so that all the products can easily be packed and stacked, and at the same time reduce packaging waste [9].

As a matter of course, the environmental aspects are also part of these important processes, including the reduction of waste during production [11]. Furthermore, improving the efficiency of packaging is an important strategic goal for the

P. Kulczycki et al. (Eds.): ITSRCP 2018, AISC 945, pp. 85–96, 2020.
https://doi.org/10.1007/978-3-030-18058-4_7

organizations considering the aspects of sustainability and economy [7]. Legislation has also forced companies to rethink their packaging operations [6].

The functions of packaging in general can be classified as follows [4]:

- Product and environment protection (physical, safety, natural deterioration, waste reduction)
- Logistics containment and handling (unit, bulk, pallet, containers)
- Information (symbol, logo, description)

Packaging systems can have different levels: primary, secondary and tertiary packaging (Fig. 1).

Primary packaging is the main package that holds the product that is being processed [8]. The aim of **secondary packaging** (it is also called transport or distribution packaging) is to preserve the product on its way from the point of manufacture to the customer. It includes the shipping container, the internal protective packaging and any utilizing materials for shipping. It does not include packaging for consumer products (primary packaging) [16]. **Tertiary packaging** combines all of the secondary packages for example into one pallet [8].

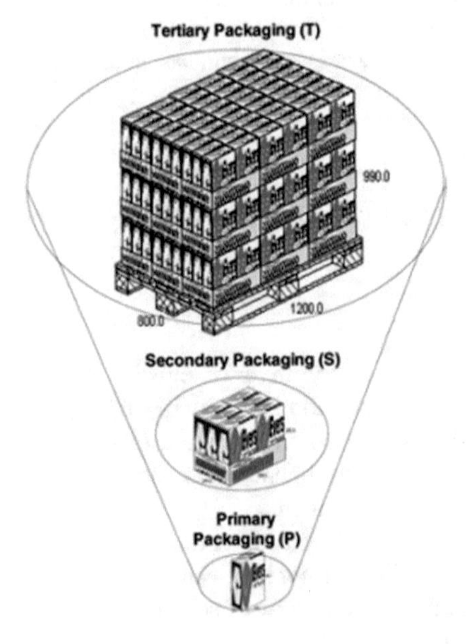

Fig. 1. Packaging system levels [8]

The functions of **transport packaging** are:

- Containment (basic purpose, supplying use value to products)
- Protection (ensuring integrity and safety of the contents and occasionally also protecting the environment from the product)
- Performance (transportation, handling, storing, selling and use of the product)
- Communication (identification of the contents and informing about package features and requirements) [16].

The most important actors of an industrial **packaging supply chain** (PSC) are suppliers, assembly factories and packaging collectors (if returnable packaging is used) [13]. Packaging producers are also important, but choosing the right packaging for the product belongs to the competency of the factories, their suppliers, or both together.

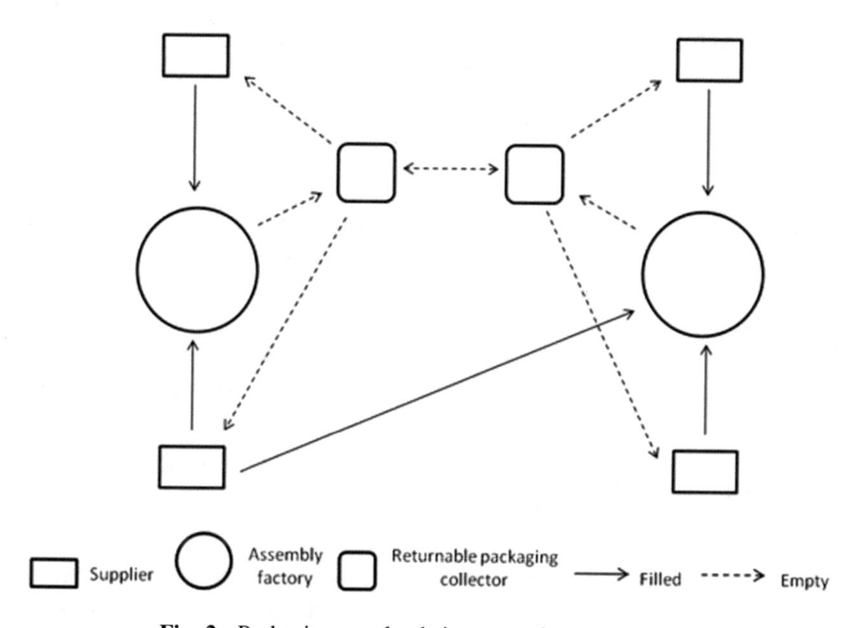

Fig. 2. Packaging supply chain – open loop system [14]

In the field of logistic packaging (industrial transportation, or even consumer packaging) the companies take decisions in order to determine the optimal packaging expenses. This decision-making situation practically means a choice between one-way (disposable) and reusable (returnable) packaging systems [1, 2]. The former is only suitable for one use as far as reusable containers and packaging are loaded with products and shipped to the destination, then the empty container is sent back to the supplier, refilled with products and this cycle is repeated over and over again as a closed-loop system. In case of an open-loop system reusable packaging is collected at a centralized return handling center, where it is cleaned, stocked, and distributed for refilling [16]. In the second case the packaging is not necessarily returning back to the initial partner who filled it (Fig. 2).

The main problem with one-way packaging is the waste created after the usage while relative production cost is lower. On the other hand, transportation and maintaining cost is a relevant issue in case of returnable packaging [2, 10]. Returnable packaging has been frequently used for example in the US automotive industry in order to reduce waste, costs, transport damages and for enabling JIT deliveries [15].

Managing returnable packaging systems requires more than just inverse transportation. The cleaning and maintenance of containers, as well as the storage and the administration are also involved in the process [3].

In most cases distance decides if the packaging comes back, but it also depends on the complexity of the supply chain. Transport modes also can play a noticeable role (for example road, rail or maritime transport).

2 The Fuzzy Signature Model

In this section the structure of the fuzzy signature modelling the packaging problem on hand will be proposed including the tree graph and the aggregation operations in the intermediate nodes.

In the FSig in the intermediate nodes the use of weighted arithmetic mean operations is proposed.

Based on expert knowledge three main aspects were defined when a decision has to be made about one-way or returnable packaging, thus the weight in the respective aggregation will be as follows:

- characteristics of the product which has to be packaged $(a_{1;}\ w = 3)$
- characteristics of the supply chain $(a_2;\ w_2 = 9)$ and
- external factors $(a_{3;}\ w_3 = 2)$

All leaves of the tree assume their values (μ_i) from the interval [0,1]. The values belonging to the intermediate nodes are calculated by respective functions specified to each leaf according to the logistics meaning and role. The relations among the individual descendants on the same level are determined with respective aggregations (see in Sect. 2.1). The final purpose of the model is to support the decision whether a disposable (one-way) or returnable packaging system should be used. When the final value created by the aggregation in the root (a_0) is close to 0, it should rather be one-way, if the result of a_0 is close to 1, the packaging should rather be returnable (Fig. 3).

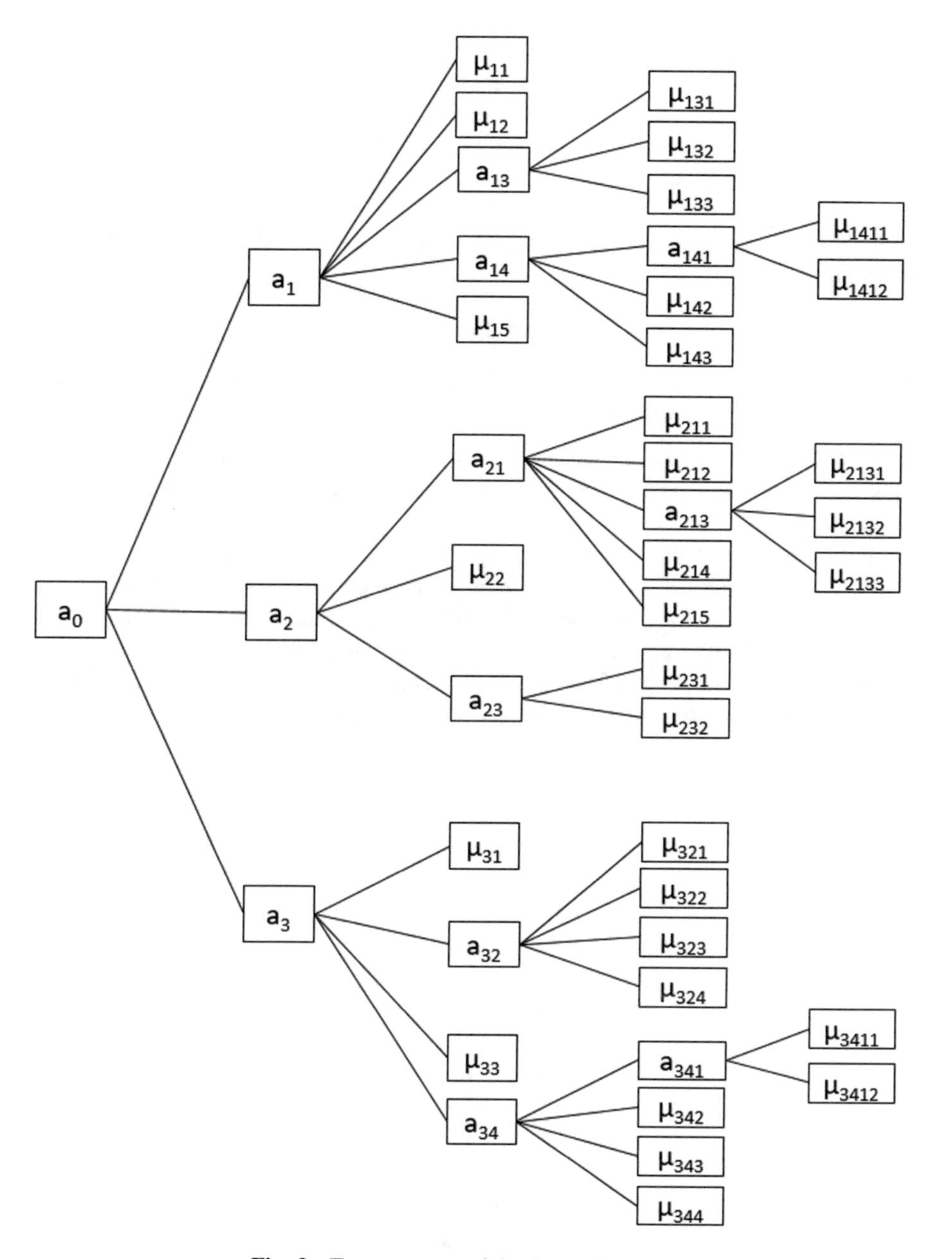

Fig. 3. Tree structure of the fuzzy signature

2.1 Definition of the Aggregation Operators

Based on the opinion of a panel of logistics experts is was decided that all aggregations are of the weighted arithmetic mean type, because the components of the individual characteristics and features are comparably importance, which may be expressed by weights of the same order of magnitude (given in Tables 1, 2 and 3).

It must be mentioned however that certain special goods with special characteristics (such as dangerous goods) may necessitate the usage of aggregations of different type, especially the minimum aggregation or some rather drastic aggregations, because of the either necessary relation of the packaging materials or very high cost of cleaning removing all remains of the material dangerous for health or the environment. The examples based on real life goods packaging technologies used by real companies have confirmed the values (Fig. 6) and the type of aggregations used in this rather complex fuzzy signature.

2.2 Defining the Main Aspects and the Weights

In the following the parent and child nodes are described in groups: weights (w_i) with aggregations of the intermediate nodes are listed in Tables 1, 2 and 3.

Product Characteristics (a_1)

This attribute represents the technical aspects of designing the right packaging for a particular product.

Production batch size and turnover are two strong aspects, but also geometrical characteristics like shape, size and weight of the product should be considered. Furthermore, physical, biological, chemical sensitivity and value of the goods also play an important role. Within physical sensitivity the mechanical and climate effects must be differentiated. The ranking and weights in the fuzzy aggregation can be seen in Table 1. ID represents the position of the aspect (as a leaf or aggregation) in the fuzzy signature.

Table 1. Ranking and aggregation weights of product characteristics and its sub-trees

1 product characteristics			
ID	*Features*	*Ranking*	*Weights*
11	batch size	1-2.	9
12	turnover	1-2.	9
13	geometrical	3.	7
14	sensitivity	4-5.	2
15	value	4-5.	2
1.3 geometrical characteristics			
ID	*Features*	*Ranking*	*Weights*
131	shape	1.	4
132	size	2-3.	1
133	weight	2-3.	1
1.4 sensitivity			
ID	*Features*	*Ranking*	*Weights*
141	physical	1.	6
142	biological	2.	3
143	chemical	3.	2
1.4.1 physical			
ID	*Features*	*Ranking*	*Weights*
1411	mechanical	1.	6
1412	climate	2.	4

Supply Chain (logistics) Characteristics (a_2)

This attribute represents the logistics aspects of designing the right packaging for the given product. Transportation represents the main part of logistics activities in the supply chain (SC), but material handling and IT support are also essential for an effectively working SC. As it was mentioned earlier, transportation distance is highly important, as well as the volume of goods delivered at the same time. From the packaging point of view environmental circumstances during transportation (temperature, vibration and humidity) evidently influencing the decision. Quality of the infrastructure and the transport modes used (modality) should also not be left out of consideration. Material handling means all operations related to handling of the goods in the supply chain, including warehousing, unloading, uploading and transshipments. The organizational level of truck loads (full truck load – FTL, less than truck load – LTL) is also significant when deciding about returnable packaging because the complexity of the task is growing with the number of participants in the process. The ranking and aggregation weights in the fuzzy sub model can be seen in Table 2.

Table 2. Ranking and aggregation weights of supply chain characteristics and its sub-trees

2 supply chain characteristics			
ID	*Features*	*Ranking*	*Weights*
21	transportation	1.	8
22	IT support	2.	7
23	material handling	3.	2
2.1 transportation			
ID	*Features*	*Ranking*	*Weights*
211	distance	1.	8
212	volume	2.	7
213	impacts	3.	5
214	infrastructure	4.	3
215	modality	5.	1
2.3 material handling			
ID	*Features*	*Ranking*	*Weights*
231	transshipment	1.	8
232	FTL/LTL	2.	2
2.1.3 environmental impacts			
ID	*Features*	*Ranking*	*Weights*
2131	temperature	1.	4
2132	vibration	2.	3
2133	humidity	3.	2

External Factors (a_3)

This attribute represents the external conditions, regulations and legal aspects. The degree of cooperation among the participants is a very important aspect and it fundamentally determines the possibility of using returnable packaging systems. Environmental effects cannot always be clearly expressed and considered enough in corporate practice, but they are necessary to be built in the model. These are the following: quantity of raw materials and energy consumed in production, CO_2 emission while return transportation, effective vehicle utilization and pool size (it means the total quantity of returnable packaging devices circulating in the system to ensure the stabile operation). The ranking and weights in the fuzzy model can be seen in Table 3.

Table 3. Ranking and aggregation weights of the external factors and its sub-trees

3 external factors			
ID	*Features*	*Ranking*	*Weights*
31	cooperation	1.	8
32	regulations	2.	3
33	legal	3-4.	2
34	environmental effects	3-4.	2
3.2 regulations			
ID	*Features*	*Ranking*	*Weights*
321	environmental	1-4.	2
322	health	1-4.	2
323	benefits	1-4.	2
324	standards	1-4.	2
3.4 environmental effects			
ID	*Features*	*Ranking*	*Weights*
341	production related	1	10
342	CO_2 emission	2	8
343	pool size	3	7
345	effective vehicle utilisation	4	5
3.4.1 production related			
ID	*Features*	*Ranking*	*Weights*
3411	raw material	1.	10
3412	energy	2.	8

All membership functions are a variant of the triangular or trapezoidal membership functions (see e.g. Fig. 4).

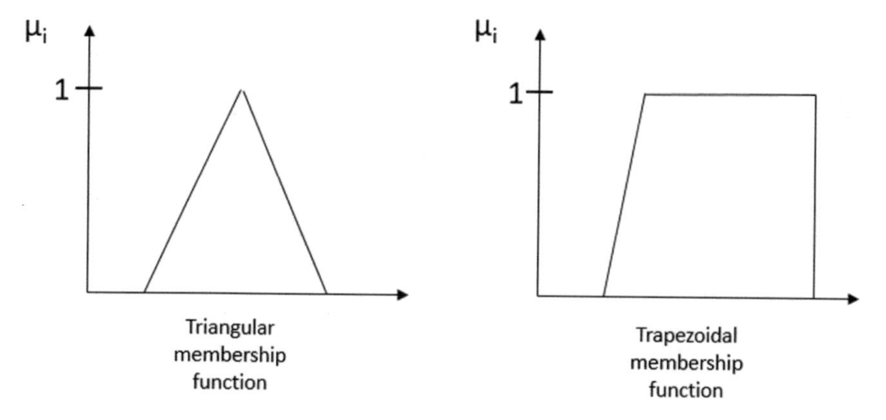

Fig. 4. Membership function examples

3 Application of the Model

In this chapter two real life examples (an industrial and a customer packaging) will be shown in order to illustrate the applicability of the proposed model.

3.1 Case Study 1

The products considered are automotive engines (CKD) transported from Europe to two different destinations in India and China. The finished engines are sensitive products therefore special (wooden) crates are used mainly to store and transport them in the practice. These ensure safe and reliable transport and storage. The columns of the crates are usually collapsible in order to save space while returning back as empty packaging transportation. There are posts inside the crate which are supposed to keep the engine in place, but these can be also collapsed (Fig. 5).

Fig. 5. Returnable packaging used for overseas CKD transport

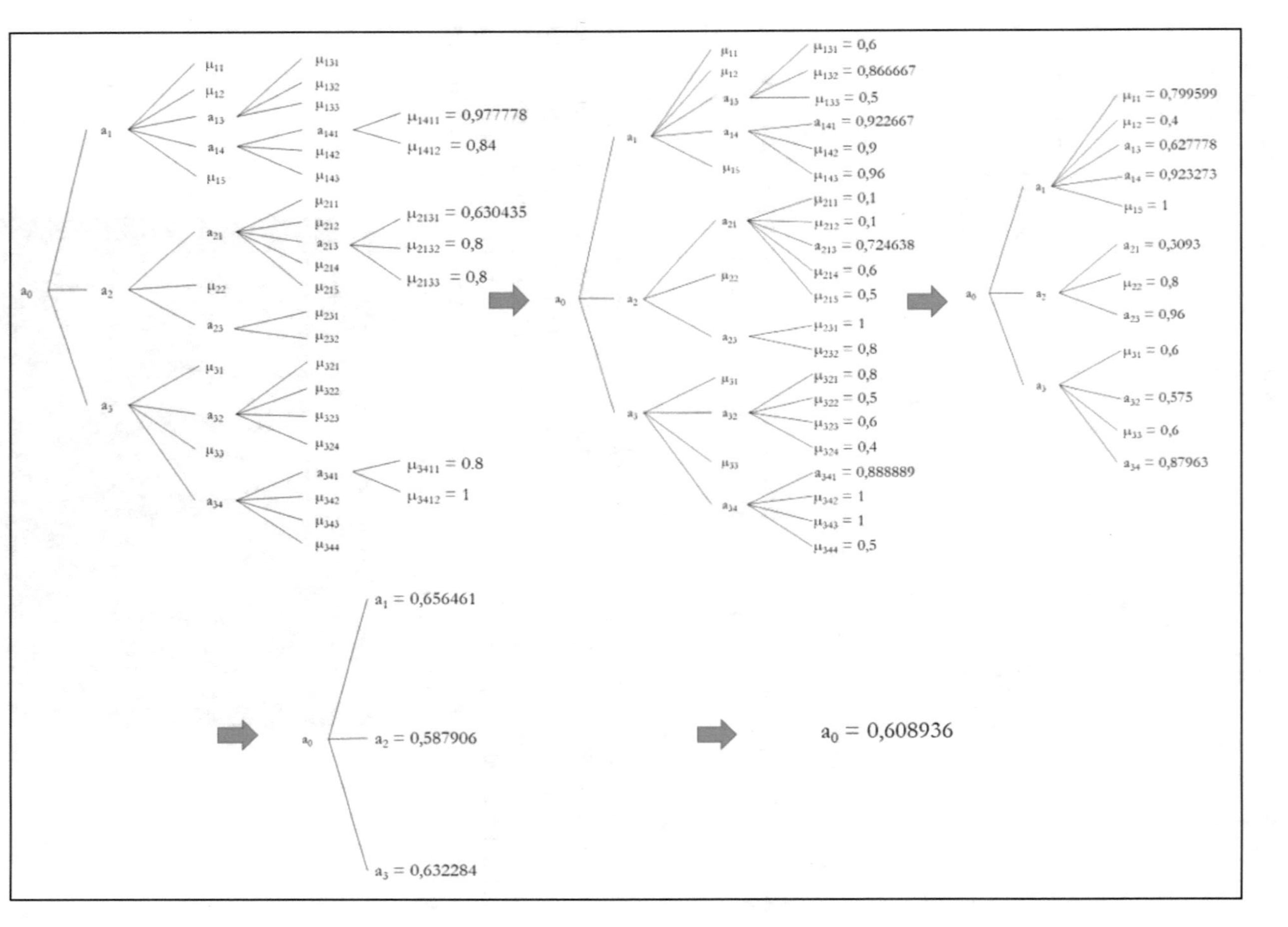

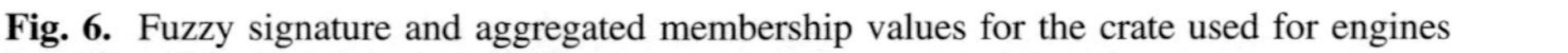
Fig. 6. Fuzzy signature and aggregated membership values for the crate used for engines

The program counts every entity according to the fuzzy signature. The values of all leaves (μ_i) are fuzzy numbers calculated from membership functions and all values of the parents are calculated according to an aggregation and their weights. The characteristics of the engine crate are described by logistics experts. The result of the whole signature is calculated by all aggregations according to Tables 1, 2 and 3 is $a_0 = 0.608936$ (see Fig. 6).

According to the aspects considered in the model the packaging system should rather be returnable. This is in full accordance with the practice of the automotive engine factory providing data for the case study.

Some of the sub sub-models result in $\mu < 0.5$ which should induce one way packaging but these values are compensated by all three sub-models, so the final result and the aggregated final membership degree all suggest returnable packaging.

3.2 Case Study 2

The packaging problem considered here is a well known customer packaging: pet bottles. The methodology is the same like in the case of the wooden crates. The result is calculated by the model above described is $a_0 = 0.519534$. Although reusable plastic bottles had issues in the practice and they are barely used today, according to the aspects considered in the model this packaging system should also rather be returnable.

4 Conclusions and Future Work

In both case study examples the use of fuzzy signature models led to overall membership degrees, supporting the use of the type of packaging that is in accordance with expert domain knowledge. In the first case $\mu_0 = 0{,}609 > 0{,}5$ indicating returnable packaging. In the second case $\mu_0 = 0{,}519 \approx 0{,}5$ which value indicates that both solutions are justifiable. Indeed, in the logistics practice both returnable pet bottles and one-way solutions have been used. As a conclusion we may state that the model proposed is basically suitable for decision support, and it is expected that further refinement of the model may lead to generally applicable technique.

Further research is going on, towards refining the model, especially separating customer and industrial products; and modelling the environmental issues in more detail.

The primary version of this paper was presented at the 3rd Conference on Information Technology, Systems Research and Computational Physics, 2–5 July 2018, Cracow, Poland [17].

Acknowledgement. The authors would like to thank to EFOP-3.6.1-16-2016-00017 1 'Internationalisation, initiatives to establish a new source of researchers and graduates, and development of knowledge and technological transfer as instruments of intelligent specialisations at Széchenyi István University' for the support of the research.

References

1. Böröcz, P.: Analysing the functions and expenses of logistics packaging systems. In: Proceedings of FIKUSZ 2009 Symposium for Young Researchers, pp. 29–39 (2009)
2. Böröcz, P., Földesi, P.: The application of the game theory onto the analysis of the decision theory of logistic packagings. Acta Technica Jaurinensis **1**(2), 259–268 (2008)
3. Böröcz, P., Singh, S.P.: Measurement and analysis of vibration levels in rail transport in central Europe. Packag. Technol. Sci. **24**, 121–139 (2016)
4. Böröcz, P., Mojzes, Á.: The importance of packaging in logistics. Transpack **8**(2), 28–32 (2008)
5. Chan, F.T.S., Chan, H.K., Choy, K.L.: A systematic approach to manufacturing packaging logistics. Int. J. Adv. Manuf. Technol. **29**(9–10), 1088–1101 (2005)
6. European Commission: Packaging and Packaging Waste, http://ec.europa.eu/environment/waste/packaging/index_en.htm. Accessed 26 May 2016
7. Gnoni, M.G., Felice, F., Petrillo, A.: A multi-criteria approach for strategic evaluation of environmental sustainability in a supply chain. Int. J. Bus. In-sights Trans. **3**(3), 54–61 (2011)
8. Hellström, D., Saghir, M.: Packaging and logistics interactions in retail supply chains. Packag. Technol. Sci. **20**(3), 197–216 (2007)
9. http://www.mjspackaging.com/blog/what-are-the-best-packaging-solutions-for-automotive-packaging/. Accessed 05 Oct 2016
10. Mojzes, Á., Böröcz, P.: Decision support model to select cushioning material for dynamics hazards during transportation. Acta Technica Jaurinensis **8**(2), 188–200 (2015)
11. Smith, A.D.: Green supply chain management and consumer sensitivity to greener and leaner options in the automotive industry. Int. J. Logistics Syst. Manag. **12**(1), 1–31 (2012)
12. Stock, J.R., Lambert, D.M.: Strategic Logistics Management. McGraw-Hill Higher Education, New York (2001)
13. Vörösköi, K.: Packaging perspectives in automotive supply chain management. In: Euroma Conference, Edinburgh (2017)
14. Vörösköi, K., Böröcz, P.: Framework for the packaging supply chain of an automotive engine company. Acta Technica Jaurinensis **9**(3), 191–203 (2016)
15. Witt, C.E.: Are reusable containers worth the cost? Mater. Handling Manag. **55**(7), 75 (2000)
16. Yam, K.L.: The Wiley Encyclopedia of Packaging Technology. Wiley, USA (2009)
17. Vörösköi, K., Fogarasi, G., Buruzs, A., Földesi, P., Kóczy, T.L.: Hierarchical fuzzy decision support methodology for packaging system design. In: Kulczycki, P., Kowalski, P.A., Łukasik, S. (eds.) Contemporary Computational Science, p. 140. AGH-UST Press, Cracow (2018)

Image Enhancement with Applications in Biomedical Processing

Małgorzata Charytanowicz[1,2(✉)], Piotr Kulczycki[1,3], Szymon Łukasik[1,3], and Piotr A. Kowalski[1,3]

[1] Centre of Information Technology for Data Analysis Methods, Polish Academy of Sciences, Systems Research Institute, Newelska 6, 01-447 Warsaw, Poland
{mchmat,kulczycki,slukasik,pakowal}@ibspan.waw.pl
[2] Electrical Engineering and Computer Science Faculty, Lublin University of Technology, Nadbystrzycka 36A, 20-618 Lublin, Poland
m.charytanowicz@pollub.pl
[3] Faculty of Physics and Applied Computer Science, Division for Information Technology and Biometrics, AGH University of Science and Technology, Mickiewicza 30, 30-059 Cracow, Poland
{kulczycki,slukasik,pakowal}@agh.edu.pl

Abstract. The images obtained by X-Ray or computed tomography (CT) may be contaminated with different kinds of noise or show lack of sharpness, too low or high intensity and poor contrast. Such image deficiencies can be induced by adverse physical conditions and by the transmission properties of imaging devices. A number of enhancement techniques in image processing may improve the quality of the image. These include: point arithmetic operations, smoothing and sharpening filters and histogram modifications. The choice of the technique, however, depends on the type of image deficiency. In this paper, the primary aim is to propose an efficient image enhancement method based on nonparametric estimation so as to enable medical images to have better contrast. To evaluate the method performance, X-Ray and CT images have been studied. Experimental results verify that applying this approach can engender good image enhancement performance when compared with classical techniques.

Keywords: Image enhancement · Contrast stretching · Nonparametric estimation · Numerical algorithm · X-ray images

In medical diagnostics, image enhancement seems to be most demanding with regard to maintaining a high-quality presentation of all relevant details. Herein, the multi-aspect, often ambiguous and highly subjective character of the image, the human visual system and the observer's experience, will introduce a great deal of difficulties into the choice of the assessment of medical imagery quality. The principle objective is to modify image attributes to make it more suitable for

P. Kulczycki et al. (Eds.): ITSRCP 2018, AISC 945, pp. 97–106, 2020.
https://doi.org/10.1007/978-3-030-18058-4_8

the visibility of structures and objects within the body and improve perception and interpretability of the information held within the image. This is directly related to the structure of the digital image, the applied imaging methods and the employed acquisition technology. Visibility within the various anatomical regions and tissue compositions is determined by a combination of the following factors: contrast, blur, signal-to-noise ratio, artefacts and spatial distortions [12, 15]. The aforementioned contrast originates within the human body as some form of physical contrast and is transformed into visible contrast in the image. Therefore, it constitutes the foundation image quality characteristic. The ability to adjust and optimize the contrast for maximum visibility is one of the major advantages of digital images.

The current rapid growth in computational power allows the use of nonparametric methods in elaborating new techniques for image processing. In the paper [16], an adaptive image enhancement method based on kernel regression and local homogeneity is proposed. First, the image is filtered via kernel regression. Local homogeneity computation is then introduced as this offers adaptive selection about further smoothing. The overall effect of this algorithm includes noise reduction and edge enhancement. The approach is considered to have better performance than other filter methods.

Another filter class is based on the nonparametric estimation of the density probability function in a sliding filter window, and is proposed in [11]. The study is addressed to the problem of impulsive noise removal in multichannel images. The obtained results show that the proposed technique excels over the standard methods currently applied.

Apart from enhancement methods, more advanced procedures based on nonparametric estimation have been elaborated. A new method of edge detection based on kernel density estimation is presented in [9]. In this algorithm, pixels in the image with minimum value of density function are treated as edges. Extensive experimental evaluations has revealed that the edge detection method is significantly a competitive algorithm and the experimental results may be extended to real problems in the field of medical imaging.

Moreover, the paper [2] presents an image segmentation approach based on the gradient clustering algorithm [3,7,8]. This was used for detecting objects structure that have been studied by way of computed tomography. The study results have shown the adequacy of using nonparametric kernel estimation theory.

In our research, the main aim is to propose an alternative method of enhancing and optimizing the contrast characteristics by way of using kernel density estimation. The method has been implemented and tested on gray-scale medical images, and it shows a significant improvement in the output images.

1 Nonparametric Density Estimation

Let (Ω, Σ, P) be a probability space. Let further a real random variable $X : \Omega \rightarrow R$ be given, with a distribution characterized by the density function f.

The corresponding kernel estimator $\hat{f} : R \to [0, \infty)$ calculated using experimentally obtained values for the m-element random sample

$$x_1, x_2, \ldots, x_m, \tag{1}$$

in its basic form is defined as

$$\hat{f}(x) = \frac{1}{mh} \sum_{i=1}^{m} K\left(\frac{x - x_i}{h}\right), \tag{2}$$

where $m \in N \backslash \{0\}$, the coefficient $h > 0$ is called a smoothing parameter, while the measurable function $K : R \to [0, \infty)$ of unit integral $\int_R K(x)\mathrm{d}x = 1$, symmetrical with respect to zero and having a weak global maximum in this place, takes the name of a kernel.

In practical applications, it is recommended to individualize the bandwidth h on particular kernels $K\left(\frac{x-x_i}{h}\right)$, hence improving the quality of estimator (2). This relies on introducing the local bandwidth factors $s_1, s_2, \ldots, s_m$, for the parameter h, defined as

$$s_i = \left(\frac{\hat{f}(x_i)}{\tilde{s}}\right)^{-c} \quad \text{for } i = 1, 2, \ldots, m, \tag{3}$$

where $\hat{f}$ is the kernel estimator in its basic form (2), $\tilde{s}$ means the geometrical mean of the numbers $\hat{f}(x_1), \hat{f}(x_2), \ldots, \hat{f}(x_m)$ and $c \in [0, \infty)$ denotes the sensitivity parameter. Based on indications for the optimization criteria, the standard value $c = 0.5$ is usually suggested.

Finally, the kernel estimator with adaptive bandwidths $h_i = hs_i$ takes the following formula:

$$\hat{f}(x) = \frac{1}{mh} \sum_{i=1}^{m} \frac{1}{s_i} K\left(\frac{x - x_i}{hs_i}\right). \tag{4}$$

Specifying the kernel estimator of a density function f, gives a natural description of the distribution of X, and allows the estimator of the cumulative distribution function, denoted hereinafter as $\hat{F} : R \to [0, 1]$, to be found from the relation

$$\hat{F}(x) = \int_{-\infty}^{x} \hat{f}(u)\mathrm{d}u, \tag{5}$$

where $\hat{f}$ denotes the kernel density estimator (4). Denoting the primitive of a kernel K as $I : R \to [0, 1]$, that is

$$I(x) = \int_{-\infty}^{x} K(u)\mathrm{d}u \tag{6}$$

the kernel estimator of the distribution function can be expressed as

$$\hat{F}(x) = \frac{1}{m} \sum_{i=1}^{m} I\left(\frac{x - x_i}{hs_i}\right). \tag{7}$$

The quality of the estimation depends on the choice of the kernel K and the calculation of the bandwidth h. This is made most often by way of established optimization criterions. Fortunately, the choice of the kernel form has no significant meaning and thanks to this, it becomes possible to take into account the primarily properties of the estimator obtained. The standard normal kernel is one of the most commonly ones used in practice. The Gaussian kernel function takes the form

$$K(x) = \frac{1}{\sqrt{2\pi}} e^{-\frac{x^2}{2}}. \tag{8}$$

For this kernel, the estimated density function is smooth and includes the derivatives of all the orders. On the other hand, if the primitive of the kernel K is needed the Cauchy kernel can be recommended. The Cauchy kernel is given by the formula

$$K(x) = \frac{2}{\pi} \frac{1}{(1+x^2)^2} \tag{9}$$

and its primitive is brought about through applying the rule

$$I(x) = \frac{1}{\pi} \left(\frac{x}{1+x^2} + \arctan(x) + \frac{\pi}{2} \right). \tag{10}$$

The shape of the estimated density function depends strongly on the smoothing parameter h that has been chosen for the density estimation. Small values of this parameter lead to spiky density estimates that show spurious features whereas too large values create over-smoothed estimates that hide structural features. A frequently used bandwidth selection procedure, called the "cross-validation method", is based on optimization criteria wherein h is chosen to minimize the function $g : (0, \infty) \to R$, defined as

$$g(h) = \frac{1}{m^2 h} \sum_{i=1}^{m} \sum_{j=1}^{m} \widetilde{K} \left(\frac{x_j - x_i}{h} \right) + \frac{2}{mh} K(0), \tag{11}$$

where $\widetilde{K}(x) = K^{*2}(x) - 2K(x)$ whilst K^{*2} denotes the convolution square of the function K, that is

$$K^{*2}(x) = \int_R K(u) K(x-u) \mathrm{d}u. \tag{12}$$

Tasks concerned with the choosing the kernel and calculating the smoothing parameter, as well as the additional procedures bringing about improvements in the quality of the estimator obtained are found in [5,6,10,13].

2 Image Enhancement Using Kernel Density Estimation

In order to introduce the image enhancement method based on kernel density estimation, theoretical aspects will be shown first.

2.1 Theoretical Aspects

Suppose that the intensity levels of an image to be enhanced will be treated as a random variable x, and assume for a moment that x has been normalized to the interval $[0, 1]$. Herein, the value equal to 0 denotes the lowest intensity level and the value equal to 1 denotes the highest intensity level. Moreover consider the transformation of the form

$$y = T(x),\ x \in [0, 1] \tag{13}$$

that gives an intensity level y for every value x of the source image. Assume also, that $y \in [0, 1]$ and the transformation $y = T(x)$ is monotonically increasing and single-valued in that interval. This guarantees the increasing order from the lowest to the highest intensity in the output image and the existing inverse transformation $x = T^{-1}(y)$ for $y \in [0, 1]$.

Assume that f_x and f_y denote the probability density functions of the random variables x and y respectively. Thus

$$f_y(y) = \left|\frac{\mathrm{d}x}{\mathrm{d}y}\right| f_x(x). \tag{14}$$

Furthermore, let the transformation function be of the form [4]

$$y = T(x) = \int_0^x f_x(u)\mathrm{d}u. \tag{15}$$

The transformation T is, hence, the cumulative distribution function of x. Assume for simplicity that T is differentiable. Thus

$$\frac{\mathrm{d}y}{\mathrm{d}x} = \frac{\mathrm{d}T(x)}{\mathrm{d}x} = \frac{\mathrm{d}}{\mathrm{d}x}\left(\int_0^x f_x(u)\mathrm{d}u\right) = f_x(x) \tag{16}$$

and

$$f_y(y) = \left|\frac{\mathrm{d}x}{\mathrm{d}y}\right| f_x(x) = 1 \text{ for } y \in [0, 1]. \tag{17}$$

The function f_y is a probability density function, and, therefore, it must be equal to 0 for every $y \notin [0, 1]$. Hence, for the transformation function given in the formula (15), the resulting density function f_y is always uniformly distributed, regardless of the form of function f_x.

2.2 Proposed Methodology

Let an image of width W and height H be defined as a two-dimensional function L which gives each pixel a nonnegative value $L(x, y)$, where x, y are spatial coordinates for $x = 0, 1, \ldots, W-1$, $y = 0, 1, \ldots, H-1$. The amplitude of $L(x, y)$ at any pair of coordinates (x, y) is called the intensity level of the image at that point. Thus the digital image may be represented as a table of discrete values $L(x, y)$ for simplicity, denoted as l_{xy}:

$$\begin{bmatrix} l_{00} & l_{10} & \dots & l_{W-1,0} \\ l_{01} & l_{11} & \dots & l_{W-1,1} \\ \dots & \dots & \dots & \dots \\ l_{0,H-1} & l_{1,H-1} & \dots & l_{W-1,H-1} \end{bmatrix}. \tag{18}$$

The nonparametric approach to estimating the probability density function from the observed data will then be applied for the intensity levels occupied by the image pixels.

As was noted in Sect. 2.1, this makes it possible to develop a transformation function that can achieve high dynamic range based only on the information available in the distribution of the aforementioned intensity levels. Consider the data set containing $m = WH$ elements l_{ij} for $i = 0, 1, \dots, W-1$, $j = 0, 1, \dots, H-1$, drawn for simplicity in the sequence

$$l_1, l_2, \dots, l_m. \tag{19}$$

Let the intensity levels $l_1, l_2, \dots, l_m$ be represented by discrete values in the range $[0, L-1]$, where L denotes the number of possible intensity levels. Using the methodology presented in Sect. 1, the auxiliary kernel estimator (4) of the elements (19), will be created:

$$\hat{f}(l) = \frac{1}{mh} \sum_{i=1}^{m} \frac{1}{s_i} K\left(\frac{l - l_i}{hs_i}\right). \tag{20}$$

Herein, the transformation function T will be defined as the cumulative distribution function of the kernel density estimator multiplied by $(L-1)$ and will take the form

$$T(l) = \text{floor}\left((L-1)\int_0^l \hat{f}(u)\text{d}u\right), \tag{21}$$

where L denotes the number of possible intensity levels and the function floor() rounds down to the nearest integer. The transformation function given in the formula (21) is uniformly distributed on the interval $[0, L-1]$, thus it has the effect of transforming the intensity levels distributions such that they are distributed more uniformly. Hence, it will flatten the intensity level distribution, and in doing so, will finally enhance the contrast in the image.

Finally, the formula to enhance the contrast of the image, normalized to the interval $[0, L-1]$, will be defined using the form

$$l'_i = \text{floor}\left((L-1)\frac{\hat{F}(l_i) - \hat{F}_0}{1 - \hat{F}_0}\right), \tag{22}$$

where the kernel estimator of the cumulative distribution function can be expressed as

$$\hat{F}(l) = \frac{1}{m} \sum_{i=1}^{m} I\left(\frac{l - l_i}{hs_i}\right), \tag{23}$$

and the function $I(l) = \int_{-\infty}^{l} K(u)\mathrm{d}u$, while the value F_0 is the minimum nonzero value of the cumulative distribution function (23).

2.3 Implementation Remarks

The proposed methodology for image enhancement, based on nonparametric kernel estimation, requires the construction of the kernel estimator of the cumulative distribution function. Therefore, the kernel K is assumed here to be in the Cauchy form (9) for which the primitive is given by the formula (10), as this is convenient for further calculations.

The output values l'_i of the source intensity levels l_i on the output image can be computed very quickly using the lookup table LUT, which assigns an output value to every possible input value l_i, $i = 1, 2, \ldots, m$ by means of the formula (22). As a result, the processed image is obtained by mapping each pixel with the intensity level l_i in the source image via a corresponding pixel with the appropriate intensity level $l'_i = \mathrm{LUT}(l_i)$.

Moreover, because of ongoing research into the computer implementation of the algorithm, the parameters appearing in the formulae (20)–(22) are effectively calculated by convenient numerical procedures. Due to the kernel estimation methodology the whole procedure can be naturally extended to the multidimensional case when full-color images are processed.

It is also worth mentioning that the presented approach overcomes the rigidity of arbitrary assumptions of the form of a density function.

3 Results and Discussion

Figures 1, 2 and 3 display the three monochrome images with 296 $\times$ 236 pixels and 256 gray levels of good, low and enhance contrast, respectively. The third image shows the results of performing the transformation (22) on the second, low-contrast image. Herein, it can be seen that the transformation has produced an output image that has nearly a uniform density estimation. Here too, it shows significant improvement, yet, as expected, a significant visual difference is also produced in the kernel density estimator. This comes about because intensity levels span the full spectrum of the gray scale (in comparison with the low-contrast image), thus, the transformation has significant effect on the visual appearance.

These results well illustrate the power of the proposed transformation. They also reveal the effectiveness of the kernel estimator-based enhancing method of making medical images consistent for efficacious detection and interpretation.

Along with the visual quality assessment, the PSNR [14] measure was computed based on the pixel difference between the good contrast image (Fig. 1) and the output images obtained using each of the following enhancing contrast methods for the low-contrast image (Fig. 2): histogram stretching, histogram equalization and the proposed procedure.

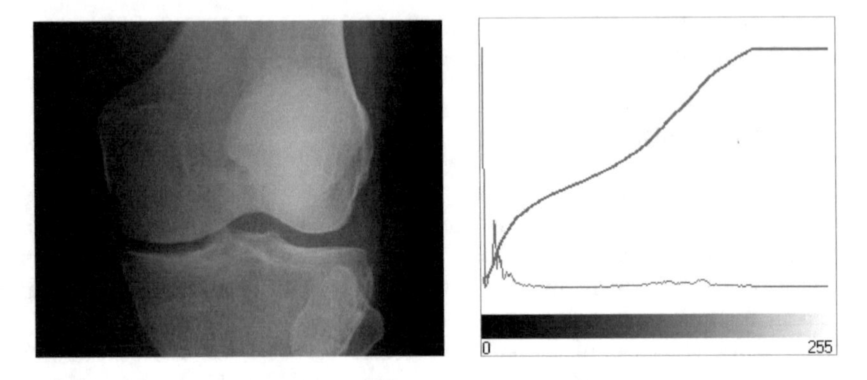

Fig. 1. The good-contrast image with its corresponding kernel density estimator (thin line) and cumulative distribution estimator (thick line).

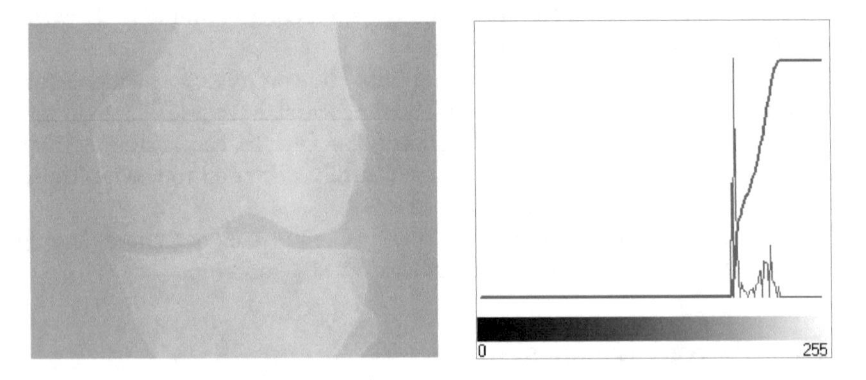

Fig. 2. The low-contrast image with its corresponding kernel density estimator (thin line) and cumulative distribution estimator (thick line).

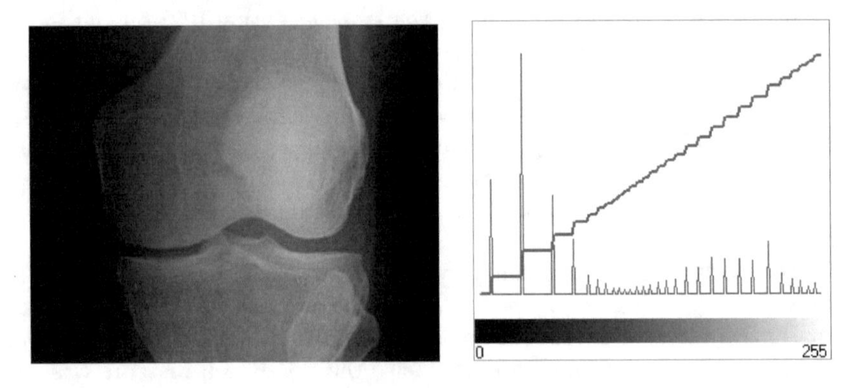

Fig. 3. The enhancement image with its corresponding kernel density estimator (thin line) and cumulative distribution estimator (thick line).

Table 1. The PSNR results obtained for enhancing contrast methods.

Method	PSNR
Histogram stretching	19.56
Histogram equalization	16.40
Proposed method	15.79

Based on obtained results (Table 1), the proposed technique generated very good image quality. Moreover, the kernel density methodology used here, through the additional procedures improving the quality of the estimation, may be useful for adapting the final results to the requirements of the problem under consideration.

The preliminary version of this paper was presented at the 3rd Conference on Information Technology, Systems Research and Computational Physics, 2–5 July 2018, Cracow, Poland [1].

4 Summary

The probability density function estimation is a fundamental task applicable within the various methods employed in data analysis. Here, an estimate of the intensity levels distribution is derived from an image and gives a natural description of its characteristics. The applied transformation allows the production of processed image with a uniform intensity levels distribution, and, as a result, generates a significant improvement in its quality. Due to the advantages of it being very intuitive, conceptually simple and flexible, this innovative technique can be easily applied in order to elaborate upon current image processing practices. What is more, the kernel density estimation method allows the use of additional procedures that will improve the quality of the estimator, as well as the fitting of the model to practical requirements. For the first, we recommend modifying the smoothing parameter and the linear transformation [5, 10], while for the second, the boundary support method [5, 13]. These procedures may serve as the basis for developing more advanced applications of the proposed methodology wherein the specified properties of the intensity distribution of the processed image are needed.

References

1. Charytanowicz, M., Kulczycki, P., Łukasik S., Kowalski, P.A.: Image enhancement with applications in biomedical processing. In: Kulczycki, P., Kowalski, P.A., Łukasik, S. (eds.) Contemporary Computational Science, p. 54. AGH-UST Press, Cracow (2018)
2. Charytanowicz, M., Kulczycki, P.: An image analysis algorithm for soil structure identification. In: Filev, D., Jabłkowski, J., Kacprzyk, J., Popchev, I., Rutkowski, L., Sgurev, V., Sotirova, E., Szynkarczyk, P., Zadrożny, S. (eds.) Information Technologies in Biomedicine, pp. 681–692. Springer, Cham (2014)

3. Charytanowicz, M., Niewczas, J., Kulczycki, P., Kowalski, P.A., Łukasik, S., Żak, S.: Complete gradient clustering algorithm for features analysis of X-ray images. In: Pietka, E., Kawa, J. (eds.) Information Technologies in Biomedicine, pp. 15–24. Springer, Heidelberg (2010)
4. Gonzalez, R.C., Woods, R.E.: Digital Image Processing. Prentice Hall, New Jersey (2007)
5. Kulczycki, P.: Estymatory jadrowe w analizie systemowej. WNT, Warszawa (2005)
6. Kulczycki, P.: Kernel estimators in industrial applications. In: Prasad, B. (ed.) Soft Computing Applications in Industry. Springer, Berlin (2008)
7. Kulczycki, P., Charytanowicz, M.: A complete gradient clustering algorithm formed with kernel estimators. Int. J. Appl. Math. Comput. Sci. **20**, 123–134 (2010)
8. Kulczycki, P., Charytanowicz, M., Kowalski, P.A., Łukasik, S.: The complete gradient clustering algorithm: properties in practical applications. J. Appl. Stat. **39**, 1211–1224 (2012)
9. Pereira, O., Torre, E., Garcés, E., Rodriguez, R.: Edge detection based on kernel density estimation. In: Proceedings of the 2017 International Conference on Image Processing, Computer Vision, and Pattern Recognition, IPCV 2017, pp. 1–24. CSREA Press (2017)
10. Silverman, B.W.: Density Estimation for Statistics and Data Analysis. Chapman and Hall, London (1986)
11. Smolka, B., Budzan, S., Lukač, R.: Nonparametric design of impulsive noise removal in colour images. J. Med. Inform. Technol. **7**, 3–14 (2004)
12. Sprawls, P.: Optimizing medical image contrast, detail and noise in the digital era. Med. Phys. Int. J. **2**, 128–133 (2014)
13. Wand, M.P., Jones, M.C.: Kernel Smoothing. Chapman and Hall, London (1994)
14. Wang, Z., Bovik, A.C., Sheikh, H.R., Simoncelli, E.P.: Image quality assessment: from error visibility to structural similarity. IEEE Trans. Image Process. **13**, 600–612 (2004)
15. Wojnar, L., Majorek, M.: Komputerowa analiza obrazu. Fotobit Design, Warszawa (1994)
16. Yang, Y.-Q., Zhang, J.-S., Huang, X.-F.: Adaptive image enhancement algorithm combining kernel regression and local homogeneity. Math. Probl. Eng. **2010**, 1–14 (2010)

Efficient Astronomical Data Condensation Using Fast Nearest Neighbors Search

Szymon Łukasik[1,2(✉)], Konrad Lalik[1], Piotr Sarna[1], Piotr A. Kowalski[1,2], Małgorzata Charytanowicz[2,3], and Piotr Kulczycki[1,2]

[1] Faculty of Physics and Applied Computer Science, AGH University of Science and Technology, Kraków, Poland
{slukasik,pkowal,kulpi}@agh.edu.pl
[2] Systems Research Institute, Polish Academy of Sciences, Warsaw, Poland
{slukasik,pakowal,mchmat,kulpi}@ibspan.waw.pl
[3] Faculty of Electrical Engineering and Computer Science, Lublin University of Technology, Lublin, Poland
m.charytanowicz@pollub.pl

Abstract. Analyzing astronomical observations represents one of the most challenging tasks of data exploration. It is largely due to the volume of the data acquired using advanced observational tools. While other challenges typical for the class of Big Data problems - like data variety - are also present, datasets size represents the most significant obstacle in visualization, and subsequent analysis. The paper studies efficient data condensation algorithm aimed at providing its compact representation. It is based on fast nearest neighbor calculation using tree structures and parallel processing. The properties of the proposed approach are preliminary studied on astronomical datasets related to the GAIA mission. It is concluded that introduced technique might serve as a scalable method of alleviating the problem of data sets size.

Keywords: Big Data · Astronomy · Data reduction

1 Introduction

Last decades are characterized by unprecedented burst of new data being generated in various fields of science and engineering. In essence, it can be useful to generate new insights, lead to discoveries and improve standard of living. However the toolbox of contemporary data science – though broad and reinforced by unconventional methods of artificial intelligence – does not contain algorithms, which cope well with the challenges of, so called Big Data. This term encompasses a set of problematic properties of data sets stored in present-day computer systems. Besides the obvious obstacle of data volume, variety – which relates to diverse data types and structure of the dataset, velocity – which refers to the speed of new data generation and veracity – which corresponds to data quality/uncertainty, can be named [10].

P. Kulczycki et al. (Eds.): ITSRCP 2018, AISC 945, pp. 107–115, 2020.
https://doi.org/10.1007/978-3-030-18058-4_9

Astronomy, among other fields of science, is nowadays strongly affected by the Big Data problems. It is due to the fact that it currently possess a wide-area of data acquisition tools. In reality the sizes of catalogs of astronomical objects reach petabytes, and they may contain billions of instances described by hundreds of parameters [13]. Even the seemingly simple task of visualizing such datasets becomes a serious challenge. In such case, along with significant computing power, sophisticated data preprocessing algorithms are required.

The aim of this paper is to provide efficient method of data condensation, that is selecting the most representative objects in data, in a way to preserve data density. Such data prototypes can later be used for visualization, as well as for other data mining procedures. For this purpose we propose a modified fast density-based multiscale data condensation algorithm [14]. Our variant utilizes approximate nearest-neighbor technique, together with parallel processing. We preliminary evaluate our approach on astronomical dataset related to the GAIA mission [1]. The preliminary version of this contribution was presented at the 3rd Conference on Information Technology, Systems Research and Computational Physics, 2–5 July 2018, Cracow, Poland [20].

The paper is organized as follows. First, in the next Section, we provide methodological preliminaries – introducing data reduction and examples of related techniques in astronomy, as well as the problem of efficient nearest neighbors calculation. The third Section overviews proposed approach and it is followed by the discussion on the first results obtained for real astronomical data, which was included in Sect. 4. Finally, general remarks regarding characteristic features of introduced approach and planned further studies are under consideration.

2 Methodological Preliminaries

2.1 Data Reduction and Its Use in Astronomy

Data preprocessing techniques used in astronomy ought to deal with large datasets – also in real-time mode. It is a consequence of rapid development of new instruments and new data gathering schemes. It effectively means that the volume of the data generated doubles every year [19]. The practical illustration of this problem is the amount of objects captured by sky surveys across the last 50 years, as demonstrated in Table 1.

Consequently data reduction is typically introduced as close as possible to the instrumentation level, i.e. at signal/image processing phase. Its goal is to bring down the size of transferred data. Such reduction in most of cases involves removing noise, signatures of the atmosphere/instrument and other contaminating factors [9,18].

When the object-based data is already available its reduction is performed typically with sampling methods [16]. Uniform sampling techniques with or without replacement are the most widely used approach – also in astronomy (e.g. see [7] or [17]). Stratified sampling – as the one preserving the ratio of objects present in different classes – is also used (e.g. [2]). As the alternative to statistical

Table 1. Selected sky surveys (as reported in [13])

Survey	Institution	Number of objects	Type	Time frame
Hipparcos	European Space Agency	0.12M	Optical	1989–1993
Tycho-2	European Space Agency	2.5M	Optical	1989–1993
DPOSS	Caltech	550M	Optical	1950–1990
2MASS	Univ. of Massachusetts, Caltech	300M	Near-IR	1997–2001
Gaia	European Space Agency	1000M	Optical	2013–
SDSS	Astrophysical Research Consortium	470M	Optical	2000–
LSST	LSST Corporation	4000M	Optical	2019–

sampling more sophisticated procedures employing probabilistic modeling could be considered. Literature of the subject contains at least one example of such technique – in [22] the dataset is clustered into hyper-balls with predetermined radii. Each of them is associated with a kernel and a weight, in such way that this mixture is exposing the local data distribution.

Paragraph above was describing general, i.e. problem independent, methods of data reduction. As astronomy relies heavily on advanced visualization many procedures of data reduction were developed to deal with the problem data abundance in visual analytics [11]. They are mainly based on creating new data context containing only selected data points – which makes data visualization less complex. Selection of such reduced set is performed either manually (as in [6]) or using distance from the observer. For more extensive overview of data reduction schemes in astronomy, along with use-case examples one could refer to [13].

Here we will discuss and use general data condensation technique proposed by Mitra et al. [14]. It was already positively evaluated for elementary data reduction tasks present in preprocessing of astronomical sky surveys [13]. It relies on finding iteratively points with the closest k-nearest neighbor (the distance to which is denoted by r_k) and then adds it to the reduced dataset, which is initially empty. At the same time other points lying within a disc of radius $2 * r_k$ are eliminated (not included in the reduced set E). Listing below describes the main steps of the data reduction algorithm.

It can be seen that the algorithm requires utilizing both, nearest neighbor search and so called radius search – finding points situated in the hyper-sphere of given radius.

2.2 Approximate Nearest Neighbors

Finding nearest neighbors in the data constitute a very important issue – as a plethora of data mining algorithms are being built on this component. It includes outlier detection [4], classification [12] and clustering [5]. That is why

Algorithm Density-Based Data Condensation Algorithm

1: Denote $B = \{x_1, x_2, \ldots, x_N\}$ as the initial dataset. Set condensation ratio k
2: Prepare empty reduced dataset E
3: For each data point $x_i \in B$ we calculate the distance $d_k(x_i)$ to its k nearest neighbor
4: Pick the point j with the smallest value of $d_k(x_j)$, i.e. $x_j = \underset{i=1,\ldots,N}{\operatorname{argmin}} d_k(x_i)$
5: Insert x_j into reduced dataset E, denote the distance to its k neighbor as r_k.
6: Remove all points from B which are situated within $2 * r_k$ from x_j
7: Repeat steps 3-6 until B is not empty

fast k-nearest neighbor calculation is crucial to data analysis – especially when it is performed on large datasets.

Naive (or so-called brute-force) k-nearest neighbor search involves the following steps:

Algorithm Brute-force k-nearest neighbor search

1: Calculate all pairwise distances $d(x_i, x_j)$, $i, j = 1, \ldots N$.
2: For each query point x_i sort distances in the ascending order.
3: For each query point x_i find a set of k closest neighbors

The time complexity of such procedure – taking into account that the neighborhood is to be identified for all query points – is $O(N^2) + (N^2 \log N)$. It becomes prohibitive for most practical applications [3].

Three major classes of algorithms have been identify to speed-up the process of locating nearest neighbors: kd-trees or other tree-based partitioning structures, hashing techniques and neighboring graph approaches. The first are based on building a hierarchical structure partitioning the data recursively, e.g. along the dimension of maximum variance. Examples of approaches based on this paradigm include the use of kd-trees [8] and vp-trees [23]. The second approach is based on to hashing points in a similarity preserving way, i.e., by putting it them in the buckets grouping similar items (as in Locality Sensitive Hashing [24]). The example of the thirds strategy can be found in [21] where a random k-NN graph approximation, updated in each step of the algorithm is being used.

For more extensive discussion on fast nearest-neighbor strategies one could refer to [15]. In subsequent part of the paper we will discuss how fast nearest neighbor calculation, based on kd-trees, can be included within the parallel scheme of data reduction strategy – to make its execution feasible, even for large datasets.

3 Proposed Approach

Proposed scheme of efficient data condensation involves at first building tree structure – using standard kd-tree algorithm. It should be then distributed among nodes which are subsequently used for k-nearest neighbor calculation. Each node is responsible for locating set of nearest neighbors for its assigned part of the dataset. The search process itself is again parallelized at multi-core level. It is followed by sequential radius search – locating points situated within $2 * r_k$ from the one with the smallest value of $d_k(x_j)$ and removing them from the dataset. Evidently each removal operation requires the update of kd-tree data structure. The process is repeated until the initial dataset is completely pruned. The algorithmic summary of the whole process is presented below.

Algorithm Data condensation with parallel nearest neighbors calculation based on kd-trees

1: Denote $B = \{x_1, x_2, \ldots, x_N\}$ as the initial dataset. Set condensation ratio k
2: Prepare empty reduced dataset E
3: Build kd-tree for B.
4: Distribute tree structure among *proc* nodes
5: Assign part of the dataset to each node
6: Execute k-nn search in parallel on each *proc* node, gather results
7: Pick the point j with the smallest value of $d_k(x_j)$, i.e. $x_j = \operatorname{argmin}_{i=1,\ldots,N} d_k(x_i)$
8: Insert x_j into reduced dataset E, denote the distance to its k neighbor as r_k.
9: Execute sequential radius search
10: Remove all points from B which are situated within $2 * r_k$ from x_j
11: Update kd-tree
12: Repeat steps 5-11 until B is not empty

It can be observed that parellization is used at the most time consuming step of locating nearest neighbors. It is not needed at the stage of radius search – as it is executed only once for one point – in each data pruning step. In subsequent part of the paper we will evaluate this approach and study its properties.

4 First Results

To evaluate the performance of proposed data condensation approach we have set up an experiment, involving reducing dataset size for a portion of GAIA data release 1 snapshot [1]. We have used only spatial information – namely transformed 3D coordinates of astronomical objects.

We examined first the impact of neighborhood size k on the resulting reduced dataset size, assuming that the initial dataset had 250 000 objects. It can be seen that the reduction provides very compact representation of analyzed snapshot – even for small number of neighbors we obtain only few percent of the initial sample (Fig. 1).

Secondly, we have studied the running times for the algorithm with varying value of condensation ratio k. The dataset size was set to 250 000 elements. One node with 4 threads running concurrently was used for this experiment.

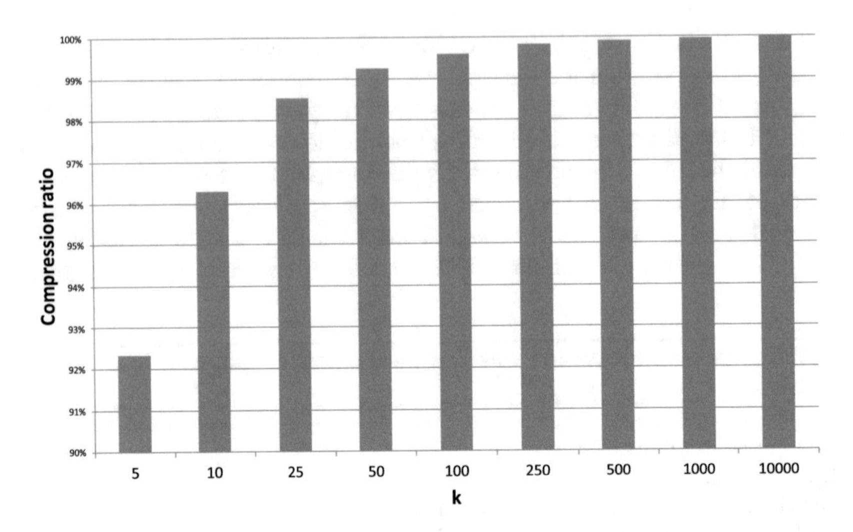

Fig. 1. Compression ratio for varying value of k

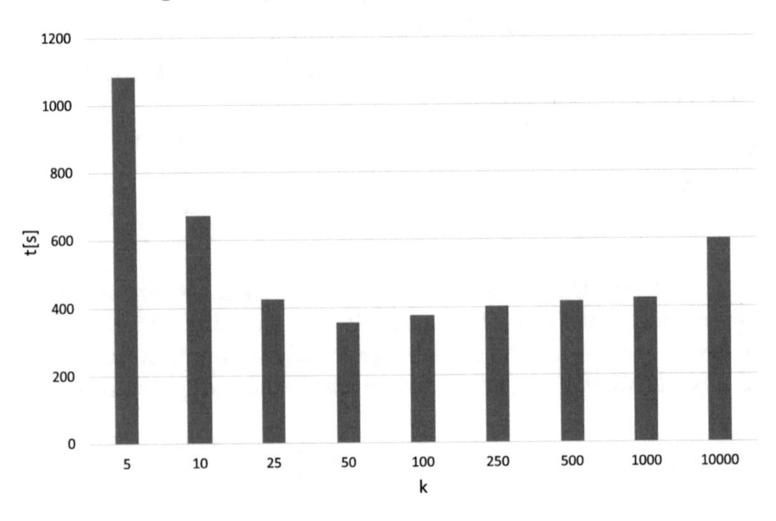

Fig. 2. Running times for varying value of k

The analysis of the results obtained at this stage (presented on Fig. 2) leads us to conclusion that the profit resulting from parallelization within one node increases with the size of k. This is due to the frequent need of synchronization, with small values of the number k, which reduces the efficiency of parallel processing. Furthermore, higher values of k result in more intensive reduction, requiring less steps of data pruning. It was observed however that for large number of neighbors the impact of identifying neighboring points becomes more tangible – and it consequently deteriorates algorithm's time performance.

Finally we have also studied the speed-up obtained for parallel calculation of nearest neighbors for 100 000 objects using 4 nodes – with one parallel thread running on each of them. It was established that parallelization is more beneficial for high number of neighbors. While the k increases, speed-up becomes asymptotically linear. The growing overhead of communication and synchronization operations was not yet observed (Fig. 3).

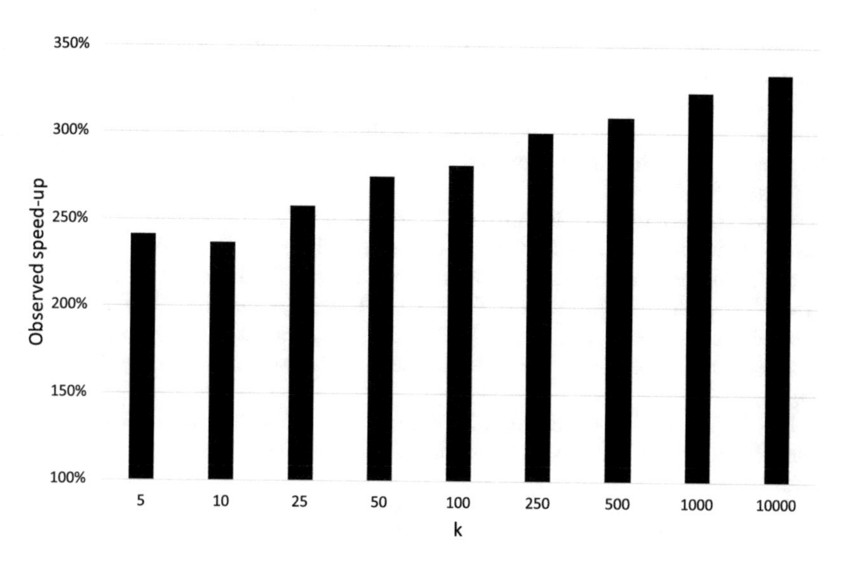

Fig. 3. Observed speed-up for 4 nodes with one thread running vs single node configuration

5 Conclusion

The paper presented an application of fast k-nearest neighbor search – based on kd-trees and parallel processing – in data condensation strategies. It was established during preliminary experiments that using improved nearest-neighbor strategies may allow to tackle large astronomical datasets. It is due to the use of efficient data representation and parallelization of data reduction scheme.

Further work in this area will concern evaluating performance of data condensation with varying error of nearest-neighbor approximation. The impact of parallelization will be studied more extensively. Finally, it is planned to investigate other methods of nearest-neighbor approximation e.g. techniques based on random kd-trees or hierarchical clustering.

Acknowledgments. This work was partially financed (supported) by the Faculty of Physics and Applied Computer Science AGH UST statutory tasks within subsidy of Ministry of Science and Higher Education.

The study was also supported in part by PL-Grid Infrastructure.

References

1. GAIA mission. https://www.cosmos.esa.int/gaia. Accessed 20 Aug 2018
2. Abraham, S., Philip, N.S., Kembhavi, A., Wadadekar, Y.G., Sinha, R.: A photometric catalogue of quasars and other point sources in the Sloan Digital Sky Survey. Mon. Not. R. Astron. Soc. **419**, 80–94 (2012)
3. Arefin, A.S., Riveros, C., Berretta, R., Moscato, P.: GPU-FS-kNN: a software tool for fast and scalable kNN computation using GPUs. PLoS ONE **7**(8), e44000 (2012)
4. Breunig, M.M., Kriegel, H.-P., Ng, R.T., Sander, J.: LOF: identifying density-based local outliers. In: Proceedings of the 2000 ACM SIGMOD International Conference on Management of Data, SIGMOD 2000, pp. 93–104. ACM, New York (2000)
5. Bubeck, S., von Luxburg, U.: Nearest neighbor clustering: a baseline method for consistent clustering with arbitrary objective functions. J. Mach. Learn. Res. **10**, 657–698 (2009)
6. Burgess, R., Falcao, A.J., Fernandes, T., Ribeiro, R.A., Gomes, M., Krone-Martins, A., de Almeida, A.M.: Selection of large-scale 3D point cloud data using gesture recognition. In: Camarinha-Matos, L., Baldissera, T., Di Orio, G., Marques, F. (eds.) Technological Innovation for Cloud-Based Engineering Systems: Proceedings of the 6th IFIP WG 5.5/SOCOLNET Doctoral Conference on Computing, Electrical and Industrial Systems, DoCEIS 2015, pp. 188–195. Springer, Cham (2015)
7. Dutta, H., Giannella, C., Borne, K., Kargupta, H.: Distributed top-k outlier detection from astronomy catalogs using the DEMAC system. Chapter 47, pp. 473–478. SIAM (2005)
8. Eastman, C., Weiss, S.F.: Tree structures for high dimensionality nearest neighbor searching. Inf. Syst. **7**(2), 115–122 (1982)
9. Freudling, W., et al.: Automated data reduction workflows for astronomy. The ESO Reflex environment. Astron. Astrophys. **559**, A96 (2013)
10. Grandinetti, L., Joubert, G., Kunze, M., Pascucci, V.: Big Data and High Performance Computing. Advances in Parallel Computing. IOS Press, Amsterdam (2015)
11. Hassan, A., Fluke, C.J.: Scientific visualization in astronomy: towards the petascale astronomy era. PASA Publ. Astron. Soc. Austral. **28**, 150–170 (2011)
12. Li, L., Zhang, Y., Zhao, Y.: k-nearest neighbors for automated classification of celestial objects. Sci. China Ser. G **51**(7), 916–922 (2008)
13. Łukasik, S., Moitinho, A.A., Kowalski, P.A., Falcão, A., Ribeiro, R.A., Kulczycki, P.: Survey of object-based data reduction techniques in observational astronomy. Open Phys. **14**, 64 (2016)
14. Mitra, P., Murthy, C.A., Pal, S.K.: Density-based multiscale data condensation. IEEE Trans. Pattern Anal. Mach. Intell. **24**, 734–747 (2002)
15. Muja, M., Lowe, D.G.: Scalable nearest neighbor algorithms for high dimensional data. IEEE Trans. Pattern Anal. Mach. Intell. **36**(11), 2227–2240 (2014)
16. Pal, S.K., Mitra, P.: Pattern Recognition Algorithms for Data Mining. CRC Press, Boca Raton (2004)
17. Rocke, D.M., Dai, J.: Sampling and subsampling for cluster analysis in data mining: with applications to sky survey data. Data Min. Knowl. Disc. **7**(2), 215–232 (2003)
18. Schirmer, M.: THELI: convenient reduction of optical, near-infrared, and mid-infrared imaging data. Astrophys. J. Suppl. Ser. **209**, 21 (2013)
19. Szalay, A., Gray, J.: The world-wide telescope. Science **293**(5537), 2037–2040 (2001)

20. Łukasik, S., Lalik, K., Sarna, P., Kowalski, P.A., Charytanowicz, M., Kulczycki, P.: Efficient astronomical data condensation using approximate nearest neighbors. In: Kulczycki, P., Kowalski, P.A., Łukasik, S. (eds.) Contemporary Computational Science, pp. 55–56 (2018)
21. Wang, D., Shi, L., Cao, J.: Fast algorithm for approximate k-nearest neighbor graph construction. In: 2013 IEEE 13th International Conference on Data Mining Workshops, pp. 349–356, December 2013
22. Wang, X., Tino, P., Fardal, M.A., Raychaudhury, S., Babul, A.: Fast Parzen window density estimator. In: 2009 International Joint Conference on Neural Networks, pp. 3267–3274, June 2009
23. Yianilos, P.N.: Data structures and algorithms for nearest neighbor search in general metric spaces. In: SODA, vol. 93, pp. 311–321 (1993)
24. Zhang, Y.-M., Huang, K., Geng, G., Liu, C.-L.: Fast kNN graph construction with locality sensitive hashing. In: Blockeel, H., Kersting, K., Nijssen, S., Železný, F. (eds.) Machine Learning and Knowledge Discovery in Databases, pp. 660–674. Springer, Heidelberg (2013)

Similarity-Based Outlier Detection in Multiple Time Series

Grzegorz Gołaszewski(✉)

Division for Information Technology and Systems Research, Department of Applied Informatics and Computational Physics, Faculty of Physics and Applied Computer Science, AGH University of Science and Technology, Kraków, Poland
Grzegorz.Golaszewski@agh.edu.pl

Abstract. Outlier analysis is very often the first step in data preprocessing. Since it is performed on mostly raw data, it is crucial that algorithms used are fast and reliable. These factors are hard to achieve when the data analysed is highly dimensional, such is the case with multiple time series data sets. In this article, various outlier detection methods (distance distribution-based methods, angle-based methods, k-nearest neighbour, local density analysis) for numerical data are presented and adapted to multiple time series data. The study also addresses the problem of choosing an appropriate similarity measure (L-p norms, Dynamic Time Warping, Edit Distance, Threshold Queries based Similarity) and its impact on efficiency in further analysis. Work has also been put into determining the impact of an approach to apply these measures to multivariate time series data. To compare the different approaches, a set of tests were performed on synthetic and real data.

1 Introduction

In most applications of time series data analysis, focus is put on finding novelties and abrupt changes in the temporal context of a particular time series itself. We have at our disposal a lot of methods for detecting such phenomena as regression modelling. A different problem is treating each time series as a single data element and trying to find within such elements those which are abnormal. There are many studies to find outliers within numerical data points, but in therms of time series data (and other highly specific data types), the methods used are mostly simple and not too sophisticated. There is a reason for this, with complex data types, even simple operations like comparing two objects to each other can have a high computational cost, so combining it with high cost analysis algorithms may be tempting but impossible in terms of actual usage. This study is an attempt to compromise numerical data outlier detection methods with the complex data similarity measures which time series are.

In Sect. 2, are presented a number of measures are presented that can be used to determine the similarity level between time series. The L-p norms are commonly used, since these have a computational complexity of $O(n)$, so it is fast

P. Kulczycki et al. (Eds.): ITSRCP 2018, AISC 945, pp. 116–131, 2020.
https://doi.org/10.1007/978-3-030-18058-4_10

and reliable in most cases. However, there are assumptions that the compared series are the same lengths which is a huge downside of them. Another popular approach is the use of elastic methods such as those originating from speech recognition, dynamic time warping and edit distance methods (like EDR and ERP). These methods are free to use with various length series at the cost of $O(n^2)$ complexity. A different approach is mapping data to the threshold-crossing time interval plane introduced in TQuEST, and then treating the time series as set of spatial points in order to find the similarity value. This method allows a different view of the data since it focuses more on frequency and points in time of crossing certain thresholds instead on comparing the time series value by value, which makes this method more noise-proof.

In Sect. 3, various method of finding outliers in numerical datasets and their adaptations are presented to be used with multiple time series datasets. A simple distance from the average value followed by a T-value test is easy to apply to numerical data, where "average value" is well defined. In time series comparison, we cannot average all time series to have a single mean value. Instead, there is shown a method of finding an "average-like" time series to which other comparisons are made. Another approach to identifying outliers is to find which part of data impacts the most in terms of changing variance in similarities between particular time series. Distance based methods are also discussed. The well known k-nearest neighbour algorithm is used almost everywhere to find similar data points and its use in time series outlier search is easy to apply. Local density methods allows identifying high density clusters of data, which makes outliers easy to spot.

Section 4 is concerned with the comparison test of different approaches to find outliers in data sets with using different similarity measures. The synthetic data generation algorithms CBF and Two-pat were used as benchmark data sets. As natural data, a Japanese Vowels set was used. As the returned results were the number of outliers found in relation to the number of outliers in test set and the number of false outliers found.

Section 5 contains a summary of results and conclusions on choosing the right outlier detecting method and the impact of the similarity measure on the efficiency on those methods.

2 Similarity Measures

In most cases in time series analysis, there is the assumption that time is one discrete dimension, while the rest of the dimensions are behavioural attributes. Throughout the article, a single n-dimensional time series of length m is $X = [X_1, X_2, ..., X_m]$, where $X_i = (x_i^{(1)}, x_i^{(2)}, ..., x_i^{(n)}, t_i)$, where t corresponds to the time attribute and $x^{(j)}$ corresponds to the attribute in j-th dimension, with the time series Y and Z accordingly. To keep it simple, there is the assumption that within the j-th behavioural dimension there is a defined metric $d^{(j)}$ or the whole time series can be compared using a n-dimensional measure d, which does not necessarily have to be metric.

2.1 L-p Norms

The simplest and most popular similarity measure are L-p norms defined as

$$L_p(X,Y) = (\sum_{i=1}^{m} d(X_i, Y_i)^p)^{1/p}.$$

If p = 1, the metric simplifies to

$$L_1(X,Y) = \sum_{i=1}^{m} d(X_i, Y_i)$$

and is called the Manhattan metric, another widely used L-p norm is $p = 2$ Euclidean metric, which can be geometrically interpreted as the distance between two points in space. In fact, in the case of $n = 1$, the whole time series are treated as data points in m-dimensional euclidean space. Advantages of L-p measures are the fact that these are metrics and have $O(m)$ computational cost. The biggest drawback of L-p measures is that these are lock-step measures which means that those can be used only with time series data of the same length and with the same time discretization.

2.2 Elastic Measures

To address the problem of comparing time series with warped time dimension, a number of the methods were introduced with most popular dynamic time warping procedure first proposed by Sakoe and Chiba [25] in speech recognition problems and then adapted to compare time series data by Berndt and Clifford [6]. The whole procedure is based on the optimization of the Euclidean or Manhattan distance within the time series stretched along the time dimension. The algorithm can be shown recursively as

$$DTW_{X,Y}(X_i, Y_j) = \delta(X_i, Y_j) + min \begin{cases} DTW_{X,Y}(i-1, j) \\ DTW_{X,Y}(i, j-1) \\ DTW_{X,Y}(i-1, j-1) \end{cases} .$$

Another approach to elastic time series measurement was based on text data analysis, where the defined Levenshtein measure as the fewest number of letter deletions, inputs and substitutions that allows the transformation of one word into the another. This idea was introduced into time series analysis as the Edit Distance (EDR) [10], which is defined recursively as

$$EDR_{X,Y}(i,j) \begin{cases} j & \text{if } i = 0 \\ i & \text{if } j = 0 \\ EDR_{X,Y}(i-1, j-1) & \text{if } d_{edr}(X_i, Y_j) = 0 \\ min \begin{cases} EDR_{X,Y}(i-1, j-1) + d_{edr}(X_i, Y_j) \\ EDR_{X,Y}(i-1, j) + d_{edr}(X_i, \varrho) \\ EDR_{X,Y}(i, j) + d_{edr}(\varrho, Y_j) \end{cases} & \text{otherwise} \end{cases} ,$$

where d_{edr} is the edit cost defined as follows

$$d_{edr}(X_i, Y_j) = \begin{cases} 0 & \text{if } d(X_i, Y_j) \leq \tau \\ 1 & \text{if } d(X_i, Y_j) > \tau \\ 1 & \text{if } X_j \equiv \varrho \text{ or } Y_j \equiv \varrho \end{cases}.$$

Since both DTW and EDR are not meeting triangle inequality, these are not metrics. Because of this, Edit Distance with Real Penalty [9] was introduced in which edit cost is not fixed and is defined as follows

$$d_{erp}(X_i, Y_j) = \begin{cases} d(X_i, Y_j) & \text{if } X_i \neq g \text{ and } Y_j \neq g \\ d(X_i, g) & \text{if } Y_j \equiv g \\ d(g, Y_j) & \text{if } X_i \equiv g \end{cases},$$

where g is called gap and is a predefined parameter, usually set to $g = 0$, as ERP authors suggest. This allows ERP to be defined as

$$ERP_{X,Y}(i,j) \begin{cases} \sum_{k=1}^{j} d(g, Y_k) & \text{if } i = 0 \\ \sum_{k=1}^{i} d(X_k, g) & \text{if } j = 0 \\ min \begin{cases} ERP_{X,Y}(i-1, j-1) + d_{erp}(X_i, Y_j) \\ ERP_{X,Y}(i-1, j) + d_{erp}(X_i, \varrho) \\ ERP_{X,Y}(i, j) + d_{erp}(\varrho, Y_j) \end{cases} & \text{otherwise} \end{cases},$$

which meets triangle inequality.

Elastic methods in general have a computational cost of $O(m^2)$. To address the high computational cost, a number of improvements were made. Saoke and Chiba [25] introduced warping windows to bound optimization paths, not only reducing calculation time but reducing the number of pathological outcomes. Another approach to the warping window was proposed by Itakura [16]. For the problem of searching for the most similar time series, lower bounds measures were introduced which have $O(m)$ computational cost and allows calculating minimal distance between time series that can be obtained using elastic measures, the simplest was proposed by Kim [18], followed by Yi [32] and Keogh [17].

2.3 Threshold Queries Based Similarity

A different approach to comparing time series data was taken by Assfalg [4]. Each time series is first represented in a threshold-crossing time interval sequence, which can be shown on the time interval plane, this allows distance between time intervals calculating as an Euclidean distance between

$$d_{int}(t_1, t_2) = \sqrt{(t_{l,1} - t_{l,2})^2 + (t_{u,1} - t_{u,2})^2}.$$

For each time interval of both time series, the most similar time interval point of other series is found and the sum of all distances are reported as a similarity measure, as follows

$$d_{TS}(X^*, Y^*) = \frac{1}{m_X^*} \sum_{i=1}^{m_X^*} min_{j=1}^{m_Y^*} d_{int}(X_i^*, Y_j^*) + \frac{1}{m_Y^*} \sum_{j=1}^{m_Y^*} min_{i=1}^{m_X^*} d_{int}(Y_j^*, X_i^*),$$

where all parameters with star (*) are corresponding values on the time intervals plane.

Threshold Queries similarity uses a predefined threshold value. Similarity computations focus not on actual values but on time points where the series reaches certain points, making this measure have applications for example in defect detection, financial analysis or in general temporal dependency detection.

2.4 Multidimensional Time Series Data

All the mentioned similarity measures are presented in original papers in the context of comparing one-dimensional time series, but can be extended to target multidimensional time series data. This can be done with two approaches. As long as there can be defined the Φ-dimensional metric d for each time point of the time series, the methods can be used straightforwardly using this metric. Problems may occur if each dimension is non-comparable with others, which can happen if the time series contains mixed data types like numerical and categorical data, defining a single similarity metric within time points may be difficult if not impossible. Another example of non-comparable dimensions may be data with high differences in variances within dimensions. In this case, even if values within dimensions were comparable (in the example all were numerical data), it may be ineffective, because of high contrast between dimensions. This problem can be addressed by treating each dimension separately. This allows defining similarity measures for each dimension and reporting the whole time series similarity as a n-dimensional similarity vector instead of a single similarity value. This idea can easily be adapted to the n^*-dimensional (with $n^* < n$) similarity by merging connected dimensions (in example treating spatial dimensions as single similarity measure, while treating other dimensions separately). Using this approach requires adapting algorithms which are using similarity measures to work with a similarity vector instead of a single value. Doing so in outlier analysis is discussed in Sect. 3 and its effectiveness in comparison to one-dimensional similarity with n-dimensional metric is tested in Sect. 4.

3 Multiple Time Series Outlier Analysis

In this chapter, a number of outlier searching algorithms will be shown. Since in most applications, time series data cannot (or at least should not) be interpreted as an n-dimensional data point, many algorithms may not find use. Moreover, there is no assurance that time series will be the same length, and this could make many spatial based methods impossible to use. Because of this, in this article there is the assumption that all that is known about time series, beside simple statistics like length of series, is the similarity measure between each series. In the multivariate case, this assumption is also extended to the n-dimensional similarity vector.

3.1 Distance Distribution Based Method

Analysis is based only on the similarity between time series, which can be interpreted as the distance between series it will be assumed that if multiple n-dimensional objects were placed in space, based on a central limit theorem, those would have n-dimensional normal distribution. Therefore the distribution of distance from the mean of this distribution would be Chi-squared distribution of (n-1)-degree. The mean value of time series object cannot be found, but it can be expected that in data set there is the object most similar to the mean object and its average square distance from other objects should be minimal, this object would be treated as a reference object. Another problem occurs with determining m, as stated there is no assurance that all time series are of the same length. The value of m can be set to an average length of time series in the set, since it would average out the Chi-squared of different degrees. Having distribution would allow reporting the time series as the outlier if the Chi-squared distribution-based probability of distance being higher than the distance from the reference object were lower than $p = 0.27\%$, based on the well-known "three sigma rule".

3.2 Angle Based Method

This method is based on the observation that the smallest area containing all data points limited by the angle pointing at a single data point has a smaller angle for the isolated data point than for points lying in close proximity to others. Using this knowledge, the angle-based outlier factor (ABOF) can be defined as

$$ABOF(X) = Var_{\{Y,Z \in S/\{X\}\}} WCos(X, Y, Z),$$

where

$$WCos(X, Y, Z) = \frac{< X - Y, X - Z >}{\|X - Y\|_2^2 \cdot \|X - Z\|_2^2}$$

and S is the set of all time series.

The method uses spatial interpretation of points to determine vectors between which the angle is computed. To apply it to the distance-based analysis, the law of cosines may be used. This allows the cosine angle between 3 objects to be computed, pointing at one of them, so the weighted cosine would be defined as follows

$$WCos(X, Y, Z) = -\frac{d(Y, Z)^2 - d(X, Y)^2 - d(X, Z)^2}{2 \cdot d(X, Y)^2 \cdot d(X, Z)^2}.$$

This approach has a computational cost of $O(n^3)$, which may lead to it being not applicable in some cases, but since cosine is weighted by the distance between points, to speed up calculations without losing too much accuracy, the k-nearest neighbours may be used as the data set for calculating ABOF for each point instead of the whole set. The points with the lowest ABOF are reported as outliers.

3.3 K-Nearest Neighbour Analysis

This algorithm is widely used in outlier analysis [19], due to its simplicity and effectiveness. The method uses the distance from the k-most similar object as the outlier factor, with higher value implicating more outlying objects.

3.4 Local Outlier Factor

A similar idea with a different application uses the local density [8] of data objects to determine which objects may be classified as outlying. Let's assume that the distance to the k-nearest neighbour for X is known as $D_k(X)$ and the set of all points with a distance lower than $D_k(X)$ as $L_k(X)$ and define the reachability distance $R_k(X,Y)$ as

$$R_k(X,Y) = max\{d(X,Y), D_k(Y)\},$$

with the observation that $R_k(X,Y) \neq R_k(Y,X)$. If the average reachability distance $(AR_k(X))$ is calculated for each $Y \in L_k(X)$, then the local outlier factor (LOF) is defined as

$$LOF_k(X) = avg_{Y \in L_k(X)} \frac{AR_k(X)}{AR_k(Y)}.$$

3.5 Multidimensional Time Series Outlier Analysis

All the methods shown use as their base a similarity measure between time series; therefore they can be used with multidimensional time series with defined n-dimensional metric without any modifications. If similarity is defined as a n-dimensional vector, there is a need to modify this approach. This can be done in a few ways, one being aggregating the vector using the length of the vector as the single similarity value, in this case, it is recommended to use some weighted length measure like Mahalanobis distance, since different dimensions may have different variations and can be dependent on each other. Another approach is to calculate the outlier factor for each dimension separately and again return its value as the n-dimensional vector. In the outlier detection application, this method may be especially useful since the target of analysis is to find atypical objects, and aggregating multidimensional results to one value may average out the potential outlier. Therefore, to determine whether the object may be counted as an outlier, the outlier factor in the dimension where it is highest (or lowest, depending on method) in regard to variance within this dimension would be taken into account. It is worth noting that different dimensions may be taken for different objects, since each object can be atypical in other ways.

4 Experiments

In order to test the effectiveness of outlier detection methods and the impact of the similarity measure, a series of experiments were performed. 1-dimensional time series data were generated using CBF [13] and Two-pat [12] algorithms. CBF allows time series data from 3 classes to be generated, while Two-pat allows data from 4 classes to be generated.

One-Dimensional Problem
For each class in an algorithm set of 9990 objects from that class and 10 objects randomly from other classes was generated, making data sets with 1% outlying objects. The goal for the algorithm was to mark all objects from other classes as outlying and not to mark class objects. For each set class subset, results were then accumulated. For the Angle-based method, the k-nearest neighbour and the Local Outlier Factor, those which had an outlying factor greater than 3 standard deviations from the average outlying factor were reported as outlying objects. For the distance distribution method, outlying factors were computed using a Chi-squared distribution probability using Chi-squared distribution of 128° as it is the length of time series in both CBF and Two-pat sets. As the threshold probability, $p = 0.27\%$ was set.

Tables 1, 2, 3 and 4 show results for one-dimensional experiments. Both fractions of outlier objects correctly marked are shown as are the number of false outliers with regard to set size.

Table 1. CBF - fraction of outliers detected

	Angle based	Distance distribution	K-nearest neighbour	Local density
Euclidean	0.9667	0.9667	1.0	1.0
Manhattan	1.0	1.0	1.0	1.0
DTW	0.9667	1.0	1.0	1.0
EDR	1.0	0.9333	1.0	1.0
ERP	1.0	1.0	1.0	1.0
TQuEST	0.0	0.3	0.6	0.6333

Table 2. CBF - fraction of regular objects marked as outliers

	Angle based	Distance distribution	K-nearest neighbour	Local density
Euclidean	0.0	0.0	0.0	0.0
Manhattan	0.0	0.0	0.0	0.0
DTW	0.0	0.0	0.0	0.0
EDR	0.0	0.0	0.0	0.0
ERP	0.0	0.0	0.0	0.0
TQuEST	0.0	0.0030	0.00346	0.0054

Table 3. Two-pat - fraction of outliers detected

	Angle based	Distance distribution	K-nearest neighbour	Local density
Euclidean	0.0	0.0	0.0	0.1
Manhattan	0.0	0.0	0.025	0.15
DTW	0.0	0.0	1.0	0.775
EDR	0.0	0.0	0.425	0.625
ERP	0.0	0.0	1.0	0.4
TQuEST	0.0	0.025	0.1	0.0

Table 4. Two-pat - fraction of regular objects marked as outliers

	Angle based	Distance distribution	K-nearest neighbour	Local density
Euclidean	0.0	0.0	0.0	0.0
Manhattan	0.0	0.0	0.0005	0.0008
DTW	0.0	0.0	0.0096	0.0058
EDR	0.0	0.0	0.0028	0.0023
ERP	0.0	0.0019	0.0005	0.0020
TQuEST	0.0	0.0129	0.0189	0.0207

Multidimensional Problem

To generate multidimensional data using CBF and Two-pat for each object in the data set, 3 time series for CBF and 4 for Two-pat were generated, and were concatenated to create 3 and 4 dimensional time series data for CBF and Two-pat, respectively. As natural data set, Japanese Vowels [20] set were used. This has 640 12-dimensional objects of lengths 7 to 29 in 9 classes. For each class, 3 random objects from other classes were added as outliers.

Table 5. CBF multidimensional - separate dimensions - fraction of outliers detected

	Angle based	Distance distribution	K-nearest neighbour	Local density
Euclidean	1.0	1.0	1.0	1.0
Manhattan	1.0	1.0	1.0	1.0
DTW	1.0	1.0	1.0	1.0
EDR	1.0	1.0	1.0	1.0
ERP	1.0	1.0	1.0	1.0
TQuEST	1.0	1.0	1.0	1.0

In the multidimensional problem, two approaches were tested. The first being the analysis of each dimensional separately. Each object marked as outlying in any dimension was marked as being globally outlying.

Tables 5, 6, 7, 8, 9 and 10 show results for analysis of each dimension separately.

The second approach used computing similarity defined for multidimensional data. In this case, the Euclidean distance was used as distance measure for the

Table 6. CBF multidimensional - separate dimensions - fraction of regular objects marked as outliers

	Angle based	Distance distribution	K-nearest neighbour	Local density
Euclidean	0.0707	0.0	0.0	0.0
Manhattan	0.0859	0.0	0.0	0.0
DTW	0.1378	0.0	0.0	0.0
EDR	0.1589	0.0	0.0	0.0
ERP	0.2222	0.0	0.0	0.0
TQuEST	0.2222	0.0081	0.0101	0.0162

Table 7. Two-pat multidimensional - separate dimensions - fraction of outliers detected

	Angle based	Distance distribution	K-nearest neighbour	Local density
Euclidean	0.6	0.0	0.0	0.4
Manhattan	0.8	0.0	0.1	0.6
DTW	0.8	0.0	1.0	1.0
EDR	0.9	0.0	1.0	1.0
ERP	0.9	0.0	1.0	1.0
TQuEST	0.9	0.1	1.0	1.0

Table 8. Two-pat multidimensional - separate dimensions - fraction of regular objects marked as outliers

	Angle based	Distance distribution	K-nearest neighbour	Local density
Euclidean	0.0677	0.0	0.0	0.0
Manhattan	0.0778	0.0	0.0020	0.0030
DTW	0.0778	0.0	0.0394	0.0253
EDR	0.2030	0.0	0.0485	0.0323
ERP	0.2030	0.0071	0.0485	0.0389
TQuEST	0.2030	0.0485	0.1020	0.1131

Table 9. Japanese Vowels - separate dimensions - fraction of outliers detected

	Angle based	Distance distribution	K-nearest neighbour	Local density
Euclidean	0.8519	0.7778	0.9630	0.9630
Manhattan	0.8889	0.7037	0.9630	0.9630
DTW	0.48148	0.8148	0.9630	0.9259
EDR	0.1481	0.0	0.0741	0.0741
ERP	0.9259	0.7037	0.8889	0.8889
TQuEST	0.7037	0.2593	0.2963	0.2963

Table 10. Japanese Vowels - separate dimensions - fraction of regular objects marked as outliers

	Angle based	Distance distribution	K-nearest neighbour	Local density
Euclidean	0.2391	0.0188	0.1047	0.1125
Manhattan	0.2453	0.0203	0.1156	0.1234
DTW	0.0672	0.0203	0.1109	0.0984
EDR	0.0813	0.0	0.0203	0.0125
ERP	0.2516	0.0313	0.1141	0.1188
TQuEST	0.1922	0.0078	0.0453	0.0422

time series points. It is worth mentioning that this approach is significantly slower than the separate dimensions approach.

Tables 11, 12, 13, 14, 15 and 16 show results for the analysis with multidimensional similarity function.

Table 11. CBF multidimensional - multidimensional similarity - fraction of outliers detected

	Angle based	Distance distribution	K-nearest neighbour	Local density
Euclidean	1.0	1.0	1.0	1.0
Manhattan	1.0	1.0	1.0	1.0
DTW	1.0	1.0	1.0	1.0
EDR	0.0	1.0	1.0	1.0
ERP	1.0	1.0	1.0	1.0
TQuEST	0.0	0.0	0.0	0.0

Table 12. CBF multidimensional - multidimensional similarity - fraction of regular objects marked as outliers

	Angle based	Distance distribution	K-nearest neighbour	Local density
Euclidean	0.0	0.0	0.0	0.0
Manhattan	0.0	0.0	0.0	0.0
DTW	0.0	0.0	0.0	0.0
EDR	0.0	0.0	0.0	0.0
ERP	0.0	0.0	0.0	0.0
TQuEST	0.0	0.0061	0.0121	0.0192

Table 13. Two-pat multidimensional - multidimensional similarity - fraction of outliers detected

	Angle based	Distance distribution	K-nearest neighbour	Local density
Euclidean	0.0	0.0	0.2	0.2
Manhattan	0.0	0.0	0.0	0.0
DTW	0.3	0.0	0.8	0.4
EDR	0.0	0.0	0.2	0.1
ERP	0.0	0.0	0.8	0.5
TQuEST	0.0	0.0	0.0	0.0

Table 14. Two-pat multidimensional - multidimensional similarity - fraction of regular objects marked as outliers

	Angle based	Distance distribution	K-nearest neighbour	Local density
Euclidean	0.0	0.0	0.0	0.0010
Manhattan	0.0	0.0	0.0	0.0010
DTW	0.0010	0.0	0.0030	0.0020
EDR	0.0	0.0	0.0010	0.0040
ERP	0.0	0.0	0.0010	0.0030
TQuEST	0.0	0.0162	0.0141	0.0192

Table 15. Japanese Vowels - multidimensional similarity - fraction of outliers detected

	Angle based	Distance distribution	K-nearest neighbour	Local density
Euclidean	0.0	0.1481	0.6296	0.7037
Manhattan	0.0	0.1481	0.7037	0.6667
DTW	0.0	0.1481	0.6667	0.5926
EDR	0.0	0.0	0.1852	0.1852
ERP	0.0	0.1481	0.3333	0.4444
TQuEST	0.0	0.0	0.0741	0.0741

Table 16. Japanese Vowels - multidimensional similarity - fraction of regular objects marked as outliers

	Angle based	Distance distribution	K-nearest neighbour	Local density
Euclidean	0.0	0.0	0.0047	0.0031
Manhattan	0.0	0.0	0.0047	0.0031
DTW	0.0	0.0	0.0031	0.0031
EDR	0.0	0.0	0.0172	0.0172
ERP	0.0	0.0	0.0078	0.0078
TQuEST	0.0	0.0063	0.0172	0.0172

5 Discussion and Summary

In the one-dimensional problem, for CBF data all but one measure had faultless or close to faultless performace for all outlier detection methods, while for Two-pat data only DTW and ERP measures kept this performance and only for k-NN algorithm, DTW and EDR in the Local Density approach worked relatively well. Both in terms of finding outliers and in avoiding marking regular objects as outliers, the worst performance was achieved using TQuEST which could be somehow expected since this method relie on crossing a certain threshold as a recognition pattern. In both CBF and Two-pat sets, the variety between classes were based on local trend rather than global changes, which were similar for all classes.

In the multidimensional problem for the separate dimensions approach, all methods had faultless performance for CBF data, with the Angle-based measure having issues with marking regular objects as outliers. In Two-pat the same issues appeared, with the addition of the Distance Distribution method, poor performance and elastic measures had significantly higher performance than L-p norms. It is worth mentioning that TQuEST had much better performance than in the one-dimensional experiment. In the Japanese Vowels set, both L-p norms and elastic methods with the exception of EDR worked relatively similarly and again k-NN and LOF worked out better both in the detecting outlier and in the avoiding marking regular objects. In the mulitdimensional similarity approach, all but the TQuEST measures worked well for CBF data (with the exception of EDR measure in the Angle-based method), but for the Two-pat set only DTW and EDR in k-NN method allowed more than half of the outliers to be detected. For the Japanese Vowels set, results were significantly lower than in the separate dimension approach and only Euclidean, Manhattan and DTW in the k-NN and LOF methods allowed more than half of the outliers to be detected.

The experiments allow the statement that in general cases, it can be expected that elastic measures will perform better than L-p norms, which is desired since with this in mind these measures were created. Another observation may be that K-nn and Local Density (with little favor for k-NN) perform better than the Angle-based methods and Distance Distribution. This can be caused by the fact that k-NN and LOF are based on the similarity of objects and the Angle-based and Distance Distribution methods require some kind of geometric interpretation for the object which may not always be provided or have practical use.

Acknowledgment. This work was partially supported by the Faculty of Physics and Applied Computer Science of the AGH University of Science and Technology.

The primary version of this paper was presented at the 3rd Conference on Information Technology, Systems Research and Computational Physics, 2–5 July 2018, Cracow, Poland [14].

References

1. Achtert, E., Kriegel, H.P., Reichert, L., Schubert, E., Wojdanowski, R., Zimek, A.: Visual evaluation of outlier detection models. In: Kitagawa, H., Ishikawa, Y., Li, Q., Watanabe, C. (eds.) Database Systems for Advanced Applications, pp. 396–399. Heidelberg, Springer, Berlin (2010)
2. Aggarwal, C.C.: Data Mining: The Textbook. Springer, Heidelberg (2015)
3. Aggarwal, C.C.: Outlier Analysis, 2nd edn. Springer, Heidelberg (2016)
4. Aßfalg, J., Kriegel, H.P., Kröger, P., Kunath, P., Pryakhin, A., Renz, M.: Similarity search on time series based on threshold queries. In: Ioannidis, Y., Scholl, M.H., Schmidt, J.W., Matthes, F., Hatzopoulos, M., Boehm, K., Kemper, A., Grust, T., Boehm, C. (eds.) Advances in Database Technology - EDBT 2006, pp. 276–294. Heidelberg, Springer, Berlin (2006)
5. Ben-Gal, I.: Outlier Detection, pp. 131–146. Springer, Boston (2005)
6. Berndt, D.J., Clifford, J.: Using dynamic time warping to find patterns in time series. In: Proceedings of the 3rd International Conference on Knowledge Discovery and Data Mining. AAAIWS 1994, pp. 359–370. AAAI Press (1994)
7. Bouguessa, M.: Modeling outlier score distributions. In: Zhou, S., Zhang, S., Karypis, G. (eds.) Advanced Data Mining and Applications, pp. 713–725. Springer, Heidelberg (2012)
8. Breunig, M., Kriegel, H.P., Ng, R.T., Sander, J.: Lof: identifying density-based local outliers. In: Proceedings of the 2000 ACM SIGMOD International Conference on Management of Data, pp. 93–104. ACM (2000)

9. Chen, L., Ng, R.: On the marriage of lp-norms and edit distance. In: Proceedings of the Thirtieth International Conference on Very Large Data Bases, vol. 30. VLDB 2004, pp. 792–803. VLDB Endowment (2004)
10. Chen, L., Özsu, M.T., Oria, V.: Robust and fast similarity search for moving object trajectories. In: Proceedings of the 2005 ACM SIGMOD International Conference on Management of Data. SIGMOD 2005, pp. 491–502. ACM, New York (2005)
11. Ding, H., Trajcevski, G., Scheuermann, P., Wang, X., Keogh, E.: Querying and mining of time series data: experimental comparison of representations and distance measures. Proc. VLDB Endow. **1**(2), 1542–1552 (2008)
12. Geurts, P.: Contributions to decision tree induction: bias/variance tradeoff and time series classification, January 2002
13. Geurts, P.: Pattern extraction for time series classification. In: De Raedt, L., Siebes, A. (eds.) Principles of Data Mining and Knowledge Discovery, pp. 115–127. Springer, Heidelberg (2001)
14. Gołaszewski, G.: Similarity-based outlier detection in multiple time series. In: Kulczycki, P., Kowalski, P.A., Łukasik, S. (eds.) Contemporary Computational Science, p. 68. AGH-UST Press, Cracow (2018)
15. Hodge, V.J., Austin, J.: A survey of outlier detection methodologies. Artif. Intell. Rev. **22**(2), 85–126 (2004)
16. Itakura, F.: Readings in speech recognition, pp. 154–158. Morgan Kaufmann Publishers Inc., San Francisco (1990)
17. Keogh, E., Ratanamahatana, C.A.: Exact indexing of dynamic time warping. Knowl. Inf. Syst. **7**(3), 358–386 (2005)
18. Kim, S.W., Park, S., Chu, W.W.: An index-based approach for similarity search supporting time warping in large sequence databases. In: Proceedings of the 17th International Conference on Data Engineering, pp. 607–614. IEEE Computer Society, Washington, DC (2001)
19. Knorr, E.M., Ng, R.T.: Algorithms for mining distance-based outliers in large datasets. In: Proceedings of the 24th International Conference on Very Large Data Bases. VLDB 1998, pp. 392–403. Morgan Kaufmann Publishers Inc., San Francisco, CA (1998)
20. Kudo, M., Toyama, J., Shimbo, M.: Multidimensional curve classification using passing-through regions. Pattern Recogn. Lett. **20**(11), 1103–1111 (1999)
21. Kuhnt, S., Pawlitschko, J.: Outlier identification rules for generalized linear models. In: Baier, D., Wernecke, K.D. (eds.) Innovations in Classification, Data Science, and Information Systems, pp. 165–172. Springer, Heidelberg (2005)
22. Kulczycki, P., Charytanowicz, M., Kowalski, P.A., Łukasik, S.: Identification of atypical (rare) elements-a conditional, distribution-free approach. IMA J. Math. Control Inf. (2017, in press)
23. Kulczycki, P., Kruszewski, D.: Identification of atypical elements by transforming task to supervised form with fuzzy and intuitionistic fuzzy evaluations. Appl. Soft Comput. **60**(C), 623–633 (2017)
24. Petrovskiy, M.I.: Outlier detection algorithms in data mining systems. Program. Comput. Software **29**(4), 228–237 (2003)
25. Sakoe, H., Chiba, S.: Readings in Speech Recognition, pp. 159–165. Morgan Kaufmann Publishers Inc., San Francisco (1990)
26. Schubert, E., Zimek, A., Kriegel, H.P.: Local outlier detection reconsidered: a generalized view on locality with applications to spatial, video, and network outlier detection. Data Min. Knowl. Disc. **28**(1), 190–237 (2014)
27. Seo, Y.S., Bae, D.H.: On the value of outlier elimination on software effort estimation research. Empirical Software Eng. **18**(4), 659–698 (2013)

28. Shaikh, S.A., Kitagawa, H.: Top-k outlier detection from uncertain data. Int. J. Autom. Comput. **11**(2), 128–142 (2014)
29. Tang, J., Chen, Z., Fu, A.W.C., Cheung, D.W.: Enhancing effectiveness of outlier detections for low density patterns. In: Chen, M.S., Yu, P.S., Liu, B. (eds.) Advances in Knowledge Discovery and Data Mining, pp. 535–548. Springer, Heidelberg (2002)
30. Vlachos, M., Hadjieleftheriou, M., Gunopulos, D., Keogh, E.: Indexing multidimensional time-series. VLDB J. **15**(1), 1–20 (2006)
31. Yang, H., Yang, T.: Outlier mining based on principal component estimation. Acta Math. Applicatae Sin. **21**(2), 303–310 (2005)
32. Yi, B.K., Jagadish, H.V., Faloutsos, C.: Efficient retrieval of similar time sequences under time warping. In: Proceedings of the Fourteenth International Conference on Data Engineering. ICDE 1998, pp. 201–208. IEEE Computer Society, Washington, DC (1998)

Metaheuristics in Physical Processes Optimization

Tomasz Rybotycki(✉)

Centre of Information Technology for Data Analysis Methods,
Systems Research Institute, Polish Academy of Sciences, Warsaw, Poland
tomasz.rybotycki@ibspan.waw.pl
http://www.ibspan.waw.pl/~trybotyc/

Abstract. The subject of this work is applying the artificial neural network (ANN) taught using two metaheuristics - the firefly algorithm (FA) and properly prepared evolutionary algorithm (EA) - to find the approximate solution of the Wessinger's equation, which is a nonlinear, first order, ordinary differential equation. Both methods were compared as an ANN training tool. Then, application of this method in selected physical processes is discussed.

Keywords: Evolutionary algorithm · Firefly algorithm · Neural network · Wessinger's equation

1 Introduction

The aim of this paper is to explore how modern metaheuristics can affect physical computations. The case of approximating differential equations was selected as an example, as many physical processes are modeled using these kind of formulas. There are many methods (mostly iterative) of approximating differential equations, e.g. Runge-Kutta [4] or improved Euler being most popular. These are, however, inefficient, when it comes to solving equations which have fully implicit solutions such as Wessinger's equation (see [13] and [16]). In order to properly address this issue another kind of approximation method was used, namely artificial neural networks. This approach was exploited for this purpose in previous works, such as [13] and [16], this work, however, covers wider scope than aforementioned is such sense, that instead of providing comparison between ANN trained by a metaheuristic to numerical methods, it compares performance of multiple metaheuristics between themselves. Moreover instead of giving sole set of weights, that was deemed as best discovered suboptimal solution, in this work different resultant solutions are discussed. Lastly the issue of possible applications of this method to physical computations is briefly addressed.

This paper is organized as follows. In Sect. 2 neural networks are presented, as a method of approximating differential equations. The description of ANN's topology, used in experiments, is also covered in this part. Following section

P. Kulczycki et al. (Eds.): ITSRCP 2018, AISC 945, pp. 132–148, 2020.
https://doi.org/10.1007/978-3-030-18058-4_11

contains a brief introduction to metaheuristics in the context of using them as an ANN teaching tool. Two metaheuristics used in experiments are described in separate subsections. Next section consists of research methodology and experimental data. It also includes conclusions of the experiments. In the last section summary of this paper, possible areas of use in physics and future research ideas are presented.

2 Neural Networks

Neural networks are relatively old, nature-inspired technology, which, thanks to deep learning algorithms, is currently in the center of researchers attention, as it is capable of solving non-trivial tasks such as recreating a picture in selected painters style or driving a car [14]. Before that, however, ANNs were used for multiple simpler tasks such as data analysis, signal filtration and function prediction or approximation to name a few (see [12] and [23]). It's also a well known fact, that they can also be used to approximate differential equations in particular (see [15] or [3]).

There are multiple kinds of artificial neural networks. In this work, feedforward neural network (FFNN) was used for several reasons. First of all, it has been shown that this kind of ANN can be properly trained using metaheuristics, and several methods of mutations and crossovers have been proposed for EA in [10] for this exact task. Moreover, these have been shown to work well when estimating ordinary differential equations, Wessinger's equation in particular (see [16] and [13]). Lastly, the main focus of this research are metaheuristics, thus using the simplest kind of neural network was the most natural approach.

Neural networks can be summarized as a set of interconnected neurons, that are divided into layers. Neurons here are meant to be seen as a mathematical models of their biological equivalents. Many of them are known in the literature. The most common one is McCulloch-Pitts model (see [12] and [22]) and thus it's been used in this work. Each neuron is equipped with activation function A. For the purpose of performed researches, sigmoid function given by Eq. 1 has been used:

$$A(x) = Sig(x) = \frac{1}{1 + e^{\beta x}}, \tag{1}$$

where x is input value (of neuron), e is the base of natural logarithm and β is power coefficient. In this work, $\beta = 1$.

Sigmoid function was used for several reasons, one of them being that it's a standard activation function (along hyperbolic tangent). Moreover it's proven to work well for function approximation (see [13] and [16]). Lastly, thank to that function, the output signal of neuron is elegantly smoothed.

Aforementioned output value of i-th neuron is given by formula 2:

$$n_i = A\left(\sum_{j=1}^{n} w_{ij} x_j + b_j\right), \tag{2}$$

where x_j is the j-th neuron output value, b_j is j-th neuron bias, n is the number of neurons which are connected to i-th neuron and w_{ij} is the weight of the connection between i-th and j-th neuron.

There are several types of neural networks. The most basic are feedforward neural networks (FFNN) in which signal travels only forward thus neurons on i-th layer sends signal only to $(i+1)$-th layer neurons. In general there are no rules telling how to connect neurons to one another or how many connections should single neuron contain. There is, however, a specific, most common kind of FFNN, called multilayer perceptron, wherein all neurons from i-th layer are connected with each neuron from $(i+1)$-th layer (see [23]). These kind of ANNs were successfully used for approximating differential equations, thus were also used in this work (see [13,16]). General connection scheme of three layer perceptron has been shown on Fig. 1.

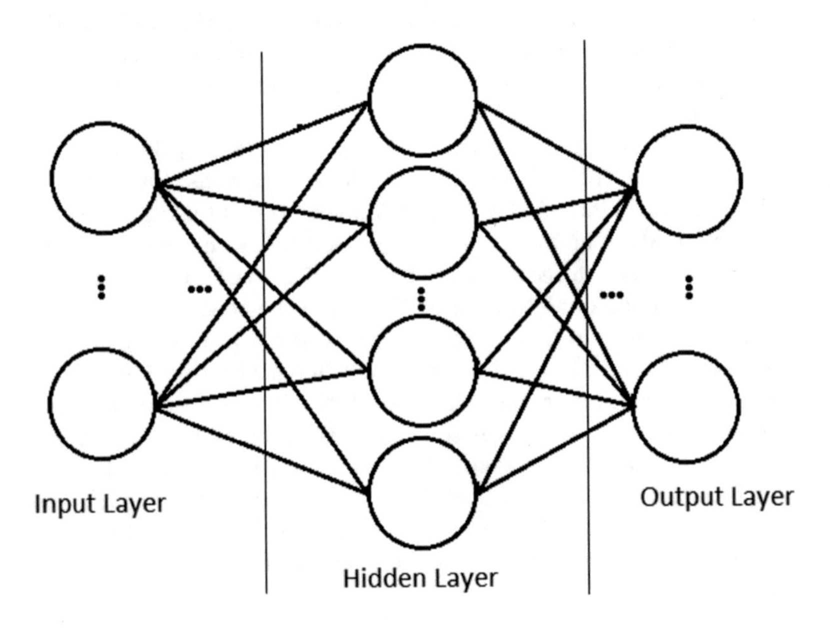

Fig. 1. Three layer perceptron.

The last important characteristic of neural networks is their topology. According to Kolmogorov's existence theorem every function of n parameters can be approximated by three-layer perceptron with $(2n + 1)$ nodes in hidden layer. It has be shown that performance of such net doesn't depend on number of hidden layers, but it strongly depends on number of neurons in hidden layer (see [1,20] and [11]). In this work, an ANN with topology 1-4-1 (meaning one input neuron, 4 neurons in hidden layer and one in output layer) was used. This topology was used for two reasons. Firstly it fulfills the condition given by Kolmogorov's theorem and secondly it has been shown in [13] and [16], that this kind of net works well for discussed task.

Another issue that has to be addressed when neural networks are considered is their training. In this work, instead of using classic approach and backpropagation algorithm, metaheuristics were used. The reason for that is that backpropagation (and other gradient based techniques), have tendencies to get stuck in local minima (see [17]). Another drawback is that gradient based methods require target function and neurons target function to be in C^1 function class (see [10]). Metaheuristics doesn't have these drawbacks. The method of using metaheuristics as training algorithms has been discussed in Sect. 4.1 of this work.

3 Metaheuristics

Metaheuristics are sets of stochastic techniques used to find suboptimal solutions for optimization problems, when deterministic approaches don't exist or are inefficient and when the problem domain is too large to use brute force approach (see [25]). Metaheuristics are divided into several groups, also including problem specific types, however, in this work only one group was considered – population-based metaheuristics. It has been shown in many works, for instance [5,10,13,16] and [9], that this type of techniques are proved to work well as training algorithms for ANN, hence selection. These methods are well known in the literature (see [18] and [25]) and their general characteristics won't be discussed further in this work. Problem specific traits, such as individuals representation or population initialization, shall be briefly discussed in this section.

The first problem that has to be addressed for population-based metaheuristics to work properly is individuals representation. In this work, an individual has to store information about weights of connections (or outputs weights) and biases of neurons (real numbers) of each neuron and connection in the net. Two options were considered – a one-dimensional vector of values and a n-dimensional vector of values, where n denotes the number of layers (excluding output layer). Although one-dimensional approach was successfully applied in previous work (see [9]) this time multi-dimensional method was used. The reasoning behind that is more intuitive applying of reproduction operators to individuals. For more details see [24].

Another important issue is the population initialization problem. It's necessary for individuals to be scattered on the domain (in the beginning), because it ensures that search space is thoroughly examined. The standard procedure for population initialization is filling representatives vector with random value (see [18]), however Montana presented more sophisticated approach to this matter, in which random values from two-sided exponential distribution are selected add appended to individuals vectors (see [9]). In this work, the latter is exploited. Another difficulty is determining population size, as it cannot be to low (for it would require many iterations) or too big (for computational reasons). In this work, fixed population size equal to 50 individuals was selected, where 10 is an arbitrary number, and 5 is the number of neurons with weighted output connections.

3.1 Evolutionary Algorithm

Evolutionary algorithm is relatively old conception loosely based on biological evolution. Because of that it's very well known in the literature. In this work, EA was prepared for a given task (ANN training) by applying specialized reproduction operators to it. Both mutation and crossover algorithms were selected from [10]. For mutation operator *node mutation* was selected and for crossover the *crossover weights* was selected. These reproductive methods proved to work best for ANN training in aforementioned work and that's the reason behind the selection. Evolutionary algorithms also contains one more important operator – the selection operator. In this work, *roulette wheel selection* was used as it's characterized by balanced selective pressure and thus eliminates the problem of algorithm stopping in the local extrema (see [18]).

3.2 Firefly Algorithm

Firefly algorithm is modern, nature-inspired metaheuristic, similar to Particle Swarm Optimization (in fact it can be reduced to PSO with it's attributes set to specific values, compare [25] and [7]). It is based on fireflies mating process. This method is described in great detail in [25], thus description of this algorithm will be reduced only to describing individuals movement. Position of firefly in $(i+1)$-th iteration is given by formula 3:

$$x_{i+1,k} = x_{i,k} + \beta_0 e^{-\gamma r_{ij}^2}(x_{i,j} - x_{i,k}) + \alpha\epsilon, \tag{3}$$

where $x_{i,k}$ denotes k-th firefly position in i-th iteration, α is step size, β_0 is base attraction of the firefly, e is a base of natural logarithm, r_{ij} is a distance between i-th and j-th firefly, ϵ is a pseudo-random vector generated during that iteration step for each firefly (meaning that it's neither global nor is it constant for each iteration) and γ is light absorption coefficient. In this work, euclidean distance, given by formula 4, was measured:

$$r_{ij} = \sqrt{\sum_{k=1}^{d}(x_{i,k} - x_{j,k})^2}, \tag{4}$$

where d is dimensions number, and $x_{i,k}$ denotes k-th attributes value of i-th agent (firefly).

There are a few additional things to mention. First of them is that fireflies adjust their position basing on position of each brighter firefly in the swarm. This means that Eq. 3 describes only a part of firefly's movement towards j-th brighter firefly. The brightest agent will in this case only move towards random direction given by vector $\alpha\epsilon$. In this nomenclature, brightness or attractiveness of a firefly can be described by a fitness function of given agent. For more details about FA see [24] and [25].

There were various reasons behind selecting firefly algorithms as an ANN training tool. First of them was that it has previously been shown, that FA can successfully train a FFNN (see [5]). Moreover, there already were similar works using *PSO* to solve the Wessinger's equation (see [13]), thus applying similar and more general algorithm seemed to be a natural approach additionally providing optional comparison data.

4 Experiments

This chapter is organized as follows. First and foremost methodology of researches is presented. In following subsection algorithms tuning method is discussed in greater detail. Lastly results of the experiments and their conclusions are presented in Subsect. 4.3.

4.1 Methodology

Wessinger's equation is given by formulas 5, 6 and 7:

$$tx^2x'^2 - x^3x'^2 + (t(t^2+1))x' - t^2x = 0, \quad 1 \leq t \leq 4, \tag{5}$$

$$x(1) = \sqrt{\frac{3}{2}} \tag{6}$$

$$x(4) = \sqrt{\frac{33}{2}} \tag{7}$$

where $x \equiv x(t)$ is an unknown function, t is parameter of this function and x' is it's first derivative.

Fully implicit solution to that equation is given by formula 8:

$$x_{imp}(t) = \sqrt{\frac{1}{2} + t^2}. \tag{8}$$

One can quickly notice, that it satisfies given boundary conditions.

In the previous work, one of the most serious problems was being unable to provide good approximate solution that would respect boundary conditions at the same time [8]. One of the ideas to tackle this issue, was dividing two separate elementary solutions – one respecting the boundary conditions, second one properly approximating the function in remaining points of the domain – and then joining these solutions into a final one. However, to properly apply numerical methods, one should first discretize domain of the problem. In this work, discrete set of values, given by formula 9 has been used:

$$T = \{t_i : 1 \leq t_i \leq 4, t_{i+1} = t_i + 0.1\}. \tag{9}$$

Similar sets were also used in [13] and [16] and proved to work well, hence the choice. Note, that boundary points has been kept in the discretized domain, to ensure, that proposed solution indeed fulfills them.

Formal representation of aforementioned two part solution would the be described by formula 10:

$$s(t, w) = B(t) + F(t, N(t, w)), \tag{10}$$

where s is solution, t is a point in which target function should be approximated, B is part of solution that satisfies boundary conditions, F is a part of solution that evaluates estimation values for the rest of the points in the domain and N is the output of ANN with given set of weights w.

In this work, same equation as in [13] has been used. It's expressed by following formula 11:

$$s(t, w) = \frac{\sqrt{66}}{6}(t-1) - \frac{\sqrt{6}}{6}(t-4) + (t-1)(t-4)N(t, w). \tag{11}$$

One may notice, that indeed, two first addends of the equation, handles the boundary condition explicitly and the last part is used to accurately approximate the rest.

The methodology of research is similar to one in [13] and [16], namely, instead of using standard backpropagation algorithm to teach ANN, a metaheuristic was used as it proves to yield better results for this kind of problem [10]. The task which metaheuristics are meant to solve is to find suboptimal values of weights and biases of neurons. These weren't used for creating ANNs (meaning considering different topologies or connection patterns), but only for determining the set of real values, that were applied to the network with selected, fixed topology as the weights and biases of neurons. In this nomenclature weights of neuron can be considered weights it's output connections.

For the fitness function, which in this case was maximized, the difference between 1 and squared errors sum for all the points of discretized domain was selected. This can be formally represented by formula 12:

$$f(a) = 1 - \frac{E(a)}{E_{max}}, \tag{12}$$

where a denotes solution represented by agent, f denotes fitness function, E denotes error and E_{max} is maximal error that occurred during search.

Fitness function in this form is not computationally complex, which is an important factor, because in population-based metaheuristics it's typical for the fitness function to be called thousands of times. Another characteristic of this function is that it's normalized and thanks to that agents performance is easier to compare. There's, however, one thing define, namely the error function E of proposed solutions.

Knowing that, in general, first order differential equations can all be expressed by formula 13:

$$f(t, x, x') = 0, \tag{13}$$

one can define the error function in the form of following formula:

$$E(a) = \sum_{t \in T} f^2\left(t, s(t, a), \frac{ds(t, a)}{dt}\right), \tag{14}$$

where the last parameter is first derivative of the solution and a denotes set of ANN weights and biases represented by given agent.

It can be noticed, that this function, in fact, satisfies all necessary requirements. The better the approximation is, the lower the error function gets. Each addend is squared in order to handle the possible negative sign (note that in case of larger errors absolute value can also be considered, however in [13] and in [16] this kind of method has been used, thus choice). Additionally, one may notice, that points scattered across whole domain are considered. The only thing left, that has to be found is t derivative of s. This derivative can be expressed by following formula 15:

$$\frac{ds(t,a)}{dt} = \frac{\sqrt{66}-\sqrt{6}}{6} + (t-4)(t-1)N' + (t-4)N + (t-1)N, \tag{15}$$

where $N' \equiv \frac{dN}{dt} \equiv \frac{dN(t,a)}{dt}$ is the t derivative of neural nets response N and the rest of symbols is identical with their previous explanation.

Using simple algebraic identities, one can formulate simplified version of this derivative as formula 16:

$$\frac{ds(t,a)}{dt} = (t^2 - 5t + 4)S' + (2t-5)S + const. \tag{16}$$

The next and last step in defining error function would be calculating t derivative of N. Knowing, that sigmoid function was selected as neuron activation function, it can be expressed by formula 17:

$$\frac{dN(t,a)}{dt} = \sum_{j=0}^{n} i_j o_j \left(\frac{1}{1+exp(-(i_j t + b_j))} - \left(\frac{1}{1+exp(-(i_j t + b_j))} \right)^2 \right), \tag{17}$$

where n is the number of neurons in the hidden layer (in this case 4), or to be more exact the number of neurons input-output weights pairs, i_j is j-th neuron input weight, o_j is j-th neuron output weight and exp is exponential function.

Moreover, before applying both metaheuristics, they had to be tuned. Tuning of the algorithms is described in next subsection. Because of stochastic nature of these methods, it was also necessary to perform experiments multiple times and consider an averaged result. Both metaheuristics were compared in terms of the convergence, dispersion, quality and variety of provided solutions.

4.2 Algorithms Tuning

To achieve necessary performance level of metaheuristics the algorithms had to be tuned adequately. Algorithm tuning is a process wherein suboptimal values of algorithm's parameters, such as number of iterations, population size and other, algorithm-specific attributes are found.

First tuned parameter was iterations number. For given population size (50) and repetitions number (10) different iteration numbers were tested (1000–10000 for EA and 500–2000 for FA) starting from the lowest. Theory claims that ANN performance should be better or the same, as the number goes up, thus the number after which the average fitness function values differed only slightly was picked as suboptimal iterations number. The numbers were equal to 7000 and 700 for EA and FA respectively. For more details and numeric data see [24]. The next step was searching for suboptimal population size. It was found by applying same repetitions number (10) for iterations number selected in previous step. For both algorithms same sizes were considered (30–70). EA and FA both performed at their best for 50 agents, just as in [10]. For more details and numerical data see [24].

Table 1. Proposed sets of firefly algorithms parameters.

i	β_0	α	γ
1	1	0.01	1
2	1	0.05	5
3	0.5	0.05	5
4	0.5	0.01	1

Finally, the algorithm specific parameters had to be selected. For EA this is mutation rate. It was lower than in [10] and was capped at 5%, which is more standard value for mutation rate (see [18]) than in [10]. As FA has not one, but three algorithm specific parameters – randomization parameter α, attractiveness variation γ and base attractiveness β_0. Knowing that for the most applications $\alpha \in [0, 1]$ and $\beta_0 = 1$ are the usual values, and that if $\beta_0 = 0$ the algorithm transforms from biased random walk to a normal walk, values for these both parameters were adequately chosen. As for γ, it's theoretical value can be selected from $[0, \infty)$, but usually it's picked from $[0.1, 10]$.

Table 2. Numerical values of errors during parameters tuning for FA.

Set\Run	1	2	3	4	5	6
1	2.49262	3.6834	4.67687	1.18447	2.72291	3.26068
2	1.33543	4.03752	2.18129	4.59007	1.76474	17.5748
3	1.04537	4.48536	2.21552	1.86416	1.08421	0.708
4	1.01501	0.99084	2.32162	3.24539	1.28454	1.7119
Set\Run	7	8	9	10	Average	
1	2.67247	3.37571	9.07496	2.04096	3.518505	
2	8.09901	7.52599	2.27193	2.04017	5.142095	
3	12.5581	1.64589	6.63027	4.30066	3.653754	
4	2.46041	2.43687	0.959687	2.8774	1.9303667	

Instead of testing performance of the algorithm for each of them separately 4 sets of attributes were proposed and tested. Respective values of each parameter for each set has been presented in Table 1.

The task, on which these set were tested, was the same task for which the algorithm was meant to be used – finding the sets of weights and biases for ANN. Average of final error values was used for verification purposes. The conclusions parameters tuning has been presented in Table 2.

Noticeably best performance was achieved by set number 4, thus this was selected.

4.3 Results

Noticing, that in typical situation, neural network training will be performed once, the accuracy of approximation was more important factor than training time. Nonetheless both of them were taken into consideration during experiments. On Figs. 2 and 3 comparison between errors and training times are presented.

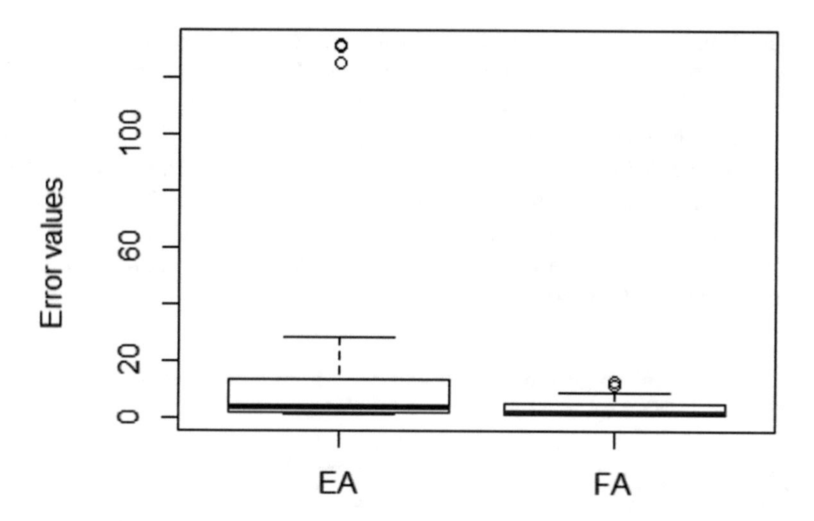

Fig. 2. ANN's errors when trained by respective algorithms.

In the Fig. 2 it can be seen, that in terms of training accuracy firefly algorithm outperformed evolutionary algorithm. Moreover, FA also is more consistent in terms of error value and hasn't shown the tendency to generate large errors. Evolutionary algorithm, on the other hand, was several times less consistent and generated sets of weights that, when applied to ANN, resulted in large error values. It is possible, that these values could be lowered with larger number of iterations, however to point of comparison was testing algorithms with values that has been found as suboptimal during algorithms tuning.

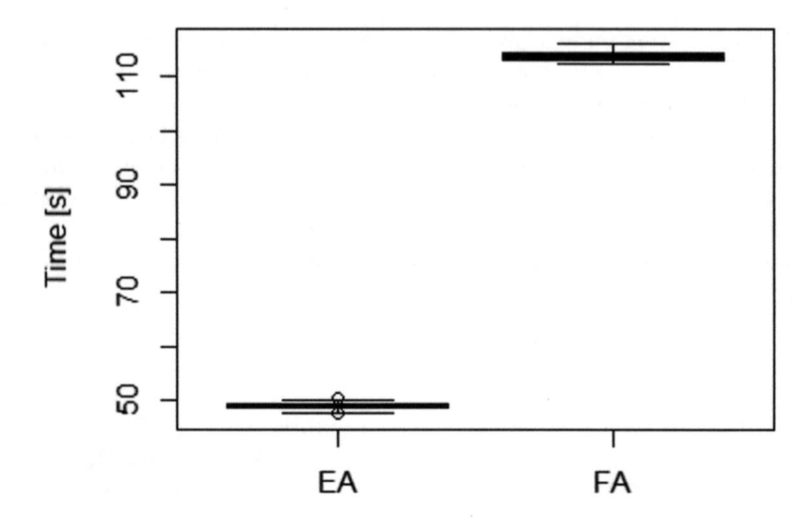

Fig. 3. Times of ANN's training when trained by respective algorithms (in seconds).

In the Fig. 3 one can see, that evolutionary algorithm outperforms fireflies in terms of training time. Despite it's 10 times bigger population size, it still manages to train ANN around 2 times faster than it's opponent. It can also be seen, that both algorithms are consistent in terms of training time – there are very few cases in which it differed significantly from the average (shown as circles on the figure). It has, however, to be noticed, that FA is much more sophisticated method and despite it's training time, it is far more accurate in terms of approximation than EA. Most important numerical data of both these experiments has been shown in Tables 3 and 4.

Table 3. The most important numerical data from EA temporal and accuracy experiments.

Name\Value	Min	1st quartile	2nd quartile	3rd quartile	Max
Error	0.933613	1.47776	3.580225	12.8494	131.457
Time [s]	47.499	48.583	48.87	49.2235	50.462

Along temporal aspects and accuracy of the algorithms a few theoretical properties of population-based algorithms were also studied. First of them was the convergence of both metaheuristics. Theory has it, that by the end of the training values of the fitness function should approach similar extreme values. The convergence of both algorithm has been shown of Figs. 4 and 5.

It can be seen, that the population converges. However, one may notice, that end values differs only slightly from ones at the beginning. That is due to fact how fitness function is normalized (with the maximal error value up to iteration).

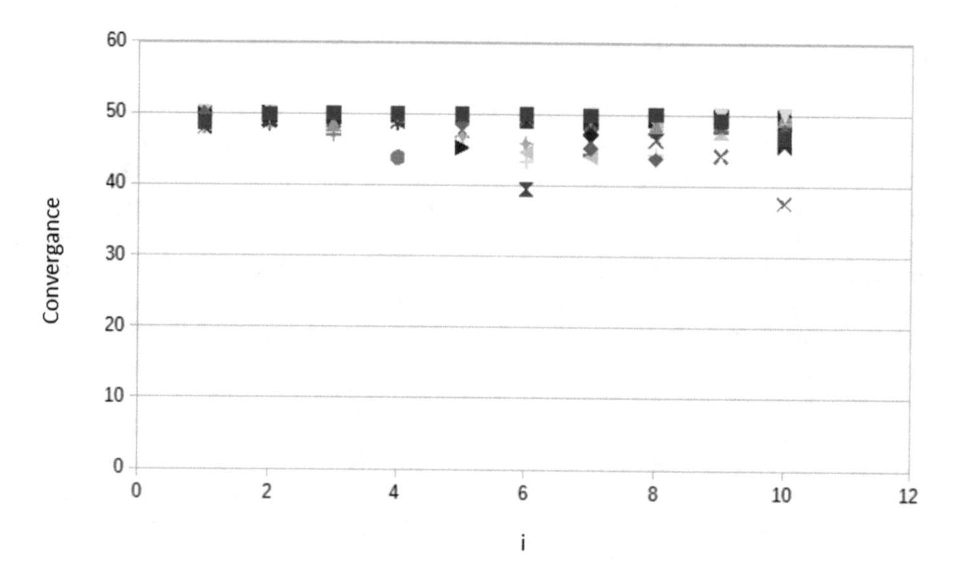

Fig. 4. The EA population convergence in $\frac{i}{10}$ fraction of training.

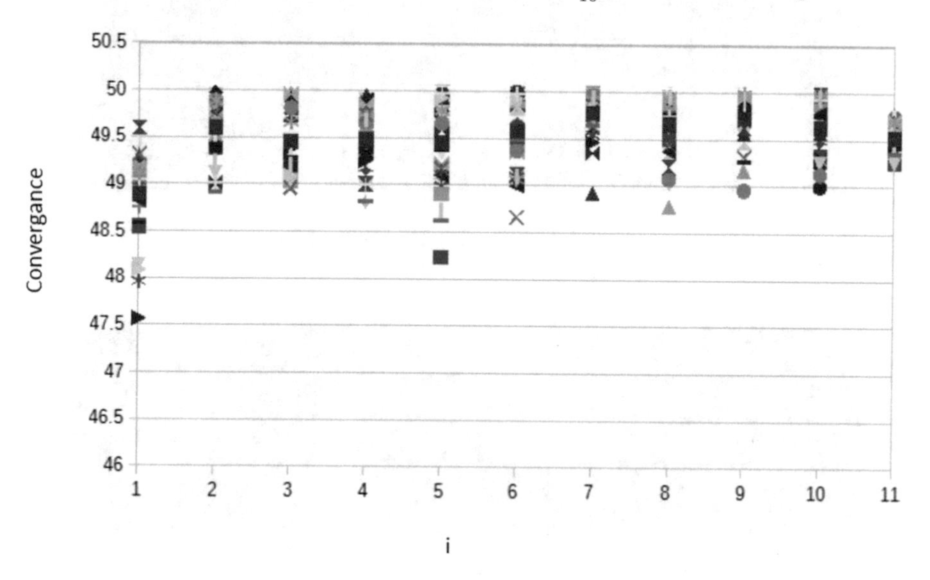

Fig. 5. The FA population convergence in $\frac{i}{10}$ fraction of training.

Table 4. The most important numerical data from FA temporal and accuracy experiments.

Name\Value	Min	1st quartile	2nd quartile	3rd quartile	Max
Error	0.803548	1.297595	2.05172	4.70623	12.7431
Time [s]	112.641	113.183	113.8315	114.624	116.1

This issue has been addressed further in [24] and is the reason why figures may seem to contradict the theory.

Another thing studied during the research was the dispersion of the population. Theoretically it should be lower by the end of the training, than it was at the beginning. To look into this issue, a standard deviation of population was measured at the start and at the end of training was computed. The researches proved once again, that theory is correct. The dispersion of population indeed lowered by the time algorithms finished finding suboptimal weights sets. It's worth noting, that EA was found to have lower dispersion than FA. Considering, that standard deviation was count based on averaged fitness function values, and that EA happened to stumble upon larger errors, this behavior was to be expected. For more details and numerical data see [24].

The other researches that were conducted during the experiments was determining variety of the suboptimal solutions provided by the algorithms. On the Figs. 6, 7 and 8 selected suboptimal sets of weights and biases, denoted further as a sets of weights, found by metaheuristics, have been presented.

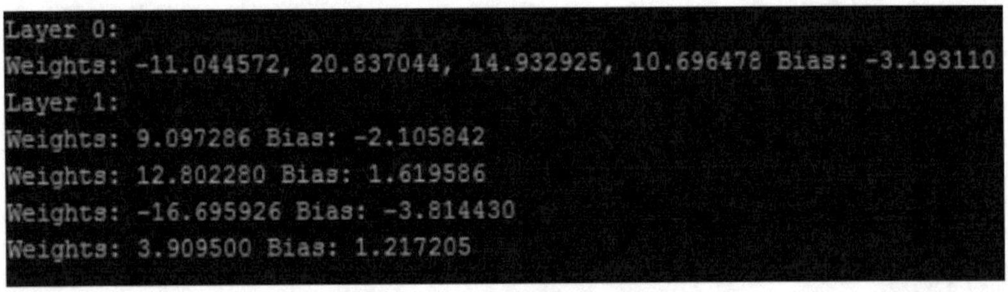

```
Layer 0:
Weights: -11.044572, 20.837044, 14.932925, 10.696478 Bias: -3.193110
Layer 1:
Weights: 9.097286 Bias: -2.105842
Weights: 12.802280 Bias: 1.619586
Weights: -16.695926 Bias: -3.814430
Weights: 3.909500 Bias: 1.217205
```

Fig. 6. First selected suboptimal set of weights.

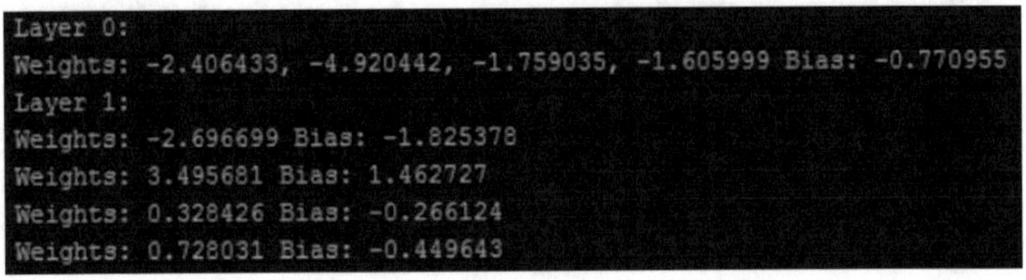

```
Layer 0:
Weights: -2.406433, -4.920442, -1.759035, -1.605999 Bias: -0.770955
Layer 1:
Weights: -2.696699 Bias: -1.825378
Weights: 3.495681 Bias: 1.462727
Weights: 0.328426 Bias: -0.266124
Weights: 0.728031 Bias: -0.449643
```

Fig. 7. Second selected suboptimal set of weights.

These figures are meant to be interpreted in the following way. Each line that starts with Layer i, denotes a new layer. Each line below represents a neuron of this layer. Each weight on position j-th describes the weight of the connection with j-th neuron of $i+1$-th layer. Weights are split by the comma. For example, in the Fig. 8, first weight 0.234483 (first under Layer 1), means, that connection of the first neuron of the hidden layer, and the only neuron of the input layer is equal to 0.234483. This neuron (connection) also has bias equal to 0.761027.

```
Layer 0:
Weights: 0.112296, 0.356483, -1.079639, -0.540388 Bias: 0.410000
Layer 1:
Weights: 0.234483 Bias: 0.761027
Weights: -0.224903 Bias: -0.517893
Weights: -0.130336 Bias: -0.425920
Weights: -0.269244 Bias: -0.941234
```

Fig. 8. Third selected suboptimal set of weights.

Analysis of Figs. 6, 7 and 8 yields important information about the form possible solutions. It's easy to notice, that most weights in each of them has different order of magnitude. This fact can be utilized in several ways. For example, knowing that suboptimal sets of values found on R^n, where n is the number of values in the set, has never exceeded a given value x (from both sides), the search domain can be reduced to e.g. $[-x, x]^n$, possibly reducing computation requirements of algorithms and increasing their accuracy (especially using fitness function proposed in this work). It's also an important fact to be aware of, that there exist many suboptimal solutions to the problem, as sometime it may be reasonable to prefer one of them over others (for a reason that was not included in fitness function). Variation analysis can also be used to study the consistency of algorithms. In this case, again, FA showed a high consistency, as solutions returned by it were similar to each other. EA, on the other hand had a tendency to provide solutions that could vary even by 2 orders of magnitude (see [24]).

Proving that ANN can do a decent job approximating differential equations, there is a huge opportunity for them to be utilized in physical computations (as this kind of processes tend to be described by differential equations very often). For instance in [2] ANNs were used for approximating nonlinear differential equations in molecular and atomic physics. It's worth noting, that in aforementioned work genetic algorithm was used to train ANN, whereas, as it was shown, FA proves to be better in terms of training accuracy. Thus applying modern metaheuristics to the same process, can optimize it even further.

There are many issues in physics that could potentially benefit from applying presented approach to solve problems related to them. E.g. in fluid dynamics, when approximating Navier-Stokes differential equation (see [6]) or in optics, when one have to deal with Klein-Gordon equation or Shrödinger equation (see [19]). There is a lack of researches proving (or disproving) the hypothesis that applying proposed technique would reduce (or simplify) computations necessary to deal with the problems that includes aforementioned or other differential equations. Further studies should tackle that issue.

For more details and numerical values of described experiments see [24].

5 Discussion

The aim of this work applying selected metaheuristics as training algorithms of ANN and then using this ANN for approximating Wessinger's equation. A comparison of both algorithms was also provided.

In most of the compared characteristics (all of them excluding time) modern metaheuristic – the firefly algorithm – proved superior to the evolutionary algorithm. In particular, it was more consistent in terms of accuracy, dispersion and convergence as well as it provided solutions with better accuracy overall. This behavior was to be expected as new algorithms are required to outperform their predecessors.

Another matter addressed in this work was subject of varying solutions. It has been shown that problem of ANN training, thus possibly also other optimization issues, can have many suboptimal solutions that differs significantly from one another. This fact can be utilized in many ways, one of which was mentioned throughout this work. In the future researches it is necessary to tackle this issue further and check the validity of authors hypothesis.

Moreover there is the problem of high maximal errors of neural net, that happened to occur in experiments and resulted in obfuscating the fitness values of individuals. They tended to be extremely high and thus the algorithm could lose the ability to distinct better solutions when near local extrema. In the future work this matter shall be addressed further, possibly by applying reduced domain (as proposed earlier in this work) or by reducing maximal possible error, either by capping maximum at some fixed value or by rejecting solutions with error higher than some constant.

The last issue, is applying proposed method to variety of physical, practical computation problems that have to deal with differential equations, including, but not limited to, formulas mentioned earlier. Future works shall focus on this matter and verify the hypothesis whether ANN trained by a metaheuristic can optimize the computation of differential equations on a level that satisfies practical, real-life, physical problems.

Acknowledgments. This work was supported by the Systems Research Institute of the Polish Academy of Sciences and is extended version of paper presented at 3rd Conference on Information Technology, Systems Research and Computational Physics, 2–5 July 2018, Cracow, Poland [21].

References

1. Malek, A., Beidokhti, R.S.: Numerical solution for high order differential equations using a hybrid neural network-optimization method. Appl. Math. Comput. **183**, 260 (2006)
2. Caetano, C., Reis Jr., J.L., Amorim, J., Lemes, M.R., Dalpino Jr., A.: Using neural networks to solve nonlinear differential equations in atomic and molecular physics. Int. J. Quantum Chem. **111**, 2732–2740 (2011)

3. Lagaris, I.E., Likas, A., Fotiadis, D.I.: Artificial neural networks for solving ordinary and partial differential equations. IEEE Trans. Neural Networks **9**, 987–1000 (1998)
4. Press, W.H., Teukolski, S.A., Vetterling, W.T., Flannery, B.P.: Numerical Recipes in C. The Art of Scientific Computing, 2nd edn. Press Syndicate of the University Press, Cambridge (1992)
5. Brajevic, I., Tuba, M.: Training feed-forward neural networks using firefly algorithms. In: Proceedings of the 12th International Conference on Artificial Intelligence, Knowledge Engineering and Data Bases, p. 156 (2013)
6. Acheson, D.J.: Elementary Fluid Dynamics. Oxford Applied Mathematics and Computing Sciences Series. Oxford University Press, Oxford (1990)
7. Kennedy, J., Eberhart, R.: Particle swarm optimization. In: Proceedings of IEEE International Conference on Neural Networks (1942)
8. Khan, J., Raja, M., Qureshi, I.: Swarm intelligence for the problems of non-linear ordinary differential equations and it's application to well known Wessinger's equation. Eur. J. Sci. Res. **34**, 514–525 (2009)
9. Montana, D.J.: Neural network weight selection using genetic algorithms. Intell. Hybrid Syst. (1995)
10. Montana, D.J., Davis, L.: Training feedforward neural networks using genetic algorithm. In: Proceedings of the International Joint Conference on Artificial Intelligence, pp. 762–767 (1989)
11. Hornick, K., Stinchcombe, M., White, H.: Multilayer feedforward networks are universal approximators. Neural Networks **2**(5), 359–366 (1989)
12. Fausett, L.: Fundamentals of Neural Networks: Architectures, Algorithms and Applications. Prentice-Hall Inc., Upper Saddle River (1994)
13. Biglari, M., Ghoddosian, A., Poultangari, I., Assareh, E., Nedaei, M.: Using evolutionary algorithms for solving a differential equation. Global J. Sci. Eng. Technol. **14**, 25–32 (2013)
14. Bojarski, M., Del Testa, D., Dworakowski, D., Bernhard, F., Flepp, B., Goyal, P., Jackel, L.D., Monfort, M., Mullar, U., Zhang, J., Zhang, X., Zhao, J., Zieba, K.: End to end learning for self-driving cars (2016)
15. Chiaramonte, M.M., Kiener, M.: Solving differential equations using neural networks (2013). http://cs229.stanford.edu/proj2013/ChiaramonteKiener-SolvingDifferentialEquationsUsingNeuralNetworks.pdf
16. Ghalambaz, M., Noghrehabdi, A.R., Behrang, M.A., Assareh, E., Ghanbarzadeh, A., Hedayat, N.: A hybrid neural network and gravitational search algorithm (HNNGSA) method to solve well known Wessinger's equation. Int. J. Mech. Aerosp Ind. Mechatron. Manuf. Eng. (2011)
17. Gori, M., Tesi, A.: On the problem of local minima in backpropagation. IEEE Trans. Pattern Anal. Mach. Intell. **14**, 76–86 (1992)
18. Engelbrecht, A.P.: Computational Intelligence: An Introduction, 2nd edn. Wiley, Chichester (2007)
19. McGurn, A.R.: Nonlinear Optics of Photonic Crystals and Meta-Materials. Morgan & Claypool Publishers, San Rafael (2016)
20. Lippman, R.P.: An introduction to computing with neural nets. IEEE ASSP Mag. **4**, 4–22 (1987)
21. Rybotycki, T.: Metaheuristics in physical processes optimization. In: Kulczycki, P., Kowalski, P.A., Lukasik, S. (eds.) Contemporary Computational Science, p. 205. AGH-UST Press (2018)
22. McCulloch, W.S., Pitts, W.: A logical calculus of the ideas immanent in nervous activity. Bull. Math. Biophys. **5**, 115–133 (1943)

23. Phan, D.T., Liu, X.: Neural Networks for Identification, Prediction and Control. Springer, London (1995)
24. Rybotycki, T.: Nowoczesne metaheurystyki w optymalizacji procesów fizycznych. Master's Thesis, University of Silesia, Sosnowiec (2016)
25. Yang, X.: Nature-inspired Metaheuristic Algorithms, 2nd edn. Luniver Press, Frome (2010)

Crisp vs Fuzzy Decision Support Systems for the Forex Market

Przemysław Juszczuk[1(✉)] and Lech Kruś[2]

[1] Faculty of Informatics and Communication, Department of Knowledge Engineering, University of Economics, 1 Maja 50, 40-287 Katowice, Poland
przemyslaw.juszczuk@ue.edu.pl

[2] Systems Research Institute, Polish Academy of Sciences, Newelska 6, 01-447 Warsaw, Poland
krus@ibspan.waw.pl

Abstract. A new concept of the multicriteria fuzzy trading system using the technical analysis is proposed. The existing trading systems use different indicators of the technical analysis and generate buy or sell signal only when assumed conditions for a given indicator are satisfied. The information presented to the trader – decision maker is binary. The decision maker obtains a signal or no. In comparison to the existing traditional systems called as crisp, the proposed system treats all considered indicators jointly using the multicriteria approach and the binary information is extended with the use of the fuzzy approach. Currency pairs are considered as variants in the multicriteria space in which criteria refer to different technical indicators. The introduced domination relation allows generating the most efficient, non-dominated (Pareto optimal) variants in the space. An algorithm generated these non-dominated variants is proposed. It is implemented in a computer-based system assuring sovereignty of the decision maker.

We compare the proposed system with the traditional crisp trading system. It is made experimentally on different sets of real-world data for three different types of trading: short-term, medium and long-term trading. The achieved results show the computational efficiency of the proposed system. The proposed approach is more robust and flexible than the traditional crisp approach. The set of variants derived for the decision maker in the case of the proposed approach includes only non-dominated variants, what is not possible in the case of the traditional crisp approach.

Keywords: Trading system · Forex · Fuzzy membership function · Multicriteria analysis

1 Introduction

In this paper, we propose an extension of the traditional trading systems based on the technical analysis using the concept of fuzziness. This concept can be implemented in a decision support system aiding the trader in making his decisions.

P. Kulczycki et al. (Eds.): ITSRCP 2018, AISC 945, pp. 149–163, 2020.
https://doi.org/10.1007/978-3-030-18058-4_12

As a field for experiments, we selected the Forex market, which is a global, decentralized market with currency pairs as basic instruments. According to the Bank for International Settlements, its average daily turnover reached $5.3 trillion in January 2014 [13]. Unlike other markets, the Forex is completely decentralized and housed electronically. It is considered as one of the largest markets in the world, where about 90% of its turnover is generated by currency speculators. Still growing number of instruments available for the trader – decision maker in the last few years make it very difficult to manually manage even the single transaction.

Three main approaches of financial data analysis are used to forecast prices on the Forex market: the technical analysis, the fundamental analysis, and the sentiment analysis. The fundamental analysis including the text-mining techniques was adapted to the stock market in [8,16], as well as for the Forex market in [15]. The sentiment analysis called also the opinion mining was presented in [4]. The technical analysis is based on the assumption, that there it is possible to predict future prices on the basis of the historical prices, in other words, that past behavior of the price has an effect on the future prices. The process of construction of such indicator can be considered as a dimension reduction [17]. In such approach, the initial data is transformed to another domain which may be simpler than the original data. Such action leads to a situation, where price and technical indicator values become an independent example of the problem.

One of the main directions of development of the trading systems based on the technical analysis consists in using various indicators which are mostly some complex formulas used for historical data to extract hidden information from price time series or to reduce some irrelevant noise. Such approach is used to identify moments to open the positions on the market. The technical analysis is the most popular tool used in the trading. Moreover, its importance is increasing over years [6].

By the trading system, we understand any system (manual or automatic), which with the use of data analyzed from the market calculates values of the market indicators. Such values are further used to generate a signal and open the position related to the selected currency pair. However, there exists a significant drawback, where all selected indicators must give the signal at the same time, thus increasing the number of market indicators leads to more seldom signals. What is obvious, even in the case, where all necessary conditions all fulfilled, there is no guarantee, that derived signal will be profitable.

There are numerous examples of fully automatic trading systems. In [12] the authors proposed to use different volatility measures as an input for the support vector machines. While in [11] ARIMA model was compared with the artificial neural network for the prediction on the Forex market. More complex systems involved the use of modern metaheuristics like Cuckoo Search Algorithm [1] or heuristic-based trading systems combining different trading rules [14] are proposed as well.

Newer works like [2] suggest, that especially very volatile markets like Forex may be very difficult to analyze. That leads to various works which limit the application of the fully automatic concept of trading systems for the decision support. There are also manual trading systems which generate signals for opening and closing positions on the market, where the final decision is made by the decision maker. One of the most significant advantages of such manual trading systems is that the decision maker may additionally apply other types of analysis on the market situations. Examples of such systems can be found in [3,5,7]. Thus one of the most important advantages of such systems is that they assure sovereignty of the decision maker.

The paper discusses decision support problems in the case of the manual trading systems. A typical system analyses data from the market, calculates technical indicators, generates buy or sell signals when a given set of rules is satisfied for a given indicator. The decision maker observing the signals can make the final decision. Still growing number of instruments available for the decision makers results in a dynamic growth of the decision space. Therefore a new approach capable to handle such difficulties is required. In this article, we investigate problems arising in the case of the traditional crisp trading systems, where the decision is made on the basis of simple binary function. The first limitation of such systems is encountered, where all initially defined rules should return the "true" value to open the position at the same time. By increasing number of rules included in the system results in reduction of the number of possibilities for the decision maker is significantly reduced and often none of the variants are considered as a promising. The traditional system disregard situations when all the indicators are very close to satisfying the assumed rules. Such situations can be in general much more promising than in the case of the signal when the rules are satisfied for only one indicator.

To cover this gap we propose to apply a multicriteria fuzzy approach. In this approach, all considered indicators are considered jointly. Each indicator is represented by a criterion in a multicriteria space of variants. The traditional strict rules are replaced by fuzzy rules. Values of the criteria are calculated by introduced membership functions. The decision maker can control the introduced fuzziness using concepts of aspiration and reservation points adopted from the reference point approach of the multicriteria analysis (see [18]). An algorithm generated non-dominated variants in the criteria space is proposed. Concepts of the domination cons applied in the algorithm ensure the high computational efficiency of the algorithm approved in experiments made on real data from the Forex market.

We present the proposed multicriteria approach for n technical indicators. The introduced fuzzy concepts are shown on the example of three indicators: the moving averages (MA), Relative Strength Index (RSI) and Commodity Channel Index (CCI). For the three indicators, the results of computation experiments on real data from the Forex market are presented and analyzed. Examples of the technical indicators are presented in Sect. 2. Section 3 introduces the concept of the traditional (crisp) trading system. Section 4 includes a detailed description of the proposed fuzzy approach along with definitions of all fuzzy membership functions included in the system. The algorithm which generates the non-dominated variants for the decision maker is proposed. Section 5 contains results of various experiments with different real-world data sets. The discussion presented along with experiments results points out the weakness of existing crisp systems and emphasizes advantages of the proposed fuzzy system.

2 Examples of Technical Indicators

In the following three equally important market indicators frequently used in the transaction systems are presented: the moving average and two oscillators. The proposed approach can, however, include any number of indicators.

The moving average equation is given as follows:

$$MA_p(t) = \frac{\sum_{i=1}^{p} price_i}{p}, \tag{1}$$

where $MA_p(t)$ is the value of the moving average for period p in time t, $price_i$ is a currency pair value for a given time i, and p is the number of included values. An example concept based on moving averages may be found in [9].

The first technical indicator belonging to the group of oscillators is the Relative Strength Index (RSI):

$$RSI_p(t) = 100 - \frac{100}{1 - \frac{avg_{gain}}{avg_{loss}}}, \tag{2}$$

where $RSI_p(t)$ is the value of the RSI indicator calculated on the basis of the last p periods in time t, avg_{gain} is the sum of gains over the past p periods and avg_{loss} is the sum of losses over the past p periods. The second oscillator to be used is the Commodity Channel Index (CCI):

$$CCI_p(t) = \frac{1}{c} \cdot \frac{price_{typical} - MA_p(t)}{\sigma(price_{typical})}, \tag{3}$$

where $CCI_p(t)$ is the value of the CCI indicator calculated on the basis of p periods in time t, $price_{typical}$ is the typical price calculated as the average value of the Close, Low and High price from a given period, σ is the mean absolute deviation and c is the constant value used for scaling the mean absolute deviation value; for $CCI_{20}(t)$ this value is equal to 0.015.

3 Crisp Trading System

Actions of the typical crisp system can be described with the use of a binary activation function. The function takes the value one when a respective condition for a technical indicator is true and takes the value zero otherwise. The signal to open a position on the market is generated only in the first case.

Let us denote these conditions for the considered indicators as $cond_{MA}$, $cond_{RSI}$ and $cond_{CCI}$. A potential BUY signal may be generated for a given currency pair when at least one of the conditions is fulfilled:

$$f_{buy} = true \text{ if } (cond_{MA_{Buy}} = true \vee cond_{RSI_{Buy}} = true \vee cond_{CCI_{Buy}} = true), \quad (4)$$

where conditions $cond_{MA_{Buy}}$, $cond_{RSI_{Buy}}$, $cond_{CCI_{Buy}}$ refer to the moving averages, RSI and CCI indicators respectively. If neither of the conditions is fulfilled the currency pair is removed from further analysis:

$$f_{buy} = false \text{ if } \neg(cond_{MA_{Buy}} = true \vee cond_{RSI_{Buy}} = true \vee cond_{CCI_{Buy}} = true), \quad (5)$$

The typical conditions used in the existing trading systems for the considered technical indicators are presented below. The condition for the moving averages takes the form:

$$cond_{MA_{Buy}} = true \text{ if } (MA_{fast}(t) > MA_{slow}(t)) \wedge (MA_{fast}(t-1) < MA_{slow}(t-1)), \quad (6)$$

where $MA_{fast}(t-1)$ is the value of the moving average from the lower period in time $t-1$, $MA_{slow}(t-1)$ is the value of the moving average from the higher period in time $t-1$. An example signal is generated if two moving averages cross each other.

In the case of the oscillators RSI and CCI, the binary activation functions are built on the basis of crossing the indicator with some predefined levels. For RSI this level will be 30. In the case of CCI it is -100. The conditions take the form:

$$cond_{RSI_{Buy}} = true \text{ if } (RSI_p(t-1) < 30) \wedge (RSI_p(t) > 30), \quad (7)$$

and

$$cond_{CCI_{Buy}} = true \text{ if } (CCI_p(t-1) < -100) \wedge (CCI_p(t) > -100), \quad (8)$$

where $RSI_p(t-1)$ and $CCI_p(t-1)$ denote respectively the values of RSI and CCI in time $t-1$.

4 Proposed Fuzzy Trading System

In the proposed system the different indicators are considered jointly and the activation conditions are fuzzy. Each currency pair is treated as a variant in a multicriteria decision space. Criteria in this space refer to particular indicators. Values of the criteria are defined by membership functions referring to particular indicators. It is assumed that the membership function for each indicator takes values in the range $\langle 0, 1 \rangle$. The membership function takes the value 1 when the value 1 is achieved by the binary activation function in the crisp approach.

In the fuzzy approach, the original signal generated in the case of crisp approach is still included. However, the situation when the conditions for a given indicator are almost satisfied, omitted in the crisp approach, can be included in the fuzzy approach with the use of the membership function.

Let each currency pair c will be treated as a variant y in the decision space $\mathbb{R}^n$; thus every variant y will be denoted as the vector of criteria $y = (y_1, y_2, ..., y_n)$, $y_i \in \langle 0, 1 \rangle$, $i = 1, 2, ..., n$, where n is the number of considered indicators. The criteria are defined by values of the membership function calculated for particular indicators.

Due to limited space, we introduce only membership functions related to the BUY signals. The membership functions for the SELL signals can be defined in a similar way. The membership function for the MMA indicator is proposed in the form:

$$\mu_{MA-BUY}(c) = \begin{cases} \frac{max - f_{low}}{max} \text{ if } (MA_{fast}(t) > MA_{slow}(t)) \wedge (f_{low} < max) \\ \wedge (MA_{fast}(t-1) > MA_{slow}(t-1)) \\ 1 \text{ if } (MA_{fast}(t) > MA_{slow}(t)) \\ \wedge (MA_{fast}(t-1) < MA_{slow}(t-1)) \\ \frac{f_{high}}{max} \text{ if } (MA_{fast}(t) < MA_{slow}(t)) \wedge (f_{high} < max) \\ \wedge (MA_{fast}(t-1) < MA_{slow}(t-1)) \\ 0 \text{ in other case} \end{cases} \tag{9}$$

where max is the maximal number of readings used in the calculations, f_{high} is a function used to count readings above the moving average with a higher period and f_{low} is a function used to count readings below the moving average with a higher period. It is assumed in the above calculations that in the case of reading without the crossover of moving averages the possibility of a trend change would rise while the present trend would continue.

The membership function defined for the RSI indicator is given as follows:

$$\mu_{RSI-BUY}(c) = \begin{cases} \frac{RSI_p(t)}{30} \text{ if } (RSI_p(t) < 30) \\ 1 \text{ if } ((RSI_p(t-1) < 30) \wedge (RSI_p(t) > 30)) \\ \vee (RSI_p(t) = 31) \\ \frac{0.9}{RSI_p(t) - 30} \cdot \alpha \text{ if } (RSI_p(t) > 31) \\ \wedge (RSI_p(t) < 50) \wedge (RSI_p(t-1) \leq 30) \\ 0 \text{ if } (RSI_p(t) > 50) \end{cases} \tag{10}$$

In the case of the CCI indicator the membership function takes the form:

$$\mu_{CCI-BUY}(c) = \begin{cases} 0 \text{ if } (CCI_p(t) < CCI_{min}) \\ \frac{CCI_p(t) - CCI_{min}}{-CCI_{min} - 100} \text{ if } (CCI_p(t) > CCI_{min}) \wedge (CCI_p(t) < -100) \\ 1 \text{ if } (CCI_p(t-1) < -100) \wedge (CCI_p(t) > -100) \\ \frac{CCI_p(t) + 50}{-50} \text{ if } (CCI_p(t) > -100) \wedge (CCI_p(t) < -50) \\ \wedge (CCI_p(t-1) > -100) \\ 0 \text{ if } (CCI_p(t) > -50) \end{cases} \tag{11}$$

where $CCI_p(t)$ is a value of the CCI indicator in the present time, CCI_{max} is the maximal considered CCI value and CCI_{min} is the minimal considered CCI value. A vector of scalar values in the range of $\langle 0; 1 \rangle$ is generated in a given time t for all of the given indicators and represents each currency pair as a variant in the multicriteria space. In this space, we made multicriteria analysis and look for the Pareto optimal (non-dominated) variants. Respective domination relations have to be introduced.

The following relations between variants are introduced in $\mathbb{R}^n$ space:

Definition 1. A variant y is at least as preferred as a variant z if each criterion of y is not worse than the respective criterion of z.

$$y \succeq z \Leftrightarrow (y_1 \geq z_1) \wedge (y_2 \geq z_2) \wedge ... \wedge (y_n \geq z_n). \tag{12}$$

Definition 2. A variant y is more preferred (better) than a variant z according to the logical formulae:

$$y \succ z \Leftrightarrow (y \succeq z) \wedge \neg(z \succeq y). \tag{13}$$

An algorithm deriving non-dominated variants is proposed. The following notions are used in the algorithm: the ideal – aspiration point u, the reservation point $x = (x_1, x_2, ..., x_n)\_where x_i \in [0, 1]$, the set of all variants Y, the set of points removed from the analysis in the algorithm Y^-, the set of points accepted for further analysis in the algorithm $Y^+ = Y \setminus Y^-$, the set of non-dominated variants ND. The aspiration point u refers to the case when BUY signals are generated for all indicators, i.e. when all the membership functions take the value 1. The reservation point x is defined by the minimum values of membership functions accepted by the decision maker.

The simplified idea of the algorithm is given below:

- **Step 0**. In this initial step, the sets Y and $ND = \emptyset$ are created. The aspiration point $u = (1, \ldots, 1)$ and the reservation point x assumed by the decision maker are fixed.
- **Step 1**. The set Y^- is generated as the set of points dominated by the reservation point and removed from further analysis. All other points belong to the set Y^+.
- **Step 2**. If there exists variant $y \in Y^+$, $y = u$, then $ND = \{y\}$. End of the algorithm.
- **Step 3**. Each variant $y \in Y^+$ is checked: if $y \in Y^-$ then it is removed from further analysis, else it is compared with the points in the set ND (it is added to ND if ND is empty). For each point $z \in ND$, if y dominates z, then z is removed from ND, y is added to ND and the set Y^- is extended by the set of point dominated by y; if y is dominated by z then y is removed from analysis, i.e. removed from the set Y^+.

The algorithm ends, when all variants from the set Y^+ are checked.

In the algorithm, the concept of domination cons is used. For each point Y added to the set ND, the set Y^- is extended using the domination cone. The successive extensions of this set of points removed from analysis assure the high computational efficiency of the algorithm.

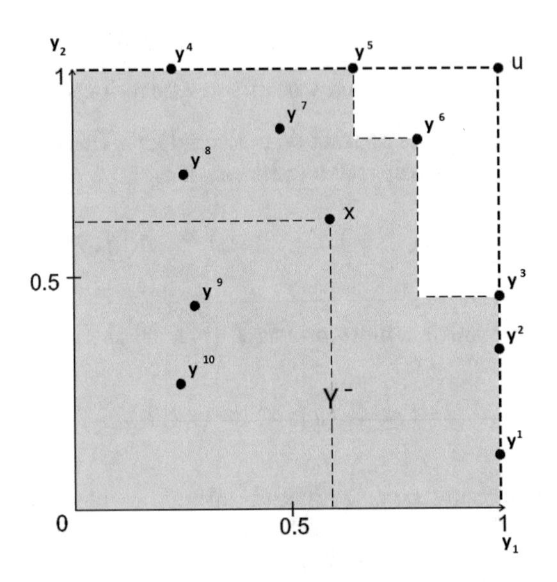

Fig. 1. An illustrative example

It has been proved that the algorithm derives all non-dominated variants in the set of variants non-dominated by the reservation point x. All other variants are eliminated from the analysis.

The algorithm has been implemented in a computer-based trading system to make experiments using real data from the Forex market.

The fuzzy approach introduces additional uncertainty which is not present in the case of the crisp approach. The decision maker selecting one of variants derived by the fuzzy trading system takes a risk that the variant can be not effective. The uncertainty and risk depend on the distance of the reservation point from the aspiration point u. In the case of the reservation point which is close to the aspiration point, the risk is lower but only a few or even not any variant can be derived by the system. The risk is greater when the distance increases but the system can derive and propose a greater number of non-dominated variants. The decision maker being aware of this decides on the positioning of the reservation point.

Figure 1 presents variants analyzed by the algorithm in a two-dimensional criteria space. The aspiration point u and the reservation point x are shown. The algorithm, from the set of all variants selects the non-dominated variants y^3, y^5 and y^6. The shadowed area represents the set of points dominated by the variants above.

The classical crisp system generates 5 signals for variants y^1, y^2, y^3, y^4, y^5 not informing, which of them are more or less promising. Let us note, that variants y^1, y^2, y^4 are eliminated and removed from analysis by the proposed fuzzy system.

5 Numerical Experiments

In this section, we present results of numerical experiments with real data from the Forex market. In the experiments, the proposed fuzzy trading system is compared with the existing crisp trading systems for the three indicators considered above. Our main motivation was to estimate the number of variants potentially interesting for the decision maker, indicated by the different trading systems. We tested 20 successive readings. By a reading, we mean a single situation on the price chart which is observed in a specific time window. The systems derived variants interested for the decision maker in every reading. We selected three different time windows (frames) corresponding to the scalping system with aggressive trading (the length of a single time window was equal to 5 min), to the intraday system (the length of a single time window was equal to 1 h) and finally to the long-term trading with the length of a single time window equal to 1 day. The overall length of the experiments in the case of the scalping system was equal to $20 \cdot 5 = 100$ min, for the intraday system the overall length of the experiments was equal to 20 h, and 20 days for the long-term trading. Information about the selected time windows can be found in Table 1. The number of variants (currency pairs) available and analyzed in every reading was always equal to 68.

Table 1. Data sets summary

	Starting date	Starting hour	Ending date	Ending hour
5 min	2017 IV 03	8.00	2017 IV 03	9.40
1 h	2017 I 02	7.00	2017 I 03	2.00
1 Day	2017 II 03	00.00	2017 III 5	00.00

Selected results of the experiments are presented in Table 2 for the 5-minute time window, in Table 3 for the 1-hour time window and in Table 4 for the largest 1-day time window. The tables present the numbers of the non-dominated variants derived by the proposed fuzzy system for four different values of the reservation point: $x = (x_1, x_2, x_3)$, $\forall_i \ x_i = 0.7$, $x_i = 0.8$, $x_i = 0.9$, $x_i = 0.95$ (the columns are marked by $x = 0.7, x = 0.8, x = 0.9, x = 0.95$ respectively). Theses results are compared to the numbers of signals generated by three versions of the crisp approach: Crisp*, Crisp** and Crisp***, wherein the Crisp* approach a signal is generated and presented to the decision maker when at least one of the conditions defined by the binary activation function is satisfied, in the second considered approach – Crisp** at least 2 conditions must be fulfilled, while in the last considered Crisp*** approach all the 3 conditions have to be satisfied. In the last case, the generated signal corresponds to the variants equal to the aspiration point u.

The Crisp* approach overproduces the number of variants proposed to the decision maker, thus selection of a single variant by the decision maker to make the trading decision may be extremely difficult. A decreasing number of criteria is observed in the case of Crisp** so that it leads to an empty set of variants derived for the decision maker. In the case of Crisp***, which corresponds to the situation, in which a variant equal to the aspiration point u should be found, even a single solution was not observed.

The fuzzy approach generates relatively small sets of non-dominated variants which are far easier to analyze by the decision maker.

We use bold font to indicate in the tables the desirable market situations when the number of variants derived by the fuzzy system for the decision maker is 4, 3 or 2. We use the italic font to indicate situations when the empty set of variants derived for the decision maker by the Crisp** method is observed, e.g. in readings 6, 7, 9, 10, 12, 13 in Table 2; see also readings 1, 3 – 6, 8 in Table 3 and readings 1, 3 – 7, 9, 13 in Table 4.

There were also situations when the proposed fuzzy approach derives a relatively large set of variants, which may be difficult to analyze by the decision maker. In the case of the 1-hour and 1-day time window, such situation is undesirable, but the decision maker has additional time to perform the analysis. While for the smaller time windows such situations need some additional extension of the proposed approach. Such an extension is planned in further works with the use of respective ranking methods.

It is crucial to understand that all variants derived for the decision maker in the case of the fuzzy approach are non-dominated, while in the case of the Crisp* approach (due to the binary activation function) many variants indicated by the system can be dominated. In the case of the crisp system, the decision maker has no information which of the generated variants is better or worse. This leads to an important observation that in the case of the crisp approach a single variant is treated as acceptable if any criterion is equal to 1. Thus, in the case of two variants, $y^1 = (0, 0.05, 1)$ and $y^2 = (0, 0.95, 1)$, both of them are treated as equally good as $(0, 0, 1)$, while in the fuzzy approach it is possible to distinguish these two variants in favor of y^2 which strictly dominates the first variant.

Similar experiments were conducted for two remaining time windows observed in Tables 3 and 4. In the case of the reservation point being far from the aspiration point $x = (0.7, 0.7, 0.7)$ the sets of generated variants often exceeded the assumed limits. The fuzzy system generates only few variants, which can be easily analyzed. In the 1-day tie window, an interesting situation could be observed in readings 4, 18 and 20, where the *Crisp*** could not deliver even a single variant while *Crisp*** generated a number of variants that greatly exceeded the analytical capabilities of the decision maker. The fuzzy approach, in turn,

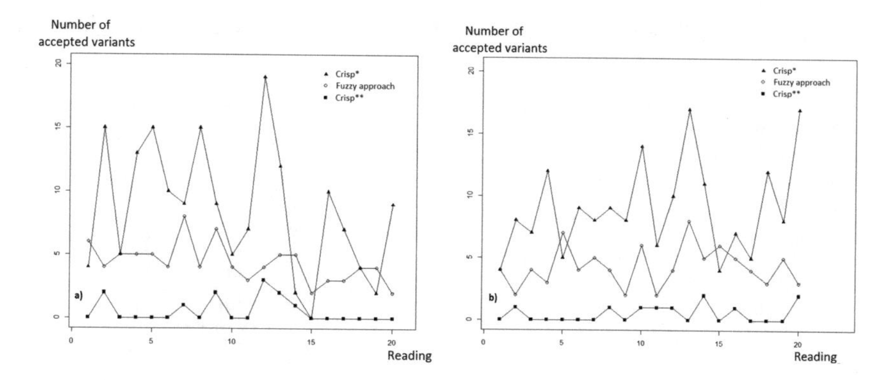

Fig. 2. (a) 1-hour time window linear chart for the fuzzy approach with $x = 0.95$, $Crisp^{*}$ and $Crisp^{**}$; (b) 1-day time window linear chart for the fuzzy approach with $x = 0.95$, $Crisp^{*}$ and $Crisp^{**}$

Table 2. Number of variants available to the decision maker for the 5-minute time window

	x = 0.7	x = 0.8	x = 0.9	x = 0.95	Crisp*	Crisp**	Crisp***
Reading 1	7	6	5	5	8	1	0
Reading 2	6	6	6	6	*16*	1	0
Reading 3	10	10	6	6	*11*	**2**	0
Reading 4	9	8	7	5	9	1	0
Reading 5	6	**4**	**3**	**3**	*12*	1	0
Reading 6	*12*	*12*	*11*	10	9	*0*	0
Reading 7	7	7	6	5	**3**	*0*	0
Reading 8	8	8	8	7	*15*	5	0
Reading 9	7	7	7	**4**	10	*0*	0
Reading 10	9	8	8	8	5	*0*	0
Reading 11	**4**	**4**	**3**	**2**	7	1	0
Reading 12	6	5	**4**	**3**	*13*	*0*	0
Reading 13	8	7	6	5	7	*0*	0
Reading 14	6	5	5	5	*12*	**2**	0
Reading 15	9	9	5	**4**	*11*	*0*	0
Reading 16	10	8	6	5	9	**2**	0
Reading 17	*11*	8	7	6	**4**	*0*	0
Reading 18	6	6	**4**	**4**	10	**3**	0
Reading 19	9	8	5	**4**	6	*0*	0
Reading 20	8	7	6	6	5	*0*	0

Table 3. Number of variants available to the decision maker for the 1-hour time window

	x = 0.7	x = 0.8	x = 0.9	x = 0.95	Crisp*	Crisp**	Crisp***
Reading 1	10	8	7	6	**4**	*0*	0
Reading 2	9	8	5	**4**	*15*	**2**	0
Reading 3	*11*	9	6	5	5	*0*	0
Reading 4	8	6	5	5	*13*	*0*	0
Reading 5	6	6	5	5	*15*	*0*	0
Reading 6	7	6	**4**	**4**	10	*0*	0
Reading 7	10	9	9	8	9	1	0
Reading 8	7	7	5	**4**	*15*	*0*	0
Reading 9	9	8	7	7	9	2	0
Reading 10	7	**4**	**4**	**4**	5	*0*	0
Reading 11	**4**	**3**	**3**	**3**	7	*0*	0
Reading 12	6	6	5	**4**	*19*	**3**	0
Reading 13	10	9	5	5	*12*	2	0
Reading 14	6	5	5	5	**2**	1	0
Reading 15	7	5	**3**	**2**	*0*	*0*	0
Reading 16	**3**	**3**	**3**	**3**	10	*0*	0
Reading 17	**4**	**4**	**3**	**3**	7	*0*	0
Reading 18	6	**4**	**4**	**4**	**4**	*0*	0
Reading 19	6	6	5	**4**	**2**	*0*	0
Reading 20	**4**	**4**	**3**	**2**	9	*0*	0

once again allowed to obtain a reasonable number of non-dominated variants in successive readings.

The obtained results are also presented in the graphical form for the fuzzy approach with $x = 0.95$, and compared to the $Crisp^{*}$ and $Crisp^{**}$ approaches. The results from Tables 3 and 4 are presented respectively in Fig. 2(a) and (b). One can easily observe disproportions in the number of variants generated by both crisp methods and a reasonable number of non-dominated variants derive from the proposed fuzzy system.

The number of solutions generated in the case of $Crisp^{*}$ fairly exceeds analytical capabilities of the decision maker, while $Crisp^{**}$ often generates no solutions at all, and the most restrictive crisp approach $Crisp^{***}$ not delivered any variants at all. The proposed fuzzy approach gives the possibility to control the number of generated variants on the basis of the risk aversion adjusted with the use of the reservation point. It may be easily extended on the trading systems with four and more indicators represented by criteria in a multicriteria space of possible decisions.

Table 4. Number of variants available to the decision maker for the 1-day time window

	x = 0.7	x = 0.8	x = 0.9	x = 0.95	Crisp*	Crisp**	Crisp***
Reading 1	9	9	7	**4**	**4**	*0*	0
Reading 2	**3**	**3**	**3**	**2**	8	1	0
Reading 3	6	6	5	**4**	7	*0*	0
Reading 4	**4**	**3**	**3**	**3**	12	*0*	0
Reading 5	7	7	7	7	5	*0*	0
Reading 6	7	6	**4**	**4**	9	*0*	0
Reading 7	8	6	6	5	8	*0*	0
Reading 8	6	5	**4**	**4**	9	1	0
Reading 9	6	6	**2**	**2**	8	*0*	0
Reading 10	9	7	6	6	*14*	1	0
Reading 11	8	5	**3**	**2**	6	1	0
Reading 12	5	**4**	**4**	**4**	10	1	0
Reading 13	9	8	8	8	*17*	*0*	0
Reading 14	8	7	6	5	*11*	2	0
Reading 15	8	6	6	6	**4**	*0*	0
Reading 16	5	5	5	5	7	1	0
Reading 17	**4**	**4**	**4**	**4**	5	*0*	0
Reading 18	6	**4**	**4**	**3**	*12*	*0*	0
Reading 19	6	5	5	5	8	*0*	0
Reading 20	**3**	**3**	**3**	**3**	*17*	2	0

6 Conclusions and Future Works

Existing trading systems based on the crisp approach have a number of disadvantages. In this article, we proposed the multicriteria fuzzy trading system including three different technical indicators. Trading rules for both: the classical crisp and the proposed fuzzy trading system were defined. A new concept of the fuzzy trading system including the possibility to generate sets of non-dominated variants derived to the decision maker was introduced as well. All concepts of trading systems were experimentally verified and tested on the limited set of technical indicators.

We experimentally verified, that proposed fuzzy trading system is capable to effectively derive Pareto-optimal variants for the decision maker. The proposed system was compared in the experiments to three versions of the crisp system: Crisp*, Crisp**, Crisp***. In contrary to the fuzzy approach, the crisp system derives very small (or even none) variants in the case of the Crisp** or number of variants is too large to be effectively handled by the decision maker – what was observed in the case of the Crisp*. The third version of the classical trading system Crisp*** was not capable to derive even single variant. One of the most important advantages of the proposed approach is that the fuzzy system is capable to derive sets of non-dominated solutions, which could be further used to develop a system for generating portfolios of variants.

Further works should include the application of methods allowing ranking of the derived variants according to preferences of the decision maker. Besides the further development of the fuzzy concept, a more robust and less computationally expensive algorithm capable to derive a set of non-dominated variants should be developed as well.

Acknowledgments. The preliminary version of this paper was presented at the 3rd Conference on Information Technology, Systems Research and Computational Physics, 2–5 July 2018, Cracow, Poland [10].

References

1. Achchab, S., Bencharef, O.: A combination of regression techniques and cuckoo search algorithm for FOREX speculation. In: Recent Advances in Information Systems and Technologies, WorldCIST 2017. Advances in Intelligent Systems and Computing, vol. 569. Springer (2017)
2. Akram, Q.F., Rime, D., Sarno, L.: Arbitrage in the foreign exchange market: turning on the microscope. J. Int. Econ. **76**(2), 237–253 (2008). https://doi.org/10.1016/j.jinteco.2008.07.004
3. Booth, A., Gerding, E., McGroarty, F.: Automated trading with performance weighted random forests and seasonality. Expert. Syst. Appl. **41**, 3651–3661 (2014). https://doi.org/10.1016/j.eswa.2013.12.009
4. Cambria, E., Schuller, B., Xia, Y., Havasi, C.: New avenues in opinion mining and sentiment analysis. IEEE Intell. Syst. **28**(2), 15–21 (2013). https://doi.org/10.1109/MIS.2013.30
5. Chen, Y., Wang, X.: A hybrid stock trading system using genetic network programming and mean conditional value-at-risk. Eur. J. Oper. Res. **240**, 861–871 (2015). https://doi.org/10.1016/j.ejor.2014.07.034
6. Gehring, T., Menkhoff, L.: Extended evidence on the use of technical analysis in foreign exchange. Int. J. Financ. Econ. **11**, 327–338 (2006). https://doi.org/10.1002/ijfe.301
7. Gottschlich, J., Hinz, O.: A decision support system for stock investment recommendations using collective wisdom. Decis. Support. Syst. **59**, 52–62 (2014). https://doi.org/10.1016/j.dss.2013.10.005
8. Hagenau, M., Liebmann, M., Neumann, D.: Automated news reading: stock price prediction based on financial news using context-capturing features. Decis. Support. Syst. **55**, 685–697 (2013). https://doi.org/10.1016/j.dss.2013.02.006
9. Holt, C.C.: Forecasting seasonals and trends by exponentially weighted moving averages. Int. J. Forecast. **20**(1), 5–10 (2004). https://doi.org/10.1016/j.ijforecast.2003.09.015
10. Juszczuk, P., Kruś, L.: Crisp vs fuzzy decision support systems for the forex market. In: Kulczycki, P., Kowalski, P.A., Łukasik, S. (eds.) Contemporary Computational Science, p. 228. AGH-UST Press, Cracow (2018)
11. Kamruzzaman, J., Sarker, R.A.: Comparing ANN based models with ARIMA for prediction of forex rates. ASOR Bull. **22**(2), 1–11 (2003)

12. Lu, C.-C., Wu, C.-H.: Support vector machine combined with GARCH models for call option price prediction. In: International Conference on Artificial Intelligence and Computational Intelligence, pp. 35–40 (2009)
13. McLeod, G.: Forex market size: a traders advantage, dailyFx, January 2014. https://www.dailyfx.com/forex/education/trading
14. Ozturk, M., Toroslu, I.H., Fidan, G.: Heuristic based trading system on Forex data using technical indicator rules. Appl. Soft Comput. **43**, 170–186 (2016). https://doi.org/10.1016/j.asoc.2016.01.048
15. Peramunetilleke, D., Wong, R.K.: Currency exchange rate forecasting from news headlines. Aust. Comput. Sci. Commun. **24**, 131–139 (2002). https://doi.org/10.1145/563932.563921
16. Schumaker, R.P., Chen, H.: Textual analysis of stock market prediction using breaking financial news: the AZFin text system. ACM Trans. Inf. Syst. **27**, 1–19 (2009). https://doi.org/10.1145/1462198.1462204
17. Wang, X., Smith, K.A., Hyndman, R.J.: Dimension reduction for clustering time series using global characteristics. In: Proceedings of the 5th International Conference on Computational Science (ICCS05) – Part III, vol. 3516, pp. 792–795 (2005)
18. Wierzbicki, A.P., Makowski, M., Wessels, J.: Model-Based Decision Support Methodology With Environmental Applications. Kluwer, Dordrecht (2000)

A Hybrid Cascade Neural Network with Ensembles of Extended Neo-Fuzzy Neurons and Its Deep Learning

Yevgeniy V. Bodyanskiy[1] and Oleksii K. Tyshchenko[1,2](✉)

[1] Control Systems Research Laboratory, Kharkiv National University of Radio Electronics, 14 Nauky Ave, Kharkiv 61166, Ukraine
yevgeniy.bodyanskiy@nure.ua
[2] Institute for Research and Applications of Fuzzy Modeling, CE IT4Innovations, University of Ostrava, 30. dubna 22, 701 03 Ostrava, Czech Republic
lehatish@gmail.com

Abstract. This research contribution instantiates a framework of a hybrid cascade neural network rest on the application of a specific sort of neo-fuzzy elements and a new peculiar adaptive training rule. The main trait of the offered system is its competence to continue intensifying its cascades until the required accuracy is gained. A distinctive rapid training procedure is also covered for this case that gives the possibility to operate with nonstationary data streams in an attempt to provide online training of multiple parametric variables. A new training criterion is examined which suits for handling nonstationary objects. Added to everything else, there is always an occasion to set up (increase) an inference order and a quantity of membership relations inside the extended neo-fuzzy neuron.

Keywords: Training procedure · Data stream · Computational intelligence · Adaptive neuro-fuzzy system · Extended neo-fuzzy neuron · Membership function

1 Introduction

Artificial neural networks [1–3] have been presently broadly applied to working out issues in the areas of Data Mining, Intelligent Control, and Image Processing due to their multipurpose fitting qualities and capabilities of training by experimental data that characterize the functioning of a studied object or a studied phenomenon.

A case is getting substantially more complicated if data come sequentially for processing in an online mode. These problems are usually scrutinized within such developing trends as Dynamic Data Mining, Video Processing, Data Stream processing, and Web Mining [4–11]. It seems comprehensible that networks which use the error backpropagation algorithm for their learning unfit completely for gainful employment in a similar situation, and it seems exceptionally justified to apply hybrid neuro-fuzzy systems of Computational Intelligence in the first place. Their output signal depends linearly on adjustable synaptic weights like radial-basis-function

P. Kulczycki et al. (Eds.): ITSRCP 2018, AISC 945, pp. 164–174, 2020.
https://doi.org/10.1007/978-3-030-18058-4_13

networks or counter-propagation networks. At the same time, these neural networks fall under the «Curse of dimensionality» which abruptly lowers their performance. In most cases, it is much more efficient to use neuro-fuzzy systems (like the Takagi-Sugeno-Kang system and ANFIS) in these tasks. Albeit there is a necessity to tune membership functions which tangles a process of their learning.

The recent interest of researchers from the area of Computational Intelligence [7–9] has been attracted to deep neural networks [12, 13] which provide a considerably much higher quality of information processing compared to conventional shallow neural networks. However, a process of deep learning in these networks happens really slow, so that data processing in an interactive computing mode by dint of the popular deep neural networks is merely impossible.

From the viewpoint of deep learning, cascade neural networks [14] look quite impressive. A process of building up cascades (layers) here can occur continuously upon reaching a required accuracy of results. That is also possible to carry out tuning the cascade network's parameters in an online manner if simplified approximating structures with an output that depends linearly on synaptic weights (that allows using fast learning procedures) are used instead of traditional elementary neurons. So, a hybrid cascade network which used neo-fuzzy neurons [15–17] as nodes was introduced in [18], and this fact allowed to improve the quality of results significantly and to implement an online process of adjusting all weights. This network is per se a deep stacked network [12, 13], albeit since a process of the zero-order fuzzy inference by Takagi-Sugeno is implemented in every layer in substance, an amount of these layers may be really high, which inherently leads to decreasing a speed of the whole system.

That is why developing a deep stacked hybrid cascade network improved approximating properties, and a high learning speed seems appropriate. In that manner, the novel nature of this network is represented by an adaptive training procedure that allows processing data with high quality when observations come to the system in an online mode.

The preliminary version of this paper was presented at the 3rd Conference on Information Technology, Systems Research and Computational Physics, 2–5 July 2018, Cracow, Poland [19].

2 A Structure of the Hybrid Cascade Neural Network with Ensembles of Extended Neo-Fuzzy Neurons

A structure of the suggested system of the Computational Intelligence is given in Fig. 1 and as a matter of fact, coincides with a topology of the hybrid cascade neural network on the grounds of an optimized pool in each cascade that was introduced for the first time in [18]. The main distinction consists of a type of used nodes and accordingly learning algorithms.

A vector $x(k) = (x_1(k), \ldots, x_i(k), \ldots, x_n(k))^T \in R^n$ (where $k = 1, 2, \ldots$ denotes the current sampled time) comes to a system's input (a receptive layer).

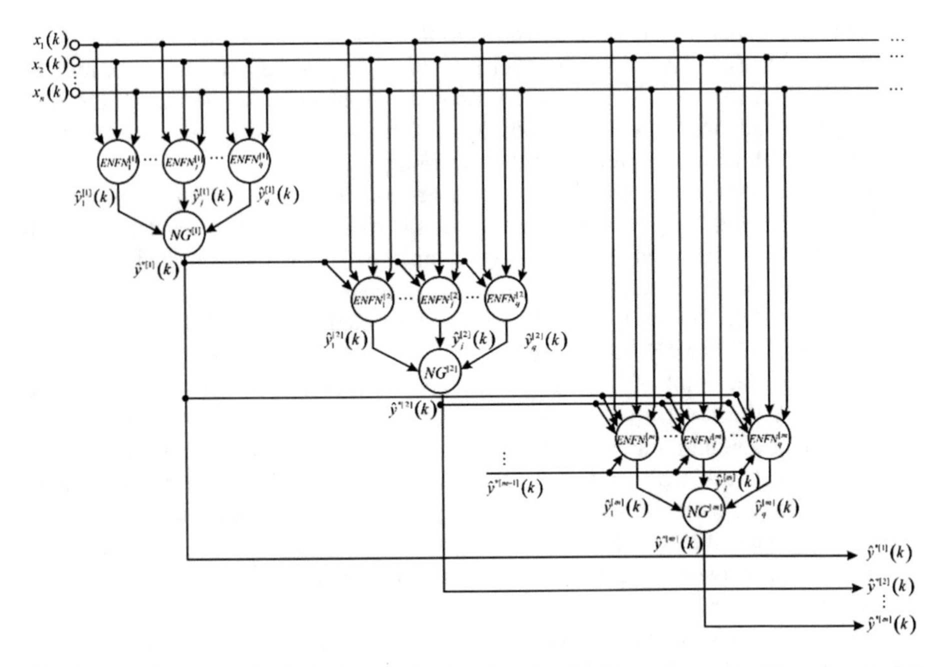

Fig. 1. A scheme of the hybrid cascade neural network driven by ensembles of extended neo-fuzzy neurons

These signals are later given to an input of each node $ENFN_j^{[m]}$ in every layer, herewith extended neo-fuzzy neurons [20] are used as nodes which are possessed of increased approximating properties as opposed to conventional neo-fuzzy neurons [15–17]. Here $j = 1, 2, \ldots, q$ stands for a number of nodes in a layer, m designates a number of a layer, wherein an amount of these layers may be growing during the learning procedure. Output signals $\hat{y}_j^{[m]}(k)$ of these nodes forming an ensemble are processed in a node of generalization $NG^{[m]}$, which synthesizes the best possible (optimal) output signal $\hat{y}^{*[m]}(k)$ of an ensemble in every layer with the help of a weighted linear combination.

If it is only a signal $x(k) \in R^n$ that arrives at an input of the first hidden layer, then it is a signal $x^{[2]}(k) = \left(x^T(k), \hat{y}^{*[1]}(k)\right)^T \in R^{n+1}$ for an input quantity of the second hidden layer, a signal $x^{[3]}(k) = \left(x^T(k), \hat{y}^{*[1]}(k), \hat{y}^{*[2]}(k)\right)^T \in R^{n+2}$ for an input of the third hidden layer. In general terms, a signal for the m-th hidden layer is $x^{[m]}(k) = (x^T(k), \hat{y}^{*[1]}(k), \hat{y}^{*[2]}(k), \ldots, \hat{y}^{*[m-1]}(k))^T \in R^{n+m-1}$. In this case, all the layers are trained in an online fashion sequentially (one after another) as a signal appears at an output of the previous layer.

3 Nodes of the Hybrid Cascade Neural Network

A prototype of the extended NFN was developed in [20] for the first time and is an ordinary neo-fuzzy neuron designed by Yamakawa, Uchino, and Miki [15–17].

The neo-fuzzy neuron is a training framework of a nonlinear nature with diversified arrival signals and a single output value to carry out the converting

$$\hat{y} = \sum_{i=1}^{n} f_i(x_i) \tag{1}$$

where x_i denotes the component i in the input vector $x = (x_1, \ldots, x_i, \ldots, x_n)^T \in R^n$ (of the dimensionality n), $\hat{y}$ designates an NFN scalar output. The NFN contains (nonlinear) synapses NS_i. Their goal is to alter the i-th component entry of x_i (in a nonlinear manner) into

$$f_i(x_i) = \sum_{l=1}^{h} w_{li}\mu_{li}(x_i) \tag{2}$$

where w_{li} indicates the synaptic weight l in the nonlinear synapse i, $l = 1, 2, \ldots, h$, $i = 1, 2, \ldots, n$; $\mu_{li}(x_i)$ stands for the membership relation l in the nonlinear synapse i which effects a fuzzification procedure of a crisp component x_i. In such a way, the NFN-performed converting may be put down like

$$\hat{y} = \sum_{i=1}^{n} \sum_{l=1}^{h} w_{li}\mu_{li}(x_i). \tag{3}$$

The fuzzy inference rule fulfilled by the identical NFN has the form

$$\begin{gathered} \text{IF } x_i \text{ IS } X_{li} \text{ THEN THE OUTPUT IS} \\ w_{li},\ l = 1, 2, \ldots, h \end{gathered} \tag{4}$$

which in turn implicates that a synapse, in fact, performs the zero-order fuzzy inference by Takagi-Sugeno [21, 22].

The NFN authors put the use of ordinary triangular expressions as membership functions to fulfill conditions of the unity partition

$$\mu_{li}(x_i) = \begin{cases} \dfrac{x_i - c_{l-1,i}}{c_{li} - c_{l-1,i}} \ \mathit{if} & x_i \in \left[c_{l-1,i}, c_{li}\right], \\ \dfrac{c_{l+1,i} - x_i}{c_{l+1,i} - c_{li}} \ \mathit{if} & x_i \in \left[c_{li}, c_{l+1,i}\right], \\ 0 & \mathit{otherwise} \end{cases} \tag{5}$$

where c_{li} signifies randomly picked out (customarily distributed on an even basis) prototypes for the membership functions in the range [0, 1], thus, expectedly $0 \le x_i \le 1$.

This pick of the membership relations feeds into the fact that the i-th element of the input term x_i makes active solely two neighboring function procedures, although their sum amounts to one which finally indicates that

$$\mu_{li}(x_i) + \mu_{l+1,i}(x_i) = 1 \tag{6}$$

and

$$f_i(x_i) = w_{li}\mu_{li}(x_i) + w_{l+1,i}\mu_{l+1,i}(x_i). \tag{7}$$

As remarked previously, the NFN's synapse NS_i executes the zero-order inference by Takagi-Sugeno only presenting the universal Wang-Mendel neuro-fuzzy system [23, 24]. It seems justified enough to enhance fitting characteristics of this computational network with the benefit of a specified constructional item said to be an "extended nonlinear synapse" (ENS_i) and to develop the "extended neo-fuzzy neuron" (ENFN) that consists of ENS_i items in place of ordinary synapses NS_i.

Scrutinizing additional variables

$$\varphi_{li}(x_i) = \mu_{li}(x_i)\left(w_{li}^0 + w_{li}^1 x_i + w_{li}^2 x_i^2 + \ldots + w_{li}^p x_i^p\right), \tag{8}$$

$$\begin{aligned} f_i(x_i) &= \sum_{l=1}^{h} \mu_{li}(x_i)\left(w_{li}^0 + w_{li}^1 x_i + w_{li}^2 x_i^2 + \ldots + w_{li}^p x_i^p\right) \\ &= w_{1i}^0 \mu_{1i}(x_i) + w_{1i}^1 x_i \mu_{1i}(x_i) + \ldots + w_{1i}^p x_i^p \mu_{1i}(x_i) \\ &\quad + w_{2i}^0 \mu_{2i}(x_i) + \ldots + w_{2i}^p x_i^p \mu_{2i}(x_i) + \ldots + w_{hi}^p x_i^p \mu_{hi}(x_i), \end{aligned} \tag{9}$$

$$w_i = \left(w_{1i}^0, w_{1i}^1, \ldots, w_{1i}^p, w_{2i}^0, \ldots, w_{2i}^p, \ldots, w_{hi}^p\right)^T, \tag{10}$$

$$\begin{aligned} \tilde{\mu}_i(x_i) = (&\mu_{1i}(x_i), x_i\mu_{1i}(x_i), \ldots, x_i^p \mu_{1i}(x_i), \\ &\mu_{2i}(x_i), \ldots, x_i^p \mu_{2i}(x_i), \ldots, x_i^p \mu_{hi}(x_i))^T, \end{aligned} \tag{11}$$

it can be put down as

$$f_i(x_i) = w_i^T \tilde{\mu}_i(x_i), \tag{12}$$

$$\hat{y} = \sum_{i=1}^{n} f_i(x_i) = \sum_{i=1}^{n} w_i^T \tilde{\mu}(x_i) = \tilde{w}^T \tilde{\mu}(x) \tag{13}$$

where $\tilde{\mu}(x) = \left(\tilde{\mu}_1^T(x_1), \ldots, \tilde{\mu}_i^T(x_i), \ldots, \tilde{\mu}_n^T(x_n)\right)^T$, $\tilde{w}^T = \left(w_1^T, \ldots, w_i^T, \ldots, w_n^T\right)^T$.

It can be marked easily that the ENFN comprises $(p+1)hn$ parametric values to be adjusted and the fuzzy inference fulfilled by each ENS_i is

$$\begin{gathered} IF\ x_i\ IS\ X_{li}\ THEN\ THE\ OUTPUT\ IS \\ w_{li}^0 + w_{li}^1 x_i + \ldots + w_{li}^p x_i^p,\ \ l = 1, 2, \ldots, h \end{gathered} \tag{14}$$

which matches the Takagi-Sugeno inference of the p-order.

The ENFN's architecture is more elementary compared to the conventional neuro-fuzzy system. This fact makes its numerical implementation easier as well.

4 A Learning Method for the Hybrid Cascade Neural Network

A training process of the system in question can be considered by an example of the j-th node in the m-th cascade described by the Eq. (13). It should be noted additionally that a one-step construction in the view of

$$E_j^{[m]}(k) = \frac{1}{2}\left(e_j^{[m]}(k)\right)^2 = \frac{1}{2}\left(y(k) - \tilde{w}_j^{[m]T}(k-1)\tilde{\mu}^{[m]}\left(x^{[m]}(k)\right)\right)^2 \tag{15}$$

was applied in [18, 25, 26] as a learning criterion; in the formula (15), $e_j^{[m]}(k)$ stands for an error at the step k, and $y(k)$ is an external reference signal.

To perform minimization of the expression (15), both the «sliding window»-based gradient procedures, as well as exponentially weighted ones based on the stochastic approximation, were employed. Although a decision quality of test cases was quite high, a convergence speed for these algorithms was insufficient in some cases.

When training the hybrid system under consideration, it arises to be more efficient to use criteria of a more general type

$$E_j^{[m]}(k) = \sum_{\tau=1}^{k} \alpha^{k-\tau}\left(e_j^{[m]}(k)\right)^2 \tag{16}$$

(here $0 \le \alpha \le 1$ is a forgetting factor) that matches the expression (15) when $\alpha = 0$ and the regular least-squares criterion when $\alpha = 1$. Minimization of the criterion (16) may be generally accomplished by applying the conventional exponentially weighted recurrent method of least squares (EWRLSM)

$$\begin{cases} \tilde{w}_j^{[m]}(k) = \tilde{w}_j^{[m]}(k-1) + \dfrac{P_j^{[m]}(k-1)e_j^{[m]}(k)\tilde{\mu}^{[m]}(x^{[m]}(k))}{\alpha_j + \tilde{\mu}^{[m]T}(x^{[m]}(k))P_j^{[m]}(k-1)\tilde{\mu}^{[m]}(x^{[m]}(k))}, \\ P_j^{[m]}(k) = \frac{1}{\alpha_j}\left(P_j^{[m]}(k-1) - \dfrac{P_j^{[m]}(k-1)\tilde{\mu}^{[m]}(x^{[m]}(k))\tilde{\mu}^{[m]}(x^{[m]}(k))P_j^{[m]}(k-1)}{\alpha_j + \tilde{\mu}^{[m]T}(x^{[m]}(k))P_j^{[m]}(k-1)\tilde{\mu}^{[m]}(x^{[m]}(k))}\right) \end{cases} \tag{17}$$

where each neuron in the cascade uses its own parameter α_j $(0<\alpha_j\leq 1)$. While processing nonstationary data streams, this approach is advantageous, since different forms of a trade-off between tracking and filtering traits of the training operation can be executed by utilizing various parameters α_j.

In addition to that, one should remember that application of the algorithm (17) makes things more confusing by the fact that it may lead to the "burst of parameters" in a covariance matrix (an exponential growth of its elements) during the learning process. This undesirable phenomenon may be prevented by choosing quite high values of the forgetting factor $\alpha_j \geq 0.95$, but tracking properties of the algorithm are lost in this case during the learning process. The "burst" can be averted by applying an exponentially weighted modification of the stochastic approximation [27] in the view of

$$\begin{cases} \tilde{w}_j^{[m]}(k) = \tilde{w}_j^{[m]}(k-1) + \left(p_j^{[m]}(k)\right)^{-1} e_j^{[m]}(k)\tilde{\mu}^{[m]}\left(x^{[m]}(k)\right), \\ p_j^{[m]}(k) = \alpha_j p_j^{[m]}(k-1) + \left\|\tilde{\mu}^{[m]}\left(x^{[m]}(k)\right)\right\|^2 \end{cases} \tag{18}$$

which is stable at any values α_j, but characterized by low processing speed.

For that matter, an optimal gradient recurrent exponentially weighted (OGREW) learning algorithm should be used instead of the formulas (17) and (18). This algorithm is a modification of the optimal adaptive identification algorithm [28] and can take on a view in this particular case

$$\begin{cases} \tilde{w}_j^{[m]}(k) = \tilde{w}_j^{[m]}(k-1) + \dfrac{\left(\bar{e}_j^{[m]}(k)\right)^2\left(r_j^{[m]}(k) - R_j^{[m]}(k)\tilde{w}_j^{[m]}(k-1)\right)}{\left\|r_j^{[m]}(k) - R_j^{[m]}(k)\tilde{w}_j^{[m]}(k-1)\right\|^2}, \\ \left(\bar{e}_j^{[m]}(k)\right)^2 = \left(e_j^{[m]}(k)\right)^2 + \alpha_j\left(\bar{e}_j^{[m]}(k-1)\right)^2, \\ r_j^{[m]}(k) = y(k)\tilde{\mu}^{[m]}\left(x^{[m]}(k)\right) + \alpha_j r_j^{[m]}(k-1), \\ R_j^{[m]}(k) = \tilde{\mu}^{[m]}\left(x^{[m]}(k)\right)\tilde{\mu}^{[m]T}\left(x^{[m]}(k)\right) + \alpha_j R_j^{[m]}(k-1). \end{cases} \tag{19}$$

We should take note of the Eq. (19) takes a form of the popular Kaczmarz-Widrow-Hoff algorithm at $\alpha_j = 0$.

An output of the neurons' signal $\hat{y}_j^{[m]}(k)$ in each layer, which forms an ensemble of cascades, are fed to inputs of generalization nodes $NG^{[m]}$, which are in a point of fact adaptive linear associators

$$\hat{y}^{*[m]}(k) = \sum_{j=1}^{q} c_j^{[m]}(k)\hat{y}_j^{[m]}(k) \tag{20}$$

moreover, its synaptic weights are calculated according to

$$\begin{cases} c_j^{[m]}(k) = \frac{\left(\sigma_j^{[m]}(k)\right)^{-2}}{\sum_{l=1}^{q}\left(\sigma_l^{[m]}(k)\right)^{-2}}, \\ \left(\sigma_j^{[m]}(k)\right)^2 = \begin{cases} (1-\alpha_j)\left(e_j^{[m]}(k)\right)^2 + \alpha_j\left(\sigma_j^{[m]}(k-1)\right)^2, \\ if\ \alpha_j \neq 1 \\ \frac{1}{k}\left(e_j^{[m]}(k)\right)^2 + \frac{k-1}{k}\left(\sigma_j^{[m]}(k-1)\right)^2, \\ if\ \alpha_j = 1. \end{cases} \end{cases} \tag{21}$$

The best in the accuracy output signal is formed for each cascade (layer).

5 Experimental Examples

Theoretical advances of our work were justified by the instrumentality of an experimental investigation depicting the forecasting issue of electric loads. It is a commonly known fact that the question of energy consumption is of the most immediate interest within the context of the everyday world. Since users' consumption keeps on being on the rise permanently, all the trends of consumption must be kept close tabs on tightly. At this point, forecasting electric loads becomes extremely important.

The data array was being gathered during 6–9 months in one of the regions of Eastern Ukraine in 2012. Generally, the data sample encompassed 6380 data points. A plurality of experiments was taken to compare performance and prediction results. In our experimental part, we used a new type of the learning criterion (16) along with the ordinary quadratic criterion (19); a changing quantity of membership functions as well as a different inference order for ENFNs. The data group was chopped into training and test data blocks. Plots of the data array in Figs. 2 and 3 illustrate the footprints of outliers stipulated by peak loads, measuring faults, and other factors. The outliers'

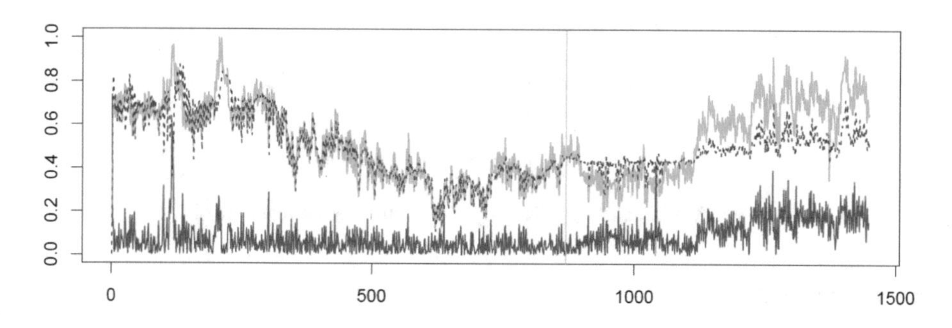

Fig. 2. A signal forecast exploited by the introduced evolving system (4 membership functions; the inference order 1)

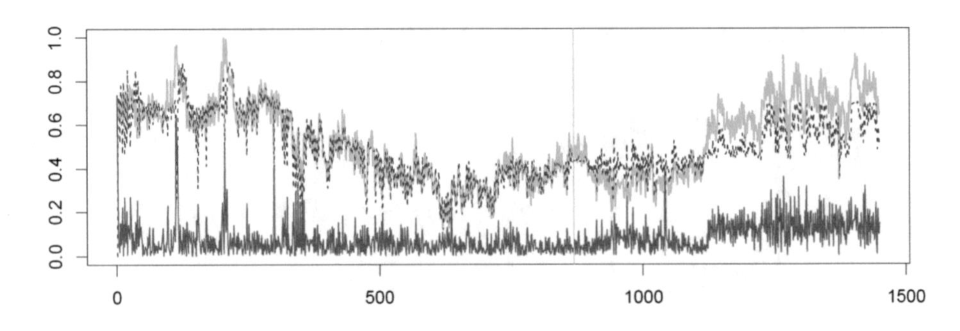

Fig. 3. A signal forecast exploited by the introduced evolving system (6 membership functions; the inference order 2)

fortuitous character is almost unpredictable and results in high prediction errors. It can be seen from Figs. 2 and 3 that a prediction quality is growing (RMSE and SMAPE are gradually falling).

The practical results prove that there is a dependency between a forecasting accuracy and a number of membership functions as well as the inference order applied (Figs. 2 and 3). There is also a relationship between a predicted fault and a number of membership functions depicted in Fig. 4.

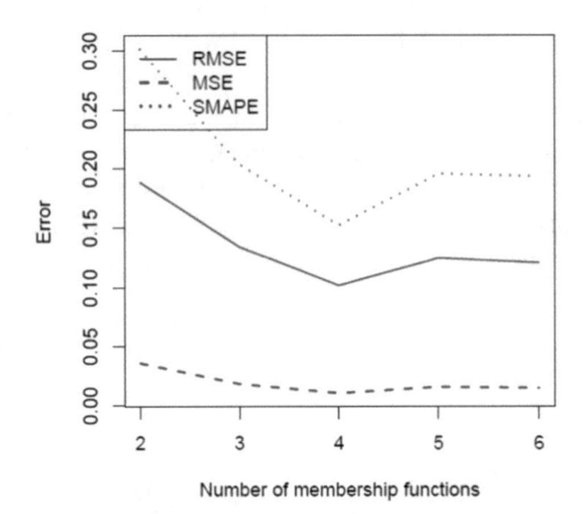

Fig. 4. A prediction fault for ENFN subject to a number of membership functions

6 Conclusion

The hybrid cascade neuro-fuzzy scheme driven by ensembles of extended neo-fuzzy neurons and an adaptive training method designated for online nonstationary data stream handling within the scope of Dynamic Stream Mining were introduced.

The suggested system is not very complicated from a calculative frame of reference by virtue of making computational processes parallel; it owns high fitted features by using ensembles of extended neo-fuzzy neurons and a high processing speed due to optimal in speed learning algorithms; and it also allows linguistic interpretation of obtained results with the benefit of fuzzy reasoning.

Acknowledgment. Oleksii K. Tyshchenko is kindly grateful for the financial assistance of the Visegrad Scholarship Program—EaP #51700967 funded by the International Visegrad Fund (IVF).

References

1. Haykin, S.: Neural Networks and Learning Machines. Prentice Hall, Upper Saddle River (2009)
2. Suzuki, K.: Artificial Neural Networks: Architectures and Applications. InTech, Hicksville (2013)
3. Hanrahan, G.: Artificial Neural Networks in Biological and Environmental Analysis. CRC Press, Boca Raton (2011)
4. Aggarwal, C.C.: A Data Mining: The Textbook. Springer, Cham (2015)
5. Delen, D.: Real-World Data Mining: Applied Business Analytics and Decision Making. Pearson FT Press, Upper Saddle River (2015)
6. Larose, D.T.: Discovering Knowledge in Data: An Introduction to Data Mining. Wiley, Hoboken (2014)
7. Kruse, R., Borgelt, C., Klawonn, F., Moewes, C., Steinbrecher, M., Held, P.: Computational Intelligence. A Methodological Introduction. Springer, Berlin (2013)
8. Bodyanskiy, Ye, Tyshchenko, O., Deineko, A.: An Evolving Radial Basis Neural Network with Adaptive Learning of Its Parameters and Architecture. Autom. Control Comput. Sci. **49**(5), 255–260 (2015)
9. Mumford, C.L., Jain, L.C.: Computational Intelligence. Springer, Berlin (2009)
10. Bifet, A., Gavaldà, R., Holmes, G., Pfahringer, B.: Machine Learning for Data Streams with Practical Examples in MOA. The MIT Press, Cambridge (2018)
11. Gama, J.: Knowledge Discovery from Data Streams. Chapman and Hall/CRC, Boca Raton (2010)
12. Goodfellow, I., Bengio, Y., Courville, A.: Deep Learning. MIT Press, Cambridge (2016)
13. Menshawy, A.: Deep Learning By Example: A Hands-On Guide to Implementing Advanced Machine Learning Algorithms and Neural Networks. Packt Publishing Limited, Birmhingham (2018)
14. Fahlman, S., Lebiere, C.: The cascade-correlation learning architecture. Adv. Neural. Inf. Process. Syst. **2**, 524–532 (1990)
15. Yamakawa, T., Uchino, E., Miki, T., Kusanagi, H.: A neo fuzzy neuron and its applications to system identification and prediction of the system behavior. In: Proceedings of 2nd International Conference on Fuzzy Logic and Neural Networks, pp. 477–483 (1992)
16. Miki, T., Yamakawa, T.: Analog implementation of neo-fuzzy neuron and its on-board learning. In: Computational Intelligence and Applications, pp. 144–149. WSES Press, Piraeus (1999)
17. Uchino, E., Yamakawa, T.: Soft computing based signal prediction, restoration and filtering. In: Intelligent Hybrid Systems: Fuzzy Logic, Neural Networks and Genetic Algorithms, pp. 331–349. Kluwer Academic Publisher, Boston (1997)

18. Bodyanskiy, Ye, Tyshchenko, O., Kopaliani, D.: A hybrid cascade neural network with an optimized pool in each cascade. Soft. Comput. **19**(12), 3445–3454 (2015)
19. Bodyanskiy, Ye., Tyshchenko, O.: A hybrid cascade neural network with ensembles of extended neo-fuzzy neurons and its deep learning. In: Kulczycki, P., Kowalski, P.A., Łukasik, S. (eds.) Contemporary Computational Science, p. 76. AGH-UST Press, Cracow (2018)
20. Bodyanskiy, Ye, Tyshchenko, O., Kopaliani, D.: An extended neo-fuzzy neuron and its adaptive learning algorithm. Int. J. Intell. Syst. Appl. (IJISA) **7**(2), 21–26 (2015)
21. Jang, J.-S.R., Sun, C.T., Mizutani, E.: Neuro-Fuzzy and Soft Computing: A Computational Approach to Learning and Machine Intelligence. Prentice Hall, New Jersey (1997)
22. Takagi, T., Sugeno, M.: Fuzzy identification of systems and its application to modeling and control. IEEE Trans. Syst. Man Cybernet. **15**, 116–132 (1985)
23. Wang, L.X., Mendel, J.M.: Fuzzy basis functions, universal approximation and orthogonal least squares learning. IEEE Trans. Neural Netw. **3**, 807–814 (1993)
24. Wang, L.X.: Adaptive Fuzzy Systems and Control. Design and Stability Analysis. Prentice Hall, Upper Saddle River (1994)
25. Bodyanskiy, Ye, Tyshchenko, O., Kopaliani, D.: Adaptive learning of an evolving cascade neo-fuzzy system in data stream mining tasks. Evol. Syst. **7**(2), 107–116 (2016)
26. Hu, Zh., Bodyanskiy, Ye.V., Tyshchenko, O.K., Boiko, O.O.: An evolving cascade system based on a set of neo-fuzzy nodes. Int. J. Intell. Syst. Appl. (IJISA) **8**(9), 1–7 (2016)
27. Otto, P., Bodyanskiy, Ye, Kolodyazhniy, V.: A new learning algorithm for a forecasting neuro-fuzzy network. Integr. Comput. Aid. Eng. **10**(4), 399–409 (2003)
28. Bodyanskiy, Ye.V., Boryachok, M.D.: Optimal control of stochastic objects under conditions of uncertainty. ISDO, Kyiv (1993)

Information Technology and Systems Research

Recurrent Neural Networks with Grid Data Quantization for Modeling LHC Superconducting Magnets Behavior

Maciej Wielgosz[1(✉)] and Andrzej Skoczeń[2]

[1] Faculty of Computer Science, Electronics and Telecommunications, AGH University of Science and Technology, Kraków, Poland
wielgosz@agh.edu.pl

[2] Faculty of Physics and Applied Computer Science, AGH University of Science and Technology, Kraków, Poland
skoczen@agh.edu.pl

Abstract. This paper presents a model based on Recurrent Neural Network architecture, in particular LSTM, for modeling the behavior of Large Hadron Collider superconducting magnets. High resolution data available in Post Mortem database was used to train a set of models and compare their performance with respect to various hyper-parameters such as input data quantization and number of cells. A novel approach to signal level quantization allowed to reduce a size of the model, simplify tuning of the magnet monitoring system and make the process scalable. The paper shows that RNNs such as LSTM or GRU may be used for modeling high resolution signals with an accuracy over 0.95 and as small number of the parameters ranging from 800 to 1200. This makes the solution suitable for hardware implementation essential in the case of monitoring performance critical and high speed signal of Large Hadron Collider superconducting magnets.

Keywords: LHC · RNN · GRU · LSTM · Signals modeling · Anomaly detection

1 Introduction

The LHC (Large Hadron Collider) is a circular proton-proton collider located at CERN (the European Organization for Nuclear Research) on Switzerland and France border. It is the largest experimental instrument which was ever built. The purpose of this huge project is verification of theories developed in elementary particle physics. The experiments build at the LHC generate a tremendous amount of scientific data which is later used in analysis and validation of the

This work was supported by the Faculty of Physics and Applied Computer Science and the Faculty of Computer Science, Electronics and Telecommunications of the AGH-UST statutory tasks within subsidy of the Ministry of Science and Higher Education.

P. Kulczycki et al. (Eds.): ITSRCP 2018, AISC 945, pp. 177–190, 2020.
https://doi.org/10.1007/978-3-030-18058-4_14

physics models regarding the history of the universe and the nature of the matter. The investigated events are so rare that the collisions must take place every 25 ns.

On the other hand the LHC itself is a unique apparatus which required development of many innovations. Therefore the LHC is a subject of many research endeavors in the fields of engineering, technology and accelerator physics. The LHC comprises many subsystems built with multitude of devices installed inside the underground tunnel. The LHC tunnel is 27 km long and located 100 m underground. The radiation level in the tunnel excludes direct intervention during operation. Therefore remote monitoring and control of each LHC device is necessary. The main engineering effort is to maximize the availability of the machine while the high safety level is guaranteed. In consequence there is a great number of data streams generated by sensors and devices depicting various subsystem conditions. Gathered data triggers an interruption of accelerator's operation immediately when dangerous anomaly is detected. Then an analysis of the event must be performed off-line by experts using PM (Post Mortem) data. The automation and improvement of on-line analysis would be very beneficiary.

The research presented in this publication is especially important for the future upgrade of the LHC and for the study of next generation circular accelerator named FCC (Future Circular Collider). The authors concentrated on a data stream coming from superconducting magnets operated at the LHC and introduced an architecture based on RNN (Recurrent Neural Networks) to model magnets behavior. The long-term authors goal is to automate the task of determining parameters of safe superconducting magnets' operation or at least reduce the necessary experts' involvement. It should be noted that specialists cannot be removed from the process of model creation, however their work can be made easier by using the proposed system.

The publication is organized as follows. Section 2 gives short view on the LHC safety system and the data used for building the RNN-based model. Section 3 provides background information about RNN. The proposed method is described in Sect. 4. Section 5 provides the results of the experiments. Finally, the conclusions of the research are presented in Sect. 6.

The preliminary version of this paper was presented at the 3rd Conference on Information Technology, Systems Research and Computational Physics, 2–5 July 2018, Cracow, Poland [1].

2 Protection of Large Hadron Collider

The LHC accelerates two proton beams travelling in opposite directions [2,3]. The particles circle the 27 km long beam pipe by 11245 times per second. Particle trajectories are formed by superconducting magnets working at a temperature of superfluid helium at about 1.9 °K. Each of the eight sectors of the LHC comprises about 154 magnets. The magnets produce a magnetic field appropriate to bend proton trajectory when they conduct an electrical current at the level of 13 kA. It means that the energy stored in one sector is about 1.2 GJ, sufficient to heat up and melt 1900 kg of copper. At collision state a separate particle has energy on the level of 7 TeV. It means that the beam of protons accumulates an energy of 360 MJ, equivalent to the energy for warming up and melting 515 kg of copper. An energy corresponding to a fraction of some 10^{-7} of the beam energy can quench a magnet when operated at full current. The quench is a phenomenon of leaving the superconducting state by a coil or a bus currying huge electric current. A huge amount of heat is released leading to catastrophic accident. The quench is a random event. The critical safety levels are, therefore, required to operate the LHC. The subsystem to protect against a consequence of this kind of event was initially built at the LHC and it is permanently maintained and developed.

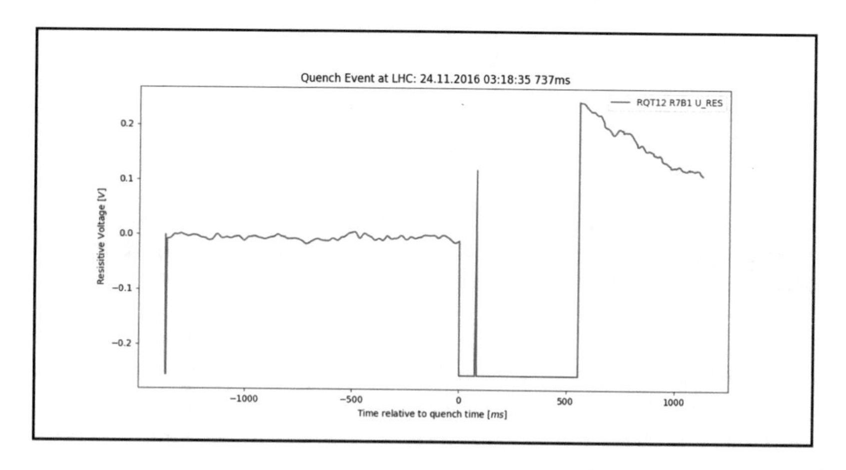

Fig. 1. The example of **post-mortem** data for one of 600 A magnet. The voltage range of the ADC is from 250 mV to −250 mV. The sampling period is 2 ms.

A system dedicated to ensure the LHC safety requirements is known as MPS (Machine Protection System) [4–6]. In general, it consists of two interlock systems: the PIS (Power Interlock System) and the BIS (Beam Interlock System).

The BIS is a superordinate system which collects signals from many sources. There are currently 189 inputs from client systems, ranging from the BLM (Beam Loss Monitor) or the FMCM (Fast Magnet Current change Monitor) to the personnel access system. However, the most important and the most complex protection subsystem is the PIS which ensures communication between systems involved in the powering of the LHC superconducting magnets. This includes the PC (Power Converters), the QPS (Quench Protection System), the UPS (Uninterruptible Power Supplies), the AUG (emergency stop of electrical supplies) and the cryogenic system. When a magnet quench is detected by the QPS, the power converter is turned off immediately. In total, there are order of thousands of interlock signals.

When a failure is detected the beams are dumped and the arrival of new particles is blocked and the trigger for data is generated. It is a request to

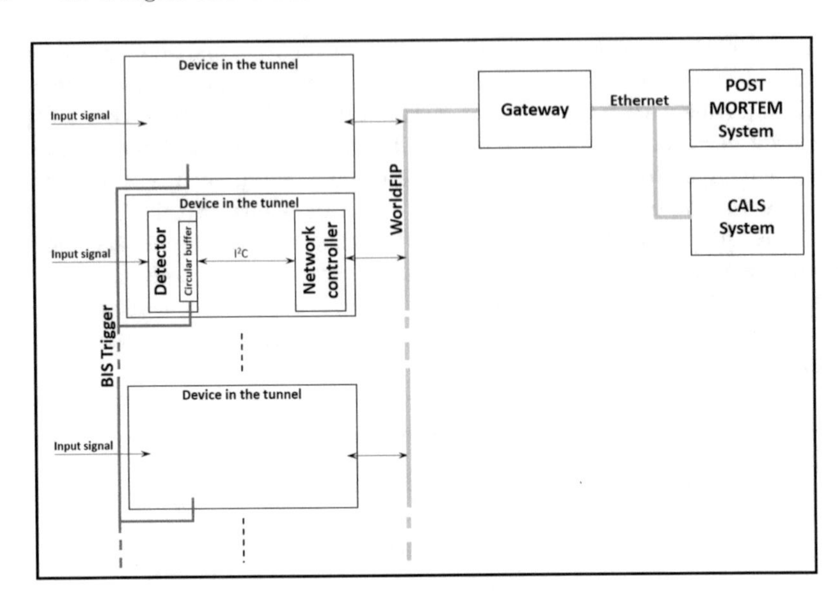

Fig. 2. The hardware path of signals from individual devices in the tunnel to the Post Mortem System. The green line marks an industrial data bus used for real-time distributed control WorldFIP (World Factory Instrumentation Protocol).

many LHC systems for providing data that were recorded locally before the failure is detected. This data gives a chance to understand reasons of the failure. Each device inside these systems, comprises a circular buffer which, at any time, serves current information about the protected component. In particular case of a quench detector, the buffer contains voltage time series acquired with a high resolution time by an ADC connected to a superconducting coil. At a trigger time, half of the buffer space is already filled with samples acquired before the event (quench) time. After the event time, the voltage samples are still recorded to fill the rest of the buffer space. Therefore, the buffer contains time series around trigger time at both sides. An example of the buffer contents is shown in Fig. 1. This kind of data is named **post-mortem** because it is recorded after the component ceased its regular activity.

The contents of the buffer is sent out by a network controller of the device over the field-bus to a gateway. Then the data is transferred to a database over Ethernet network. The data transmission path is shown in Fig. 2. There are two storage systems for data – the PM System (Post Mortem System) and the CALS (CERN Accelerator Logging Service). The first system is used during failures and requested checks and is a source of the data used in the experiments presented in this paper. The second one is used to store permanently acquired monitoring data from any device. Due to low resolution time this data is not used in this study.

The PM System is a diagnostics tool with the role of organizing the collection and analysis of transient data recorded during time interval around a failure or a request sent by any device in the MPS [7]. The main purpose is to provide a fast and reliable tool for the equipment experts and the operation crews to help them decide whether accelerator operation can continue safely or whether an intervention is required. When a failure (a beam loss or a magnet quench) happens, the individual devices' buffers are frozen and transmitted to the PM System for further storage and analysis [7–9].

The architecture of PM System is scalable, very flexible and dependable. A dynamic load of the system is balanced during data collection. Any device can dump the PM data transparently and without any additional configuration effort. The data storage is highly redundant and equipped with data consistency check. The data is stored in a file with a self-describing format know as JSON (JavaScript Object Notation). Therefore files can be processed later by any program.

Users can access the PM data by means of a specially designed REST (Representational State Transfer) API (Application Programming Interface). The aim is to serve multiple language technologies according to user preferences: Python, MATLAB, LabVIEW, C++ and Java. A user is not dependent on the data format and the file system. A direct extraction of only one signal from a big dataset is possible without the necessity of reading the entire set. The API can handle very complex queries.

3 Recurrent Neural Networks

Virtually all real world phenomena may be characterized by its spacial and temporal components. The spacial ones exist in space and it is assumed that they are stationary i.e. do not develop in time. Whereas the temporal ones unfold in time and have no spacial component. This is an idealization since there are neither pure spatial nor temporal phenomena, most of them may be described as a mixture of those two different components.

There is a well-established practice in Deep Learning applications to use FNNs (Feed-forward Neural Networks) and CNNs (Convolutional Neural Networks) to address tasks dominated by a spacial component [10]. On the contrary, data which contain more temporally distributed information are usually processed by models built around Recurrent Neural Networks. Of course, it is possible to treat time series signals as a vector of spatial values and use FNN or CNN to classify them or do some regression [11].

The voltage and current time series, which are used to train models described in this paper and make predictions, unfold in time and their temporal component is dominant. Therefore, a decision was made to use RNN networks. One of the most efficient RNN's architecture is Long Short-Term Memory [12–15].

Since its invention in 1997, the LSTM (Long Short-Term Memory) was updated and modified [16] to improve its modeling properties and reduce large computational demands of the algorithm. It is worth noting that LSTM, as opposed to a vanilla RNN [17] is much more complex in terms of the internal component constituting its cell. This results in a long training time of the algorithm. Therefore, there were many experiments conducted with simpler architectures which preserve beneficial properties of LSTM. One of such algorithms is the GRU (Gated Recurrent Unit) [18] which is widely used in Deep Learning as an alternative for LSTM. According to the recent research results, it even surpasses LSTM in many applications [19].

The field of RNN-based methods for anomaly detection is growing very fast with a progress and new discoveries in DL (Deep Learning). The basic concept of those methods uses original signal modeling.

An architecture of LSTM-based anomaly detector which incorporates both hierarchical approach and multi-step analysis was proposed in [20], capitalizing on a property of generalization which results from stacking of several RNN layers. The Gaussian distribution of an error signal - the difference between predicted and real values - was used to decide if the prediction error signifies an anomaly.

There is also a whole branch of detectors which exploit a property of inconsistent signal reconstruction in the presence of anomalies [21,22]. The authors trained the model of the autoencoder on regular data and set a threshold above which the reconstruction error is considered an anomaly. The papers deal with acoustic signals, but such an approach may be efficiently employed in other domain such as videos [23].

To comply with the CERN magnets monitoring system requirements, the detector system including RNN will need to be implemented in hardware using FPGAs (Field-Programmable Gate Arrays). The networks of the similar size as proposed for the task were already implemented for application in speech recognition and were described in [24–26]. Differences in the described approaches yield varying results, with the single iteration execution time ranging from $\sim$16 µs to 1 ms.

4 Proposed Method for Anomaly Detection

In [17], the experiments with the data obtained using Timber (the user interface to the CALS) were conducted using the setup presented in Fig. 3a, which employed RMSE measure for anomaly detection. A huge challenge in this approach is a lack of a clear reference threshold of an anomaly. In order to determine the error level, a group of experts must be consulted and it is not always easy to set one. This is due to the fact that RMSE does not always indicate anomalous behavior well enough to quantify it correctly [27].

We decided to take advantage of the experience from [17] and designed a new experimental setup which is shown in Fig. 3b. This new approach allowed to convert a regression task to the classification one, which in turn enables better anomaly quantification.

The main difference between the previously used approach and the proposed one is an introduction of a grid quantization and classification steps (see marked boxes in both Fig. 3a and b). Consequently, in the new approach the train and

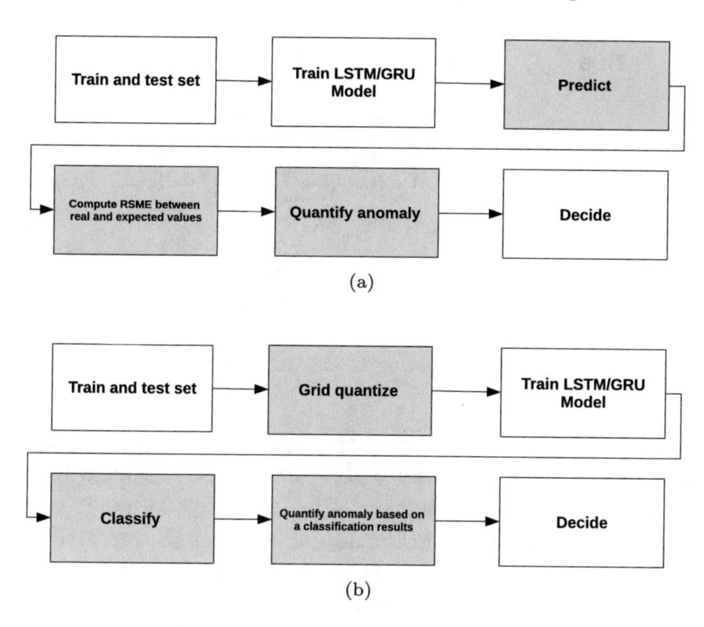

Fig. 3. Experimental setups featuring (a) RMSE (Root-Mean-Square Error) and prediction and (b) grid quantization and classification

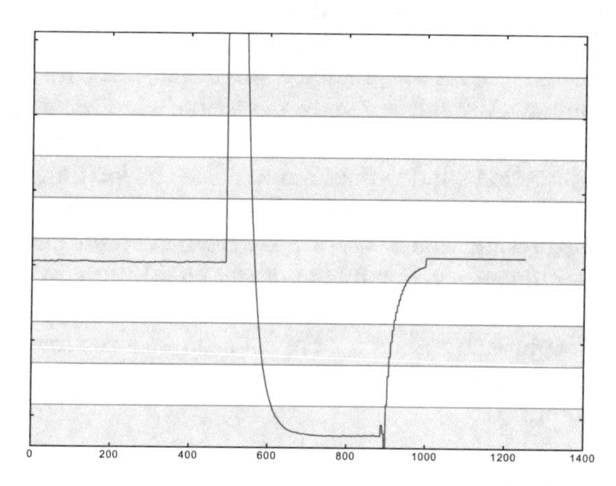

Fig. 4. Visualization of the grid size = 10

test data are brought to several categories depending on a grid size. This transformation may be perceived as a specific kind of quantization, since the floating-point data are converted to the fixed-point representation denoted as categories in this particular setup (Fig. 4). It is worth noting that the increase in the grid size leads to an increase of the resolution and it is more challenging for the classifier. Potentially, large resolution setup will demand larger model.

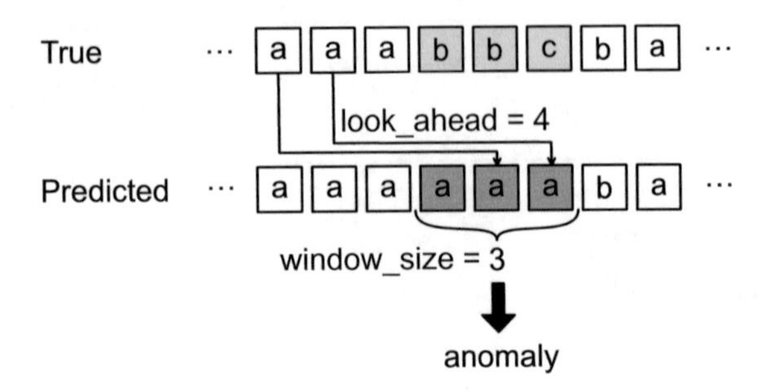

Fig. 5. Visualization of the proposed method

In order to determine if an anomaly occurs it is enough to observe the predicted categories for several time steps. Categories are predicted based on number of previous steps equal to the look_back parameter. When it turns out that the predicted category differs from the actual one over the selected time period, it means that the anomaly occurred (Fig. 5). The data expert has a much easier task in this case (when compared to the RMSE-based approach), because the only decisions required are about the grid size and the anomaly detection window, both of which are well quantifiable parameters.

The anomaly detection window is a parameter that determines how many consecutive predicted values in the signal need to differ from the true ones in order to detect an anomaly. Each predicted value that matches a true one resets the difference counter. A small anomaly detection window allows for a faster reaction time, while bigger one decreases the possibility of a false positive.

The anomaly detection window size is related to the look_ahead parameter of the model (how many time steps into the future model predicts) i.e. look_ahead value must be bigger than the window size. Such a condition is necessary in order to avoid the influence, a possible anomaly could have on values predicted within the window.

The proposed approach reliability depends on the accuracy of the model, therefore experiments on hyper-parameters influence on model performance for PM data were conducted.

5 Experiments and the Discussion

A main goal of the conducted experiments was validation of the feasibility of the application of the proposed method for detecting anomalies in PM time series of LHC superconducting magnets. An Long Short-Term Memory model was built with the Keras/Theano libraries [28].

5.1 Dataset

All the data used for the experiments were collected from CERN PM database using PM JSON API written in Python. We have collected signals from 600 A magnets current for different time series: U_{res}, U_{diff}, I_{did} and I_{dcct}.

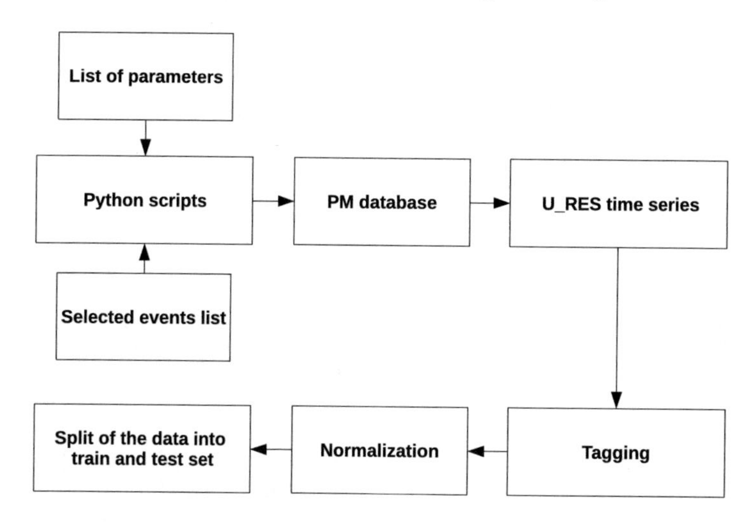

Fig. 6. The procedure for the extraction of voltage time series with selected events from the PM database using U_{res} as an example

A procedure of data extraction from the PM database is composed of several steps as presented in Fig. 6. A dedicated application and a set of parameters such as signal name and a number of time steps were used. PM database API does not allow to acquire more than one-day long signal at once. Therefore, the scripts were designed to concatenate several separate days to form a single data stream used for the experiments.

In total, 4 GB of data were collected from the database. Only a fraction of the data (≈302 MB) contained valuable information for our experiment. Consequently, we have provided a script to extract this information and keep it in separate files. As final steps, the data was normalized to [0, 1] range and split into train and test sets. Evaluation of the hyper-parameters influence on the models performance was done using a subset (≈15%) of the available data.

5.2 Quality Assessment Measures

Used as quality measures during experiments were RMSE and accuracy:

$$RMSE = \sqrt{\frac{1}{N}\sum_{i=1}^{N}(y_r^i - y_p^i)^2} \tag{1}$$

$$Accuracy = \sum_{i}^{N}\frac{Y_r^i - Y_p^i}{N}, \tag{2}$$

where: y_r^i and y_p^i are a voltage time series and its predicted counterpart, respectively, Y_r^i and Y_p^i are the true categories and ones predicted by the model and N is a dataset cardinality.

5.3 Results

This section contains all the results of the experiments conducted to validate the feasibility of the application of the presented method.

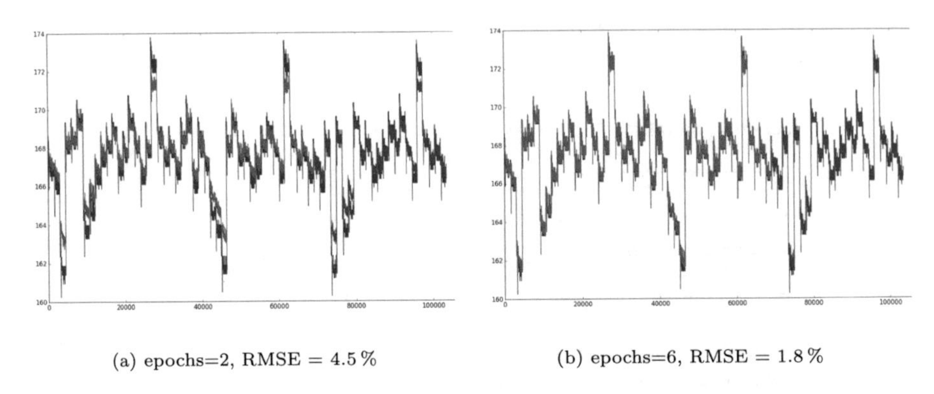

(a) epochs=2, RMSE = 4.5 %

(b) epochs=6, RMSE = 1.8 %

Fig. 7. Influence of number of epochs on model performance using simple waveforms; blue – original signal, red – train set prediction, magenta – test set prediction

The initial tests were focused on gauging the influence of number of epochs on the model performance. Figure 7a presents the results of the model for a network of 16 cells trained for two epochs, with RMSE equal to 4.5%. Increasing the amount of the epochs from two to six significantly improved the results of the model, which is reflected by Fig. 7b. Consequently, RMSE dropped to 1.5%.

Next series of experiments was conducted for different values of grid size (g), look_ahead steps (la), look_back steps (lb) as well as the number of cells (c) in the LSTM model. Batch size was fixed at 20, with number of epochs being equal to 6. The results of the experiments are presented in Table 1.

By analyzing the values of accuracy for different grid sizes, one can see that increasing the size of grid (reducing the single quantum size) leads to a deterioration of a model performance for the same parameters and the same set of data. This is the expected effect, which results from an increase in the number of categories that must be taken into consideration in the classification process while maintaining the existing network resources.

The results of the model depending on the value of the look_ahead parameter confirm that, as expected, the more forward steps are anticipated, the lower accuracy is reached, because it is more challenging for the model to predict the correct categories. This effect even deepens with increase in the grid resolution and the network size reduction – smaller net cannot handle correct classification with not enough resources available. Since the look_ahead parameter limits the anomaly detection window size, its value should be chosen carefully to allow for the best possible model accuracy while permitting a sufficiently large window size.

Table 1. Prediction accuracy of U_{res} signal using single layer LSTM; g – grid size

look_back	cells	look_ahead = 1			look_ahead = 4			look_ahead = 32			look_ahead = 128		
		g = 10	g = 40	g = 100	g = 10	g = 40	g = 100	g = 10	g = 40	g = 100	g = 10	g = 40	g = 100
1	1	0.925	0.789	0.386	0.876	0.745	0.383	0.667	0.437	0.367	0.411	0.386	0.318
	9	0.936	0.868	0.766	**0.885**	0.808	**0.710**	**0.714**	0.495	0.418	**0.543**	0.417	0.345
	17	**0.939**	**0.903**	0.767	**0.885**	0.817	**0.710**	**0.714**	**0.655**	0.540	**0.543**	0.495	**0.346**
	25	0.936	0.902	**0.771**	**0.885**	**0.818**	**0.710**	**0.714**	**0.655**	**0.541**	**0.543**	**0.496**	**0.346**
9	1	0.484	0.722	0.638	0.483	0.706	0.594	0.464	0.437	0.368	0.412	0.401	0.238
	9	0.935	0.867	0.745	0.885	0.800	0.680	0.713	0.654	0.565	0.542	0.418	0.397
	17	0.944	0.890	0.769	**0.892**	0.835	0.715	0.716	0.652	0.548	0.543	0.494	0.407
	25	**0.945**	**0.897**	**0.785**	0.891	**0.842**	**0.730**	**0.717**	**0.655**	**0.568**	**0.545**	**0.495**	**0.410**
17	1	0.853	0.717	0.542	0.483	0.691	0.530	0.464	0.433	0.368	0.479	0.402	0.319
	9	0.937	0.861	0.774	0.884	0.827	0.690	0.715	0.656	0.436	**0.546**	0.496	**0.407**
	17	0.944	0.899	0.758	0.892	0.828	0.694	0.714	0.654	0.557	**0.546**	0.496	**0.407**
	25	**0.948**	**0.900**	**0.779**	**0.893**	**0.837**	**0.719**	**0.717**	**0.669**	**0.566**	**0.546**	**0.489**	**0.407**
25	1	0.829	0.812	0.549	0.861	0.786	0.530	0.664	0.437	0.367	0.411	0.410	0.318
	9	0.936	0.871	0.766	0.884	0.821	0.679	0.713	0.643	0.440	0.548	0.495	0.351
	17	0.944	**0.897**	0.782	0.891	0.838	0.726	**0.720**	0.655	0.566	0.547	**0.500**	**0.408**
	25	**0.945**	0.894	**0.797**	**0.893**	**0.843**	**0.729**	**0.720**	**0.670**	**0.559**	**0.550**	0.494	0.404

The LSTM model performance for a different number of cells clearly shows that without enough cells the model is not able to accumulate all the training data dependencies needed to make the appropriate classification. However, using more than nine cells leads to very low improvement in the model performance. This observation leads to the conclusion that nine cells seem to be sufficient for this classification task.

It should be emphasized that the proposed method introduces a clear way to determine whether a given set of model hyper-parameters is adequate for the task (achieves required accuracy for given predetermined grid and window sizes), while giving an opportunity to simplify the architecture as much as possible. This is critical due to the fact that the size of the network significantly affect the computational complexity of training and prediction. It is of great importance also in the case of hardware implementation of LSTM and GRU networks, because of its size, which directly determines the amount of hardware resources to be used.

We also conducted experiment with the full dataset and the following architecture of the network: single layer LSTM, 128 cells, 20 epochs of training, look_back = 16, look_ahead = 128, grid = 100 and optimizer *Adam*. This resulted in a huge performance leap comparing to the results presented in Table 1. The accuracy reached almost 99.9%.

We did most of our experiments using LSTM algorithm for the sake of congruency and consistency with [17], which this paper is meant to be a continuation of, in many aspects. Nevertheless, we decided do show the comparison between GRU and LSTM performance on a sample dataset as given in Table 2.

Table 2. Comparison of GRU and LSTM performance

	GRU	LSTM
Cells	16	
Epochs	10	
Parameters	864	1152
Accuracy [%]	61.17	61.12
Training time	2 h 13 min	2 h 30 min

6 Conclusions and Future Work

This work extends existing experiments [17] using higher resolution data and more diverse models. As LHC experiments enter *High Luminosity* phase more data will be collected which raises new challenges in the maintenance of equipment.

In experiments presented in this paper a U_{res} signal was used. In the future experiments, we plan on using several signals the same time and comparing performance with the one achieved in this paper. Nevertheless, very promising results of 99% accuracy were achieved for the largest dataset of 302 MB and the following architecture of the network: single layer LSTM, 128 cells, 20 epochs of training, look_back = 16, look_ahead = 128 and grid = 100.

Another aspect worth investigating is the feasibility of implementing predictive model on FPGAs. Performing computations on a PC, works well for validation of the idea, but requirements of control systems like QPS are rather *hard real-time* which PC systems are incapable of doing.

References

1. Wielgosz, M., Skoczeń, A.: Recurrent Neural Networks with grid data quantization for modeling LHC superconducting magnets behavior. In: Kulczycki, P., Kowalski, P., Łukasik, S. (eds.) Contemporary Computational Science, p. 240. AGH-UST Press, Cracow (2018). http://itsrcp18.fis.agh.edu.pl/proceedings/
2. Brüning, O., Collier, P.: Building a behemoth. Nature **448**, 285–289 (2007). https://doi.org/10.1038/nature06077
3. Evans, L., Bryant, P.: LHC machine. J. Instrum. **3**(08), S08,001 (2008). https://doi.org/10.1088/1748-0221/3/08/S08001
4. Wenninger, J.: Machine protection and operation for LHC. CERN Yellow Report CERN-2016-002 (2016)
5. Bordry, F., Denz, R., Mess, K.H., Puccio, B., Rodriguez-Mateos, F., Schmidt, R.: Machine protection for the LHC: architecture of the beam and powering interlock system. LHC Project Report 521, CERN (2001). https://cds.cern.ch/record/531820/files/lhc-project-report-521.pdf
6. Schmidt, R.: Machine protection and interlock systems for circular machines – example for LHC. CERN Yellow Report CERN-2016-002 (2016)

7. Ciapala, E., Rodríguez-Mateos, F., Schmidt, R., Wenninger, J.: The LHC post-mortem system. Technical report LHC-PROJECT-NOTE-303, CERN, Geneva (2002). http://cds.cern.ch/record/691828
8. Lauckner, R.J.: What data is needed to understand failures during LHC operation. In: 11th Workshop of the LHC, Chamonix XI, pp. 278–283 (2001). CERN-SL-2001-003. https://cds.cern.ch/record/567214
9. Borland, M.: A brief introduction to the SDDS Toolkit. Technical report, Argonne National Laboratory, USA (1998). http://www.aps.anl.gov/asd/oag/SDDSIntroTalk/slides.html
10. Krizhevsky, A., Sutskever, I., Hinton, G.E.: ImageNet classification with deep convolutional neural networks. In: Pereira, F., Burges, C.J.C., Bottou, L., Weinberger, K.Q. (eds.) Advances in Neural Information Processing Systems 25, pp. 1097–1105. Curran Associates, Inc. (2012). http://papers.nips.cc/paper/4824-imagenet-classification-with-deep-convolutional-neural-networks.pdf
11. LeCun, Y.: Deep learning of convolutional networks. In: 2015 IEEE Hot Chips 27 Symposium (HCS), pp. 1–95 (2015). https://doi.org/10.1109/HOTCHIPS.2015.7477328
12. Graves, A.: Neural Networks. Springer, Heidelberg (2012). https://doi.org/10.1007/978-3-642-24797-2
13. Morton, J., Wheeler, T.A., Kochenderfer, M.J.: Analysis of recurrent neural networks for probabilistic modelling of driver behaviour. IEEE Trans. Intell. Transp. Syst. **PP**(99), 1–10 (2016). https://doi.org/10.1109/TITS.2016.2603007
14. Pouladi, F., Salehinejad, H., Gilani, A.M.: Recurrent neural networks for sequential phenotype prediction in genomics. In: 2015 International Conference on Developments of E-Systems Engineering (DeSE), pp. 225–230 (2015). https://doi.org/10.1109/DeSE.2015.52
15. Chen, X., Liu, X., Wang, Y., Gales, M.J.F., Woodland, P.C.: Efficient training and evaluation of recurrent neural network language models for automatic speech recognition. IEEE/ACM Trans. Audio Speech Lang. Process. **24**(11), 2146–2157 (2016). https://doi.org/10.1109/TASLP.2016.2598304
16. Greff, K., Srivastava, R.K., Koutník, J., Steunebrink, B.R., Schmidhuber, J.: LSTM: a search space odyssey. IEEE Trans. Neural Netw. Learn. Syst. **28**(10), 2222–2232 (2017). https://doi.org/10.1109/TNNLS.2016.2582924
17. Wielgosz, M., Skoczeń, A., Mertik, M.: Using LSTM recurrent neural networks for detecting anomalous behavior of LHC superconducting magnets. Nucl. Instrum. Methods Phys. Res. A **867**, 40–50 (2017). https://doi.org/10.1016/j.nima.2017.06.020
18. Chung, J., Gulcehre, C., Cho, K., Bengio, Y.: Gated feedback recurrent neural networks. In: Proceedings of the 32nd International Conference on International Conference on Machine Learning - Volume 37, ICML 2015, pp. 2067–2075. JMLR.org (2015). http://dl.acm.org/citation.cfm?id=3045118.3045338
19. Chung, J., Gulcehre, C., Cho, K., Bengio, Y.: Empirical evaluation of gated recurrent neural networks on sequence modeling. In: NIPS 2014 Workshop on Deep Learning, December 2014
20. Malhotra, P., Vig, L., Shroff, G., Agarwal, P.: Long short term memory networks for anomaly detection in time series. In: Proceedings of the 23rd European Symposium on Artificial Neural Networks, Computational Intelligence and Machine Learning, ESANN 2015, Bruges, Belgium, pp. 89–94. Presses universitaires de Louvain (2015). https://www.elen.ucl.ac.be/Proceedings/esann/esannpdf/es2015-56.pdf

21. Marchi, E., Vesperini, F., Eyben, F., Squartini, S., Schuller, B.: A novel approach for automatic acoustic novelty detection using a denoising autoencoder with bidirectional LSTM neural networks. In: 2015 IEEE International Conference on Acoustics, Speech and Signal Processing (ICASSP), pp. 1996–2000 (2015). https://doi.org/10.1109/ICASSP.2015.7178320
22. Marchi, E., Vesperini, F., Weninger, F., Eyben, F., Squartini, S., Schuller, B.: Non-linear prediction with LSTM recurrent neural networks for acoustic novelty detection. In: 2015 International Joint Conference on Neural Networks (IJCNN), pp. 1–7 (2015). https://doi.org/10.1109/IJCNN.2015.7280757
23. Chong, Y.S., Tay, Y.H.: Abnormal event detection in videos using spatiotemporal autoencoder. In: Cong, F., Leung, A., Wei, Q. (eds.) Advances in Neural Networks - ISNN 2017, pp. 189–196. Springer, Cham (2017)
24. Chang, A.X.M., Martini, B., Culurciello, E.: Recurrent neural networks hardware implementation on FPGA. CoRR abs/1511.05552 (2015). http://arxiv.org/abs/1511.05552
25. Han, S., Kang, J., Mao, H., Hu, Y., Li, X., Li, Y., Xie, D., Luo, H., Yao, S., Wang, Y., Yang, H., Dally, W.B.J.: ESE: efficient speech recognition engine with sparse LSTM on FPGA. In: Proceedings of the 2017 ACM/SIGDA International Symposium on Field-Programmable Gate Arrays (FPGA 2017), pp. 75–84 (2017). https://doi.org/10.1145/3020078.3021745
26. Lee, M., Hwang, K., Park, J., Choi, S., Shin, S., Sung, W.: FPGA-based low-power speech recognition with recurrent neural networks. In: 2016 IEEE International Workshop on Signal Processing Systems (SiPS), pp. 230–235 (2016)
27. Strecht, P., Cruz, L., Soares, C., Mendes-Moreira, J., Abreu, R.: A comparative study of regression and classification algorithms for modelling students' academic performance. In: Proceedings of the 8th International Conference on Educational Data Mining, EDM 2015, Madrid, Spain, 26–29 June 2015, pp. 392–395 (2015). http://www.educationaldatamining.org/EDM2015/proceedings/short392-395.pdf
28. Chollet, F., et al.: Keras. GitHub (2015). GitHub repository. https://keras.io/getting-started/faq/#how-should-i-cite-keras

Two Approaches for the Computational Model for Software Usability in Practice

Eva Rakovská(✉) and Miroslav Hudec

Faculty of Economic Informatics,
University of Economics in Bratislava, Bratislava, Slovakia
{eva.rakovska,miroslav.hudec}@euba.sk

Abstract. Rapid software development and its massive deployment into practice brings a lot of problems and challenges. How to evaluate and manage the existing software in an enterprise is not an easy task. Despite different methodologies in IT management, we encounter problems with how to measure usability of software. Software usability is based on user experience and it is strongly subjective. Every IT user is unique, so the measurement of IT usability has often qualitative character. The main tool for such measurement is survey, which maps her or his needs of daily work. The article comes from experimental study in the medium-sized company. It was based on the idea of using rule-based expert system for measurement of software usability in enterprises. Experimental study gave a more detailed view into the problem; how to design the fuzzy-rules and how to compute them. The article points to problems in designing a computational model of software usability measurement. Thus, it suggests a computational model, which is able to avoid the main problems arising from experimental study and to deal with the uncertainty and vagueness of IT user experience, different number of questions for each users group, different ranges of categorical answers among groups, and variations in the number of answered questionnaires. This model is based on the three hierarchical levels of aggregation with the support of fuzzy logic.

Keywords: IT software usability · Usability aggregation · Fuzzy quantifiers · Uninorm

1 Introduction

In today's competitive world, Information Technologies (IT) have a significant impact on management and efficiency of enterprises and companies. Enterprises take effort to optimize their processes and often put a lot of investments into IT. The traditional perception of IT management (HW support, network services, installation services, etc.) is rapidly changing. Information technologies are offered as services integrated into the organization to support the achievement of business goals. IT services are directly involved in an enterprise as an organic component. But, organizations and businesses often fail to manage their IT services adequately. IT management uses the standards for implementation and management IT like Information Technology Infrastructure Library [9] and Control Objectives for Information and Related Technologies [11] in a strategic

P. Kulczycki et al. (Eds.): ITSRCP 2018, AISC 945, pp. 191–202, 2020.
https://doi.org/10.1007/978-3-030-18058-4_15

level of management. However, it is not easy to control and measure the actual performance of IT services and software. Especially, small and medium-sized enterprises are not able to apply and to control IT management standards. Small and medium-sized enterprises prefer to use Balance scorecard methodology to monitor business performance and use the traditional non-financial and financial metrics also to monitoring IT performance as a part of business performance [13]. The IT metrics should be measurable, but often is problem to define metrics concerning the IT user satisfaction or IT user experience among different user groups. IT user experience comes from the qualitative and subjective opinion of users [1] such: whether is the software application response time good, whether the software availability is appropriate, whether IT services availability is appropriate, whether the software has intuitive interface, whether the software saved the user time, whether the user is able to use all functionality of software, etc. Usually the user experience is monitored with surveys and we suppose the observability and measurability are the attributes of user experience [1].

The aim of paper arises from the experimental idea to design a fuzzy-rule expert system for helping the managers to monitor the software usability as was aforementioned. So it means to monitor, which software is the most valuable in enterprise and in which department. Many enterprises use mix of software and applications, which are not compatible and sometimes are useless. We started the research with experimental study in [12] where we realized survey for users to evaluate the software usability and gain the appropriate knowledge for preparing fuzzy-rules. Surveys realized by questionnaires cope with the issues of item and unit non response and measurement error, which are far from negligible [3, 4]. This way also copes with the imbalanced number of responded questionnaires among users groups.

Thus, this research was focused on analyzing benefits and drawbacks of expressing usability by the rule-based systems and suggesting complex flexible aggregation to overcome issues in data collected by questionnaires. The preliminary version of this paper was presented at the 3rd Conference on Information Technology, Systems Research and Computational Physics, 2–5 July 2018, Cracow, Poland [15]. The paper is organized as follows: Sect. 2 gives some preliminaries and motivation for this study. Section 3 is focused on the experiential study and building a rule-based system from the collected data. Section 4 is dedicated to multilevel aggregation of questionnaires for evaluating each software tool. Finally, Sect. 5 concludes this paper and outlines future research activities.

2 Motivation and Background

The motivation for our research has arisen from the practice, where many software users are not satisfied with IT and software in their companies and the management has no real picture about the daily user needs. As we mentioned afore, the user experience reveals the connection between the user and a product and is observable and measurable. Especially, when we speak about the software user experience it is necessary to mention the software efficiency and effectivity, which are involved in the user experience. In [1] the efficiency is written as "the amount of effort required to complete the task" and effectiveness means "being able to complete the task". Software efficiency and effectiveness are usually connected with software development and software engineering. Software engineering methodologies contain standards and offer metrics for better software performance. But, to identify the real software performance as part of IT in an enterprise is highly controversial task. To increase the software performance means not only to find the hard metrics, but also try to measure software usability [1], which is based on behavior in using the software. Usability issues, as clarity of concepts, logical sequence of operations, transparency of design, quick availability of certain items, an expression of user's frustration, misinterpreting some parts and information in software etc., are often thought as purely qualitative, so they are observable and we can collect their data by questionnaires. The collected user experiences may be expressed by rule-based systems, either classical or fuzzy.

Surveys realized by questionnaires, although effective, cope with the issues of item and unit non response and measurement error, which are far from negligible [3, 4]. We have observed the same behavior (with lower extent) in the case of survey among users in a company. In this case, it causes imbalanced number of filled questionnaires (e.g. 20 filled questionnaires in one department and 12 in another). Respondents may not be careful in filling questionnaires, i.e. they can fill neighboring value in categorical answers.

In this extent, quantified aggregation in the sense of quantified summaries [18] and hybrid aggregations [19] are promising. The former may solve issues related to imbalanced response rates, whereas the latter are suitable for emphasizing highly usable software and penalizing the lowly usable due to the full reinforcement effect [5, 7].

3 Experimental Study

The aim of the section is to sketch the main problems with software usability measurement by using fuzzy rules, when we wish to measure usability among various departments regarding the users` experience. First, we started with experimental study, in which we prepared user experience survey and applied fuzzy rule-based approach for evaluating it. It may be expected, that the fuzzy-rules could be a good method, because for the answers the Likert scale is mostly used [20]. However, the fuzzy rules approach has shown the following main problems

- How to create the questionnaires on a common platform for each software in each department in the enterprise,
- How to aggregate the answers among all departments in enterprise to gain relevant and usable result.

3.1 Qualitative Software Efficiency Evaluation

Software usability in practice is often a moot question. Many enterprises use a mix of incompatible software or useless software. The main indicator of appropriate software usage should be its efficiency and usability. Software usability measurement is something what is vague, because it is measured by experiments, case-studies, best-practices and surveys. Survey is the basic method for user experience. Therefore, we started our research with experimental study in the company as was mentioned before [12]. The experiment consisted from two parts

- Qualitative software usability evaluation,
- Preparing the fuzzy rules for software usability evaluation.

The evaluation was based on user satisfaction surveys using [20]. The questionnaire was inspired by the certain methods of software usability evaluation and their combinations (System Usability Scale, Software Usability Measurement Inventory, etc.) and contained 26 questions [12]. Here, it is necessary to divide the users to groups depending on their positions in the company. We divided them to three main groups: IT managers, IT developers, Economic department/other users, and we created customized questionnaires for each group separately. Each questionnaire started with a question concerning the most frequently used software "Type the software you most frequently work with. This software will be further the subject of this questionnaire, and all questions will be asked only about it" [12]. Other questions (in total 26) mapped the user satisfaction with this software from the various points of view.

We asked for total software satisfaction; whether the software product is up to date; how users are working with the software product; whether they use the service and maintenance services; whether the software is intuitive, consistent, or whether they will prolong the license. Some of the other questions were: What is the interaction between the user and the software; whether the software sometimes ended unexpectedly; whether it is a satisfactory software language; whether it has all the necessary features; whether the user feels frustrated when working with the software or the software environment is easy to use; how often the software is working, or whether the software is slower after hours of work etc.

Table 1 shows the results from questionnaires at Economic Department of company, where the users preferred and frequently used software *S* and some parts of it (S_1, S_2). Majority of questions used answer scale from 1 to 5 (1 = disagree, 2 = rather disagree, 3 = I cannot judge, 4 = rather agree and 5 = agree) and some questions used more detailed scale (from1 to 10) for evaluating the attributes such as interaction with software. One row in Table 1 represents 26 answers collected from one user.

Table 1. Illustrative sample of answers from users' questionnaires in experimental study

S1,"","4","2","3","4","4","4","5","4","4","4","8","2","2","4","2","4","5","4","2","8" ,"4","4","4","1","8"
S2,"4","4","4","4","4","4","4","5","4","4","4","7","4","2","4","4","4","5","4","1"," 7","4","4","4","1","8
"S2","2","4","4","5","4","2","3","5","2","4","2","7","2","2","4","2","4","4","4","2", "8","4","4","5","1","8"
"S2","4","4","2","5","4","2","2","4","2","3","3","5","2","2","2","2","3","4","3","2", "6","4","4","2","2","6"
"S","2","4","4","5","5","4","2","4","2","4","5","7","2","1","2","2","4","5","4","1"," 8","3","2","2","1","8"
"S","4","4","3","4","4","4","","5","4","4","4","8","4","4","3","4","4","4","4","2","8 ","4","4","","4","7"

3.2 Fuzzy Rules Preparation for Software Usability Evaluation in MATLAB

The next step in software usability evaluation was appropriate quantification from the survey. First, we have chosen the evaluation by fuzzy-rules in MATLAB software (Fuzzy Logic Toolbox and graphical user interface), because of its good applicability for questionnaires evaluations.

Secondly, we considered 25 inputs for fuzzy rule-based system and settings as Mamdani inference: "And method = min", "Or method = max", "Implication = min", "Aggregation = max", "Defuzzification = centroid". The domain of possible answers was covered by 3 or 5 sets depending on the question type [12]. We designed the set of fuzzy-rules (Rule Editor), debug them (Rule Viewer). The system behavior (Surface Viewer) is shown on Fig. 1. Finally, we adjusted the designed rules in Rule Viewer again.

To sum up, we have produced three types of questionnaires (each type for one group), each contains 25 questions. The domain of each set of answers in the Fuzzy Logic Toolbox ranges from three to five fuzzy sets. The number of combinations for one questionnaire was as follows: 11 767 897 353 375 rules. In order to cover all the options, we would have to create as many rules as possible. After multiplying three (three questionnaires), the number would increase even more, and so we decided to take the number of rules more slowly. Another option was to divide the constructed fuzzy rule-based system into 5 parts and gradually count the results for the first 5 questions, the other 5 questions, and so on. In such a way, a whole set of rules combinations would be covered. The number of rules would be 2 583, 1 647 and 3 333

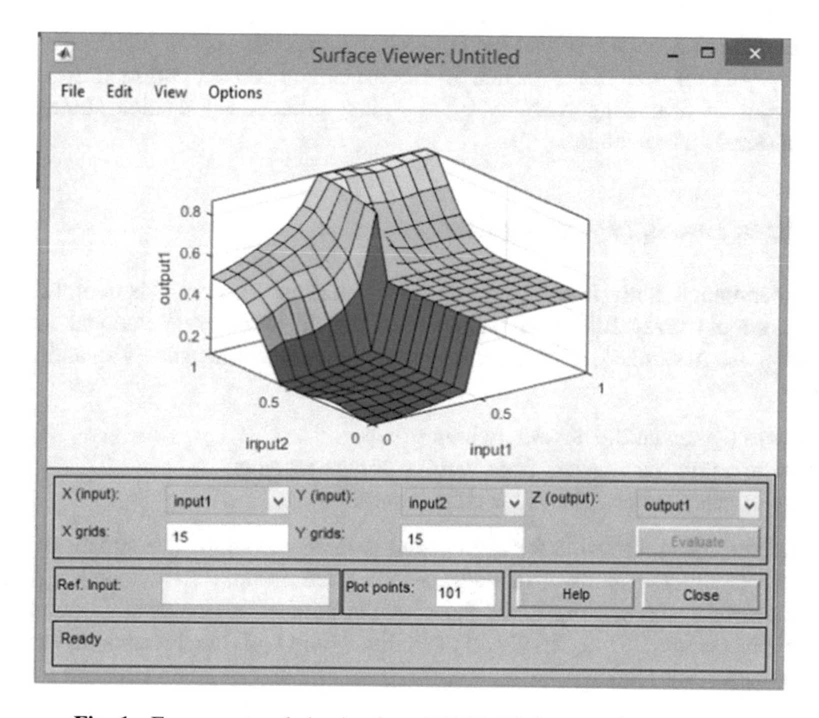

Fig. 1. Fuzzy-system behavior from MATLAB for experimental study

for the Management questionnaire. The number of rules is different due to the number of fuzzy sets in each question. So the number of combinations increases. Such a number of rules is possible to get into the system, but the computer would be exposed to intense calculations, and therefore the system response time would increase rapidly. Neither the disjunctive normal form is the solution, because there are few rules with common input parts. Finally, we decided for restriction of the number of rules. We have reduced the number of rules to 1 200 rules using the FIS matrix in MATLAB, but the number was still significant. We have not used weighted rules, because all the questions have the same importance.

3.3 Experimental Study Conclusion

Looking back on the experiment we can summarize the pros and cons of it. To sum up, the conditions were

- only three groups of users,
- the same number of questions in each questionnaire (although some questions were different depending on users group),
- only two types of question; the first type has the answer value from 1–5 and second type from 1–10,
- the number of respondents was relatively small (less than 50).

We chose a relatively well-known fuzzy-rules method for evaluating our experiment and we worked with MATLAB software, where we used settings with Mamdani inference (Sect. 3.2).

The main drawback was that the computing complexity is too high and we have difficulty with the size of the model. It was the reason to start considering the other possibilities of evaluating usability by applying different aggregation strategies for different levels of evaluation.

4 Three Levels of Aggregation

The experimental study in the previous section has shown how it is complicated to make a simple fuzzy rule-based aggregation and to avoid computational intensive activities. As mentioned before, we had some problems with survey quantification, namely:

- How to aggregate the answers within group,
- How to aggregate answers from various groups together,
- How to calculate usefulness for all software.

Therefore, the motivation for aggregation is to develop a flexible survey system in the sense of different number of questions for each group, different ranges of categorical answers among groups, and variations in number of answered questionnaires. Before proceeding, let us briefly explain the mining of the hierarchical view on questionnaires' evaluation.

4.1 The Organization of Questionnaires

Let us have n groups in the survey, G_i, $i = 1 \ldots n$ (e.g. G_1 is group of managers, G_2 is group of IT developers, G_3 is group from the economic department, etc.). The number of groups depends on the type of organization (enterprise). Let us mark each respondent as R. In each group the number of respondents is $R_{ji,}$ $i = 1 \ldots .n$, $j = 1 \ldots .m$. Each respondent fill one questionnaire, so the number of respondents is equal to the number of questionnaires. Finally, Q_{li} $l = 1 \ldots L$, where l is the number of question in the questionnaire for group i. Figure 2 sketches the aggregation within group G_1, where Aggregation 1 means aggregation of answers for one respondent. Analogously, Fig. 3 sketches the aggregation among groups in organization regarding evaluated software S_p, $p = 1 \ldots P$.

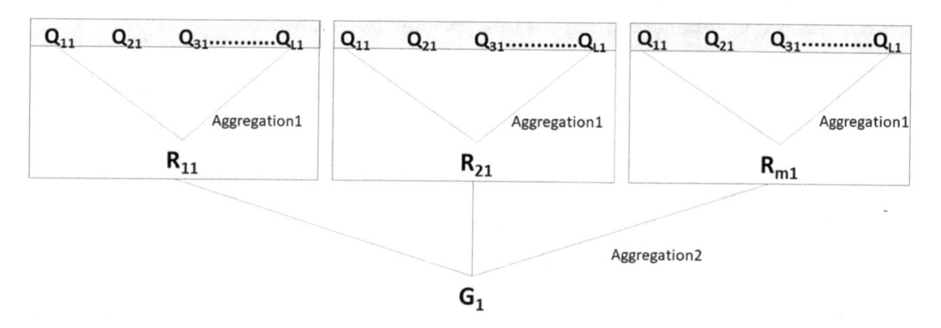

Fig. 2. Aggregation scheme within group G_1

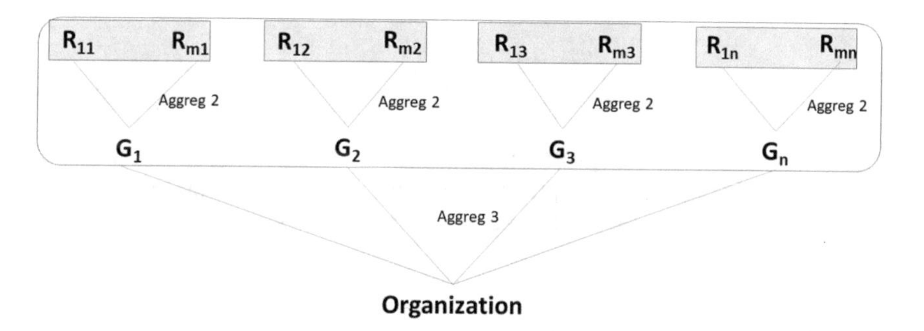

Fig. 3. Scheme of groups' aggregation within organization concerning one software

At the lowest level are questionnaires for particular groups of users, where L is the number of questions in a questionnaire for group i. Each group may have different number of categorical questions and the number of possible answers. For instance, for group G_1, possible categorical answers to a question are 1, 2, 3, 4, 5; whereas for G_2 answers are 1, 2, 3, 4, 5, 6, 7, 8, 9, 10 (where 1 is the worst, and 5 or 10 the best option, respectively). Answers might be also expressed by linguistic terms, where one number corresponds to one linguist term. Comparing such two answers is not a problem because the method suggested in [10] carries out the necessary conversions. Thus, the different number of categories or terms is not a problem. The problem is when we have different number of questions and respondents (or filled questionnaires) among groups.

Aggregation operators reduce a set of values into a unique representation or meaningful number [7]. In this direction, we have searched for suitable aggregation for the problem plotted in Figs. 2 and 3. The question is how to envelop all three levels of aggregation to find the final aggregated value for each evaluated software in order to rank them. This value should be in [0, 1] interval. In this case, it is clear that 0 is related to the worst, whereas 1 to the ideal software.

4.2 Aggregating Values for Respondents

In our study, the aggregation at the first level is the simplest one, sum of answers. The score of each questionnaire is the sum of respondent's answers

$$v(R_{ji}) = \sum_{l=1}^{L} x_l \quad (1)$$

where x_l is the answer to lth question for respondent j in ith group.

Another option may be weighted average

$$v(R_{ji}) = \sum_{l=1}^{L} w_l \cdot x_l \quad (2)$$

where w_l is weight of answer l, or arithmetic mean

$$v(R_{ji}) = \frac{1}{L}\sum_{l=1}^{L} x_l \quad (3)$$

Values of suggested aggregations (1–3) are not in the unit interval. These functions have been chosen due to their simplicity. This step is considered as the preparation step for the next ones. The illustrative data are shown in Table 2 for one group of respondents.

4.3 Aggregating Values for Groups

The main question is how to aggregate, or find a representative value for each software S_p by each group when the number of questions and possible answers is different. The possible answer is quantified aggregation by summary *most of questionnaires highly rated software S*. The classical prototype forms of Linguistic Summaries (LSs) [18] are suitable for solving this task.

Table 2. The illustrative group G_1 and its respective respondents and membership degrees to the fuzzy set *high*, which is used in the next subsection.

Respod. in G_1	Questions Q_1 (answers 1-5)					Sum of answers	Membership to set high
R1	1	2	2	3	2	**10**	**0**
R2	5	2	4	5	4	**20**	**1**
R3	2	1	4	3	4	**14**	**0.016129032**
R4	3	3	3	3	3	**15**	**0.338709677**
R5	2	5	4	1	5	**17**	**0.983870968**
R6	4	4	4	5	4	**21**	**1**
R7	3	3	3	3	3	**15**	**0.338709677**
R8	1	1	2	3	1	**8**	**0**
R9	4	4	4	4	4	**20**	**1**
R10	4	5	4	5	4	**22**	**1**
R11	4	5	3	3	5	**20**	**1**
R12	5	4	3	2	5	**19**	**1**
R13	4	4	4	5	4	**21**	**1**
R14	4	5	5	4	5	**23**	**1**
R15	4	5	4	4	3	**20**	**1**
R16	3	2	5	3	4	**17**	**0.983870968**
R17	1	2	3	1	1	**8**	**0**

In order to solve this task terms *highly rated software* and *most of* should be formalized. Regarding the latter, it is a usual relative fuzzy quantifier, which is in our case expressed as

$$\mu_Q(y) = \begin{cases} 0, & \text{for } y \leq 0.5 \\ y - 0.5 & \text{for } 0.5 \leq y \leq 0.85 \\ 1, & \text{for } y > 0.85 \end{cases} \tag{4}$$

where y is the proportion of highly rated responses.

The term *highly rated software* is content dependent on each group. From the minimal and maximal score among questionnaires (Table 2), sets *high* and *low* rates were created in the sense of [16] by uniformly covering the domain of scores [17].

The validity for the quantified aggregation is calculated as

$$v(G_i) = \mu\left(\frac{1}{ni}\sum_{j=1}^{ni} \mu_{high}(R_j)\right) \tag{5}$$

where $\frac{1}{ni}\sum_{j=1}^{ni} \mu_{high}(R_j)$ is the proportion y of questionnaires which highly rated software tool and n_i is the number of responded questionnaires in group i.

Thus, the result of aggregation is in the [0, 1] interval. Regarding the group G_1 from Table 2, the solution is 0.532.

4.4 Aggregating Values for Software Tools Among Groups

In Sect. 4.3 quantified aggregation regarding the opinion of each group has been calculated. The result of aggregation assumes value in the [0, 1] interval, which opens the path for applying a variety of aggregation functions. The four main classes of aggregation functions are [6, 8]: conjunctive, averaging, disjunctive and mixed.

When all the groups agree that a tool is effective, this tool should be emphasized by upward reinforcement, which is possible to realize by t-conorm functions. Analogously, when all the groups assign low values to the tool, it should be attenuated by downward reinforcement, which is realizable by t-norm functions. But, we need to apply the full or two-directional reinforcement. Such property is met by uninorm functions. A possible choice is 3-Π function suggested in [19].

$$v(S_p) = \frac{\prod_{i=1}^{n} x_i}{\prod_{i=1}^{n} x_i + \prod_{i=1}^{n} (1 - x_i)} \tag{6}$$

An illustrative example is shown in Table 3, where values of G_1 for S_1 are taken from the previous subsections. Other values are evaluated for the illustrative purpose. The upward reinforcement holds for S_2, whereas downward reinforcement holds for S_5. For tools S_2 and S_4, the aggregation is behaving as an averaging function.

Table 3. Aggregating of utilities by all groups for considered tools

Group/Software	G_1	G_2	G_3	G_4	Uninorm (6)
S_1	0.532	0.151	0.85	0.35	0.381533986
S_2	0.835	0.725	0.788	0.931	0.998507712
S_3	0.25	0.32	0.41	0.22	0.029828289
S_4	0.63	0.826	0.253	0.366	0.612460096
S_5	0.11	0.22	0.32	0.18	0.00358814

5 Conclusion

The evaluation of software usability in the company is not an easy task due to the different number of questions and possible (categorical) answers to the same question among groups, the different number of responded questionnaires by groups and the like. Usually, rule-based systems are used to express users view and knowledge regarding software usability. However, these rule-based systems might be very large, especially when the company has a larger number of software tools and departments or groups of users.

The next task was ranking all software tools across the company. In order to solve this problem, we have suggested three levels of aggregation for the schemes shown in Figs. 2 and 3. The aggregation on questionnaire level is realized by the sum of answers or more sophisticated functions like the weighted average. The second level of aggregation is realized by the quantified aggregation among questionnaires to cope with the aforementioned problems. The result is in the unit interval, which opens space for a large variety of aggregation operators in the third level of aggregation. In this step, the suitable functions are uninorms due to their full reinforcement property.

For future research, we would like to examine other aggregation functions for all levels and to realize further experiments regarding computational effort. The suggested aggregation model may be applicable to some tasks in companies concerning customer satisfaction, customer experience, marketing evaluation etc. Based on the suggested aggregation model it is possible to develop a valuable decision support system.

Acknowledgements. This paper is part of a project VEGA No. 1/0373/18 entitled "Big data analytics as a tool for increasing the competitiveness of enterprises and supporting informed decisions" by the Ministry of Education, Science, Research and Sport of the Slovak Republic.

References

1. Albert, W., Tullis, T.: Measuring the User Experience, Collecting, Analyzing, and Presenting Usability Metrics (Interactive Technologies), 2nd edn. Elsevier, Waltham (2013)
2. Bandarian, R.: Evaluation of Commercial potential of a new technology at the early stage of development with fuzzy logic. J. Technol. Manag. Innov. **2**(4), 73–85 (2007)
3. Bavdaž, M.: Sources of measurement errors in business surveys. J. Official Stat. **26**(1), 25–42 (2010)
4. Bavdaž, M, Biffignandi, S., Bolko, I., Giesen, D., Gravem, D., Haraldsen, G., et al.: Final report integrating findings on business perspectives related to NSIs' statistics. Deliverable 3.2., Blue-Ets Project (2011). http://www.blue-ets.istat.it/fileadmin/deliverables/Deliverable3.2.pdf. Accessed June 2016
5. Beliakov, G., Pradera, A., Calvo, T.: Aggregation Functions: A Guide for Practitioners. Springer, Heidelberg (2007)
6. Calvo, T., Kolesárová, A., Komorníková, M., Mesiar, R.: Aggregation operators: properties, classes and construction methods. In: Calvo, T., Mayor, G., Mesiar, R. (eds.) Aggregation Operators. New Trends and Applications, pp. 3–104. Physica-Verlag, Heidelberg (2002)
7. Detyniecki, M., Fundamentals on aggregation operators. In: Proceedings of the AGOP 2001, Asturias (2001)
8. Dubois, D., Prade, H.: A review of fuzzy set aggregation connectives. Inf. Sci. **36**, 85–121 (1985)
9. Greiner, L., White, S.K.: What is ITIL? Your guide to the IT Infrastructure Library, in digital magazine CIO from IDG. https://www.cio.com/article/2439501/itil/infrastructure-it-infrastructure-library-itil-definition-and-solutions.html. Accessed 28 Mar 2018
10. Herrera, F., Martíez, L.: A model based on linguistic 2-tuples for dealing with multigranular hierarchical linguistic contexts in multiexpert decision-making. IEEE Trans. Syst. Man Cybern. Part B Cybern. **31**, 227–234 (2001)
11. ISACA, Service IT governance professionals, COBIT5, an ISACA framework. http://www.isaca.org/cobit/pages/default.aspx. Accessed 28 Mar 2018
12. Králiková, L.: Testovanie efektívnosti softvéru v podnikovej praxi z hľadiska užívateľov = Software effectiveness testing in business practice from a users` perspective, (in Slovak), Master Thesis, University of Economic in Bratislava (2017)

13. Pavlík, L.: Metrics for Evaluating Information Systems, Posterus, portál pre odborné publikovanie, ISSN; 1338-0087 http://www.posterus.sk/?p=18957. Accessed 21 Mar 2018
14. Rakovská, E.: Projekty znalostného manažmentu ako súčasť metodiky Balanced scorecard = Knowledge management projects as a part of Balanced scorecard methodology In: Eduard Hyránek, E., Nagy, L., Výsledky riešenia končiacich grantových úloh VEGA 1/0261/10, 1/0872/09, 1/0384/10, 1/0415/10: zborník vedeckých statí. EKONÓM, Bratislava (2011)
15. Rakovská, E., Hudec, M.: Two approaches for the computational model for software usability in practice. In: Kulczycki, P., Kowalski, P.A., Łukasik, S. (eds.) Contemporary Computational Science, p. 21. AGH-UST Press, Cracow (2018)
16. Ruspini, E.: A new approach to clustering. Inf. Control **15**, 22–32 (1969)
17. Tudorie, C., Qualifying objects in classical relational database querying. In: Galindo, J. (ed.) Handbook of Research on Fuzzy Information Processing in Databases. Information Science Reference, pp. 218–245. Hershey (2008)
18. Yager, R.R.: A new approach to the summarization of data. Inf. Sci. **28**, 69–86 (1982)
19. Yager, R.R., Rybalov, A.: Uninorm aggregation operators. Fuzzy Sets Syst. **80**, 111–120 (1996)
20. QP-Quality Progress, the official publication of ASQ. Allen, I.E., Seaman, Ch.A.: Likert Scales and Data Analyses (2007). http://asq.org/quality-progress/2007/07/statistics/likert-scales-and-data-analyses.html. Accessed 29 Mar 2018

Graph Cutting in Image Processing Handling with Biological Data Analysis

Mária Ždímalová[1(✉)], Tomáš Bohumel[1], Katarína Plachá-Gregorovská[2], Peter Weismann[3], and Hisham El Falougy[3]

[1] Slovak University of Technology in Bratislava, Radlinského 11, 810 05 Bratislava, Slovak Republic
zdimalova@math.sk

[2] Institute of Experimental Pharmacology and Toxicology, Slovak Academy of Science, Bratislava, Slovak Republic
plachakaterina@gmail.com

[3] Institute of Anatomy, Faculty of Medicine, Comenius University of Bratislava, Bratislava, Slovak Republic

Abstract. In this contribution we present graph theoretical approach to image processing focus on biological data. We use the graph cut algorithms and extend them for obtaining segmentation of biological cells. We introduce completely new algorithm for analysis of the resulting data and sorting them into three main categories, which correspond to the certain type of biological death of cells, based on the mathematical properties of segmented elements.

Keywords: Graph cuts · Segmentation · Cell analysis · Apoptosis · Necrosis · Computer morphometry

1 Introduction

Graph theory has many real applications, in economy, [3,4], in image processing [11], didactic, chemistry and many others. We consider specially its applications in image processing.

Image processing and pre-processing are well-known and suitable tools for handling with different types of data, [1,8,11]. In biology and medicine the main aim is to simplify the representation of the obtained images for users. These methods are very useful for handling medical and biological data as well. There are a lot of approaches for image processing and pre-processing, e.g. trash-holding, graph cutting, level sets methods and many others. For our purposes we focus on a segmentation of objects, especially in our work on finding specific cells. In this concept the aim of segmentation is to distribute an image into regions and thus simplify its representation. Two types of regions are considered, object regions and background regions. The output of the segmentation is a binary image with extra information held representing "object" and "background" segments. We consider "objects" all cells of interest and the background

P. Kulczycki et al. (Eds.): ITSRCP 2018, AISC 945, pp. 203–216, 2020.
https://doi.org/10.1007/978-3-030-18058-4_16

the rest of the image. The main goal of the segmentation is to simplify and change the representation of an image in the more meaningful and easier way for next analyses.

In our case we deal with biological real data. We focus on the following analysis. Parkinson's disease, Alzheimer disease and amyotropical lateral sclerosis are major neurodegenerative disease of adults, which are characterized by age-related neurodegeneration of many neural structures. Although, in some forms of the mentioned diseases, heredity plays a role, the mechanism of cell death is not totally understood. For the purpose of precise analysis of programmed cell death in neurodegenerative disease research, we have created a computer software to detect apoptosis/necrosis on histological preparations. We tested its functionality on histological preparations of the brains gained from Wistar (7 days old), which we exposed to hypoxic - ischemic insult for 12 and 36 h. We used cryostat coronal cuts of 30 um width colored by cresyl violet. We tested pyramid cells in the region CA1 of hipocamp.

The aim of this work was the automatically distinction, the searching and the counting of concrete types of cells, it means the apoptosis (the apoptotic cells), the nuclear morphology (hybrid state) and the necrotic cells (the necrosis) which are characterize below. Every cell from this category has its specific feature, which are size and number of burst of elements, from which these cells consist. We searched as well for presence of cell wall and many other properties, see e.g. [9].

The preliminary version of this paper was presented at the 3rd Conference on Information Technology, Systems Research and Computational Physics, 2–5 July 2018, Cracow, Poland [12].

2 Segmentation and Graph Cutting

The segmentation can be provided by many different techniques [8,14]. The segmentation in the image processing can be formulated in mathematics as a minimization problem, see [7,11]. Segmentation can work as a powerful energy minimization tool producing globally optimal solution. For segmentation we use the mathematical method called "graph cutting", see [2,5,7]. In the work we focused primarily on Ford-Fulkerson and Edmonds-Karp algorithms [5,6]. We process 2D image, which we first abstract as a graph (graph theory in mathematics) and then we try to find a maximum flow in it. After finding the maximum flow [5], we are able to segment the image. The graph cuts are used in medical and biological image segmentation following few dynamic algorithms, finding the local minimum of the energy. Compare to threshold technique, this approach gives more realistic results.

The principle of Ford-Fulkerson and Edmonds-Karp algorithms is based on increasing of the flow in the graph (the network) through the augmenting paths. The algorithm progress while any augmenting path can be found. When there is no augmenting path available, the algorithm ends and the maximum flow is reached. The value of the maximum flow equals the sum of the capacities of the "minimal cut" edges.

Minimal cut is the result of the graph cut algorithms (mentioned above) applied. The simplified explanation of finding the minimal cut is the process of pushing flow (imaginary units) from the source vertex named s to tank vertex named t through the graph consisting of the vertices and edges while possible. Once the process is finished and there is no capacity of the edges to transport any other flow, the minimal cut can be found as the union of such edges.

In the image segmentation process the pixels of any 2D image can be abstracted into the graph vertices and the graph used in the theoretical mathematics can be constructed. After that the graph cut algorithms can be applied, the minimal cut can be found and finally the image can be represented by the objects and the background.

3 Program and User Manual

For such purposes mentioned in the paragraph above we created the software with built in "graph cutting" algorithms, for handling medical and biological data, which are in our case output images from microscope. The advantage of this method is that it can provide global segmentation as well as local and we are also able to detect the edges, which represent to the boundaries between cells. In our specific case we need to provide the pre-processing of the images as well. We applied Gaussian kernel for pre-processing of the first input data from electronic microscope. The main benefit of this software is its complexity. It is able to do pre-processing of the images, segmentation of the image, which in our special case means finding the corresponding cells, and finally counting and categorization of the cells. Once the process of the image segmentation, cell counting and categorization is done, the user can manually adjust image output (change the categories of marked cells, etc.) and use numerical output for the statistics.

The process of the image segmentation and cell categorization consists of the following steps:

1. Data gathering from the microscope.
2. Pre-processing of the data.
3. Software initialization and specific input image loading.
4. Setting up the object and background pixels of the image.
5. Image segmentation process.
6. Setting up the parameters for each of the cell types.
7. Counting, color marking and categorization of the cell types.
8. Manual adjustment of the marked cells done by user.
9. Output saving (image and numerical data).

The application of the image segmentation described above is used to identify three different types of cells presented in the images produced by the microscope which are the apoptosis, the nuclear morphology and the necrosis. Once the segmentation is done, we have to differentiate the segmented objects and group them together, because each of the cell types consists of many of the smaller objects. Each group of the segmented objects consists of the objects that have the characteristic sizes, distances between each other and quantities due to what they can be categorized as one of the three cell categories.

4 Application in Biological Data

The whole process of dealing with data (images) consists from three main steps: preprocessing of obtaining data, processing and post-processing of data. Pre processing is devoted to the correction and the preparation of data obtained from the microscope. Under the processing of data we understand handling with data from graph theoretical approach. Last step is devoted to the final analyzing of resulting data.

(a) Pre-processing
Pre-processing is the whole first preparation data before segmentation. Our input data are medical images from Microscope, which are necessary transform to 24 bit map and afterwards correct contrast of input data image that results images will not be too dark, too light or we can see good visibly contours. We use two methods, the normalization of the histogram and the shadding corrections.

Normalization of the Histogram: Input data (microscope images) were took by professional microscope with certain intensities. The range of intensities of the microscope are multiple times bigger, as we usual work in image processing (265 of values). Because of that such obtained images in usual format had slow contrast. It means that in histogram were the values cumulated on small interval and not in his full range of intensities, provided by given format.

Normalization of histogram corrects the distribution of intensities, and so it changes their range and this way it improves the contrast of the image. The next liner transformation transforms values of intensities of pixels from interval from the range $< L, U >$ for new values I_{new} from interval in range $< L_{new}, U_{new} >$.

$$I_{new} = (I_p - L)\frac{U_{new-L_{new}}}{U - L} + L_{new}.$$

Above mentioned method fails in case that in interval $< L, U >$ appears the only pixel with very small or very big intensity, which is faraway from the top of interval of the histogram, and so outside of the range of intensities of pixels I_p creating the biggest ratio in the histogram (image). In such case will be intervals $< L, U >$ and $< L_{new}, U_{new} >$ almost identified and the requested affect will not be reached.

It is more reliable to choose it for L a U some percentil of the original histogram. The range of the interval $< L, U >$, will be created 90% of pixels of the origin range cumulated around the top of the histogram. To pixels with very small or very big intensities (10)% will be automatically assigned new value of the intensity I_{new} equals L_{new}, or U_{new}.

Shadding Correction: The shading correction is a background correction in an image meant to compensate for irregular illumination effects. This correction is also known as unsharp masking.

(b) Processing

In this part are implemented and used all graphs algorithms used in the segmentation of image. After pre-processing and correction soft input data in requested quality, then follows their processing. The input: we normalize the images by the histogram and the shadding corrections. The output: are segmented data, which are in other process classified.

For searching of the maximal flow of the network we used Edmonds-Karp algorithm, see [5]. We choose the Edmonds-Karp algorithm for the shortest path algorithms. We implemented the algorithm with minor variations over the original one. For segmentation we used and extended the method of graph cutting, see the Sect. 5, [10] and [13]. We modified some of steps of algorithm and as well we did optimization of the fast running of algorithm.

The result of the segmentation (processing) is the image classified into two classes: object and background pixels. The fact and information, if the resulting pixel belongs after segmentation to the background or the object is important by post analysis (post-processing). Post-processing is devoted to classifying of the obtained cells. Because we needed to distinguish concrete types of cells, it was necessary to create a new algorithm for classifying obtained cells.

(c) Postprocessing: Characterization of output data-New algorithm for classification of cells

After finishing the segmentation we use the segmented image for the next analysis: the distinguishing, searching and the counting of concrete types of cells. We focuse on the distinguishing of cells according the fact which type of death the cells died. In our work we consider three specific categories:

1. the apoptosis (apoptotic cells)
2. the nuclear morphology (hybrid cells)
3. the necrosis (necrotic cells).

From mathematical and programming approach we do not distinguish categories of the cells according the type of death. So we do not analyse by subjective way, which come from our sensual experience. We need to use characteristics, which are recognizable by analysing of the image, and which are objective and which are described by the properties of corresponding categories. It means categories which are able to be measured. Every category has its typically mathematical properties.

Apoptosis: From medical point of view, the apoptosis is a regulated death of the cell on the polycells organism. The cell simply dies, (or necrosis), after that they will be created big clumps of cells, which concentrate to small number of the bigger whole (mostly 1–3 of pieces). In this process the cell wall will not be break, remains in one unit.

From mathematical point of view is the apoptosis a group of objects, which consists from bigger number of pixels. Average value of distances of the centres of these objects is continuously equal to the average value of one and half of sum of the radius of all couples of objects, see Fig. 1.

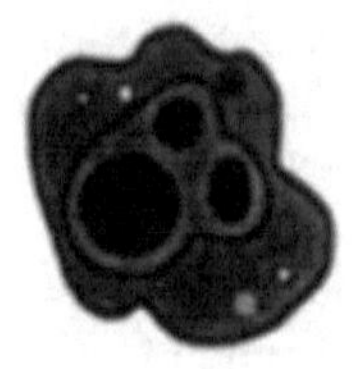

Fig. 1. Image of the apoptosis

Nuclear Morphology: The nuclear morphology is the intermediate state between the apoptosis and the necrosis, called as well apoptosis-necrosis continuum. Bigger wholes which arised by apoptosis, will finally decompose. But they are still markedly, many times bigger like by necroses. The wall of the cell remains without any break points.

The nuclear morphology (the hybrid state) is in the sense of image processing the group of more objects, multiple less sizes as are sizes of objects creating apoptosis. These appear about in the distance twice bigger as is the multiple of the sum of the radial of two objects of any objects, as is shown in Fig. 2.

Fig. 2. Image of the nuclear morphology

Necrosis: From definition the necrosis is a form of traumatic cell death. This type of death is conditional with acute injury or damage of the cell. This type of death is conditioned by acute injury or, respectively, cell damage. Death occurs by breaking the cell wall, which eventually almost completely disappears, and the inside of the cell (the clump of particles) dissolves into the space. These elements are then smaller in size, but they form larger aggregates.

From the viewpoint of the image analysis, the necrosis, see Fig. 3, is represented by miniature objects with dimensions that are several times smaller than the distances between these objects compared to the apoptosis and the morphology.

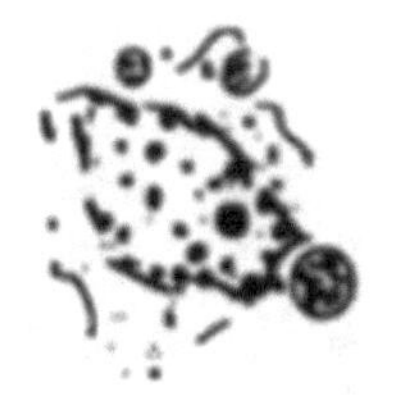

Fig. 3. Image of the necrosis

Algorithm for Analysing Types of Cells. As we mentioned before, we created a new algorithm for the analysing of cells. The algorithm is able to do the categorization of segmented cells and sum them. The algorithm consists from steps described below.

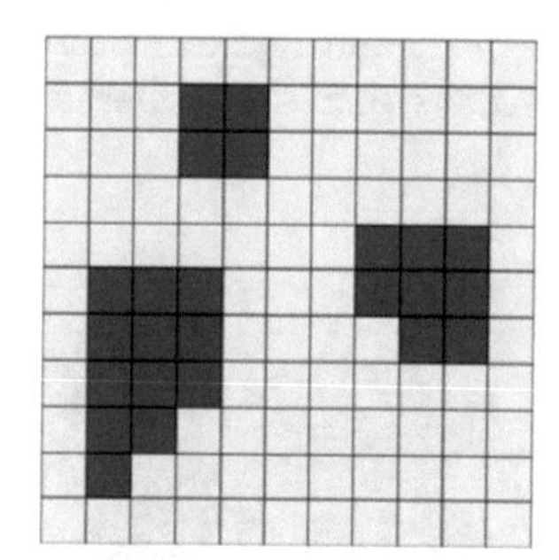

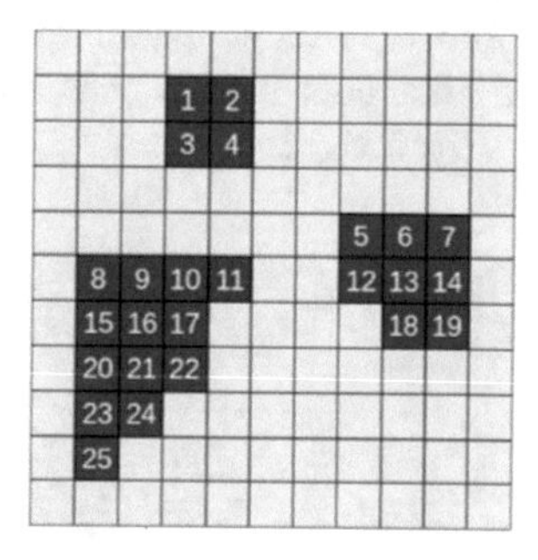

Fig. 4. Left: detection of the cells; Right: Enumerating of the objects

Detection of objects is a process where we check the whole image and count all pixels, which were denoted by segmentation as the object pixels. We number every object pixel by the number, see Fig. 4 left.

Enumerating of objects go this way: we check the whole image again and all numbered pixels will be assigned to certainly objects, which are subsequently enumerated. The result are the enumerated objects (families of objected pixels) as showed at Fig. 4 right.

Measuring of the size of objects we obtain by summing up of all pixels, which appear in the corresponding enumerated object. The size of the object is given by the number of pixels and it is one of the criterium needed by the categorization of cells, see Fig. 5 left.

Classification of object according the size we will do such way that we consider all objects (elements) which are with their size typically for given category of the cell. We just ignore all others objects.

Re-numbering will be like we again check all enumerated objects and that which we just ignored in the previous step we will not enumerate. In this step we select all objects which will be used by categorization of cells.

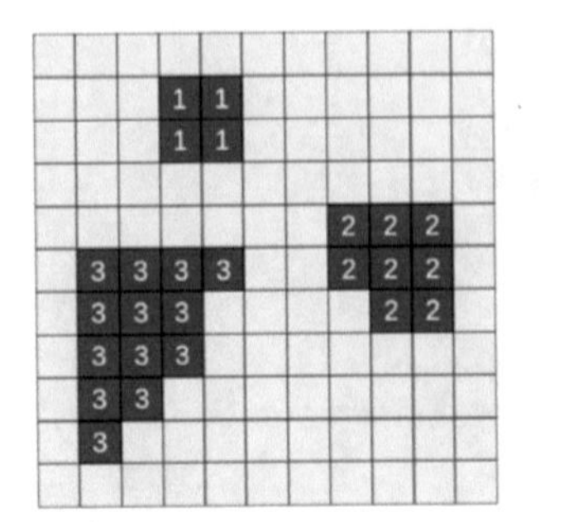

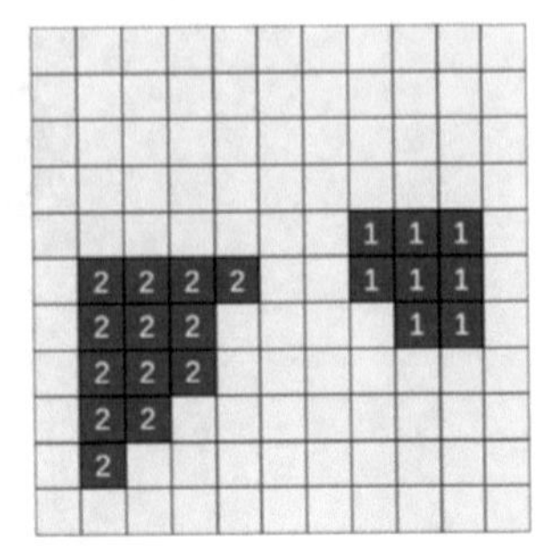

Fig. 5. Left: measuring of the cells; Right: re-numbering of the cells

Detection of the middle pixels of cells: It is possible to find the middle pixels of objects by recognizing its coordinates for each pixel that is a part of the numbered object. We then calculate the value of the mean pixel as the arithmetic mean of the pixel coordinates that belong to that concrete numbered objects, as is shown in Fig. 5 right.

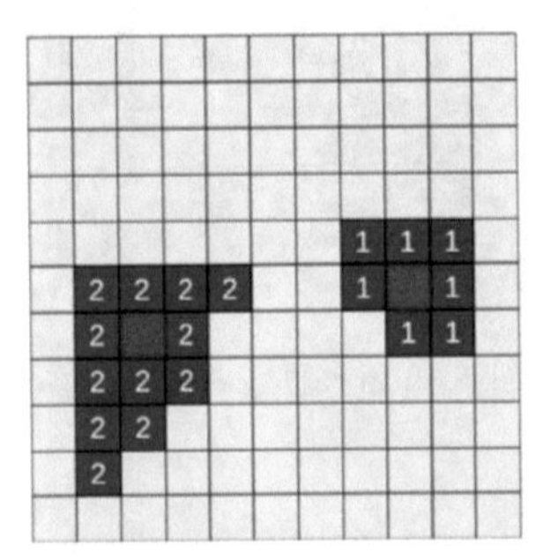

Fig. 6. Left: measuring the distances of objects; Right: detection of the middle pixels of the cells

Object distance measurement is the step in which we compare distances (in pixels) between the mean pixels of enumerated objects (elements) which are by their size typical for a given cell category. If the average pixels of the objects are in some special distance typically for a certain cell category, then the clusters of these objects will create specific cells, see Fig. 6 left.

Determination of the mean pixels of the cells is achieved by finding a mean pixel for each concrete luster-aggregate. That means, the middle pixel of the objects is the of acceptable size and at the acceptable distances. Then we calculate the average pixel value of a particular cell as the arithmetic average of the coordinates of the middle pixels of the objects in the given particle cluster of elements, see Fig. 6 right.

Counting and summing up of the cells is the last step in which we can easily calculate the middle pixels of a given cell category, that is, the number of middle pixels of apoptic, hybrid and necrotic cells.

5 Implementation in the Program

We describe how we implemented graph theory approach in implementation of the program to obtain the segmentation. The program is written in the language C. For searching of maximal flow of the network we used Edmonds-Karp algorithm. We modified some of steps of algorithm and as well we did optimization of the fast running of algorithm. We decreased mainly the number of iterations. If finally was not possible to decrease the number of iteration, we decreased the number of operations, which were done in this iteration to the minimum. Program consists from three main parts. The first one is the initialization, which describes and specifies our approach to the network. Another part is the algorithm for finding of maximal flow of the network. The last step is the processing of the resulting segmented data.

Capacity of the Edges. We count the capacity for the corresponding links (edges). That one, which connects exactly two of neighboring pixels (vertices) p and q we call N-links. That links, which connect exactly one pixel with the source s and the sink t, we call T-links, where the pixel is presented as gray cube, N-links as horizontally links, and T-links as vertical lines. More about the implementation and the transformation to the network, see [13]. We use the following notation for the counting of the capacities:

- P the set of all pixels,
- (p, q) the edge connecting neighboring pixels p and q,
- I_p the value of the intensity of the pixel p,
- M the maximal value of the intensity of the pixel (of the responsible figure),
- D the difference of the maximal and minimal value of the intensity of the pixel (of the responsible figure),
- O_{avr} the average value of the intensity of object seed pixels,
- B_{avr} the average value of the intensity of the background seed pixels,
- $S(p)$ the capacity of the edge(link) connecting the sink (the vertex s) and corresponding pixel p,
- $T(p)$ the capacity of the edge (link) connected output source (the vertex t) and concrete pixel (the vertex p),
- $N(p, q)$ the capacity of the edge (link) connected neighbors pixels p a q,
- λ the weighing constant.

The weighing constant λ determines the result of the segmentation. The note $N(p, q)$ express the relationship between intensities of p an q, $S(p)$ and $T(p)$ express the relationship between intensity values of pixels and the values O_{avr} and B_{avr}, see Table 1. More about connection between this variables and constants you can find in [10, 13].

Table 1. Table-The capacity evaluation

Type	Edge	Capacity	
N-links	(p,q)	if $(p,q) \in P$	$N(p,q)$
T-links	(s,p)	if $p \in P \setminus \{o \cup B\}$	$\lambda S(p)$
		if $p \in O$	0
		if $p \in B$	0
	(p,t)	if $p \in P \setminus \{o \cup B\}$	$\lambda T(p)$
		if $p \in O$	0
		if $p \in B$	∞

Capacity Evaluation. First we select the values $S(p)$ and $T(p)$ for pixels p, which we denoted as object seed pixels and background seed pixels. The value $S(p)$ will have the value of infinity, while we denoted the pixel p as object seed pixel, so $S(p) = \infty$. It is because $S(p)$ can not be never saturated, while it connects the source s and object pixel. It means object pixel need to be always reachable from the source s. The note $T(p)$ will have the value equals to 0, while we denoted the pixel p as object seed pixel, so $T(p) = 0$. It is because this edge needs to be saturated and output t should not be reachable from the pixel p. Similarly, but vice versa, we do similar assignments also for all background seed pixels. We guarantee this way that after finishing of segmentation will be object seed pixels surely the part of the object and background seed pixels will be part of the background.

Linear Diffusion Coefficient. Capacities of N-links and T-links are depended from intensities of the concrete pixel. Others values of capacities we count from the values of intensities of pixels as follows: Both N-line and T-line capacities depend on the intensity of the pixel. Therefore, the next capacity values are calculated from the pixel intensity values as follows [6]:

$$N(p,q) = D - |I_p - I_q|$$
$$S(p) = M - |O_{avr} - I_p|$$
$$T(p) = M - |B_{avr} - I_p|$$

It is precisely because of the character (definition) of the M and D constants that the capacities are played non-negative. In extreme cases, some capacities may be zero. Taking into account all previous claims, we assign specific capacities to specific edges in the following way as shown in the Table 1.

Non-linear Diffusion Coefficient. We suggest as well different and new approach how to give values to edges. If we want to approach the assignment of N-line capacities in a way that takes greater account of the relative intensity of pixel intensities and penalizes their differences, it is necessary to choose a

non-linear coefficient for calculating their capacity. The non-linear coefficient causes neighboring pixels with similar intensity values to have high-capacity edges (lines) and a certain drop in intensity of the neighboring pixels, and the edge-to-edge gain is almost zero. For the interpretation of next formulas the following definition is needed:
s - absolute value of intensity difference of two neighboring pixels
σ - penalizing constant
k - the penalty.

Thus, the value s is calculated as the absolute value of the difference of the two adjacent pixels in the picture, the penalizing constant is optional and affects the course of functions, especially how rapidly their first derivation changes and the penalty k is given by the formula.

TV diffusion coefficient

$$d(s) = \frac{1}{s}, s \in N$$

BFB diffusion coefficient

$$d(s) = \frac{1}{s^2}, s \in N$$

Charbonnier's diffusion coefficient

$$d(s) = \frac{1}{\sqrt{(1 + \frac{s^2}{k^2})}}, s \in N$$

Perona-Malik's diffusion coefficient

$$d(s) = \frac{1}{1 + (\frac{s}{k})^2}, s \in N$$

$$d(s) = e^{-(\frac{s}{k})^2}, s \in N$$

Weickert's diffusion coefficient

$$d(s) = \begin{Bmatrix} 1 & s = 0 \\ 1 - e^{\frac{-3.31488}{(s/K)^8}} & s > 0 \end{Bmatrix}, s \in \mathbb{N}$$

Comparison of Diffusion Coefficients. The diagram below on Fig. 7 shows the diffusion coefficients of the diffusion coefficients, depending on the intensity differences from 0, 255, at the selected value as follows: Linear coefficient - blue, TV coefficient - red, BFB coefficient - purple, Charbonnier coefficient - yellow, Perona-Malik coefficient - green, (-black), Weickert coefficient - orange.

Using of different types of coefficients help us to obtain better segmentation and better results. From the programming point of view, the implementation can be divided into three main parts, which can also be further elaborated. They are as follows: marking procedure, path reconstruction, the distribution of vertices.

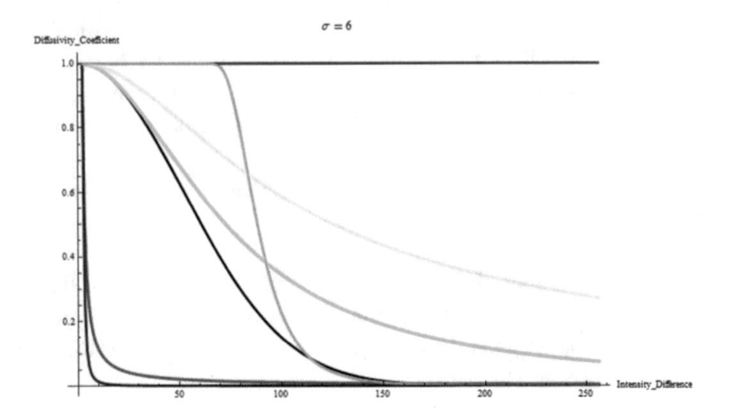

Fig. 7. Comparison of diffusion coefficients

6 Results

We used the application of the image segmentation, described above, to identify three different types of cells presented in the images produced by the microscope. They are the apoptosis, the nuclear morphology and the necrosis. When we finished the segmentation, Fig. 8, we are supposed to differentiate segmented objects and group them together. Each of the cell types consists of many of the smaller pieces like objects. Each of this segmented objects consists of other object which have the corresponding sizes, distances between each other and quantities due to they can be categorized to main three categories. Below, there is a part of the input and the output image, Fig. 9, with marked apoptosis (red), nuclear morphology (blue) and necrosis (green).

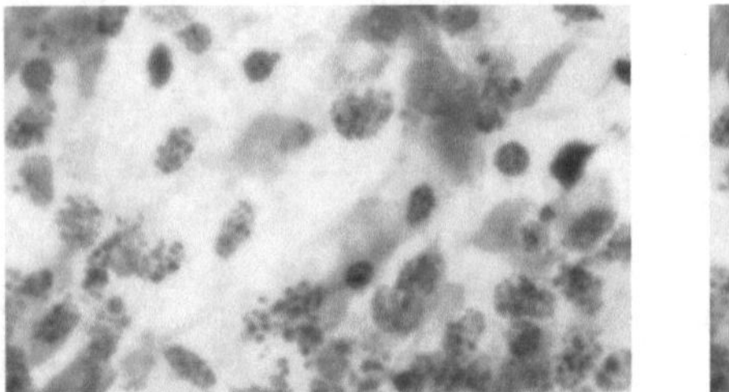
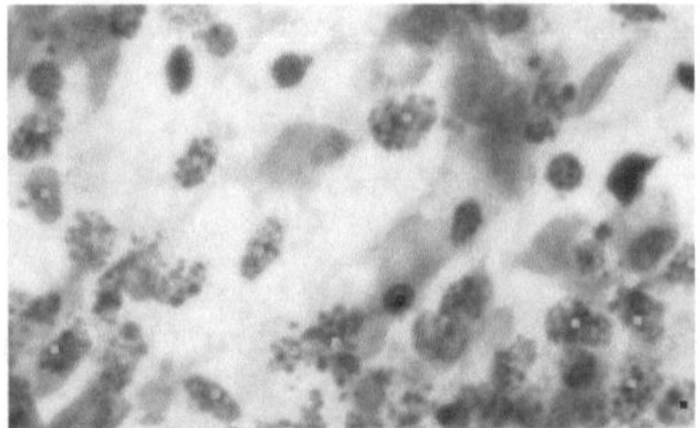

Fig. 8. Left: input: original data; Right: output: data after the segmentation

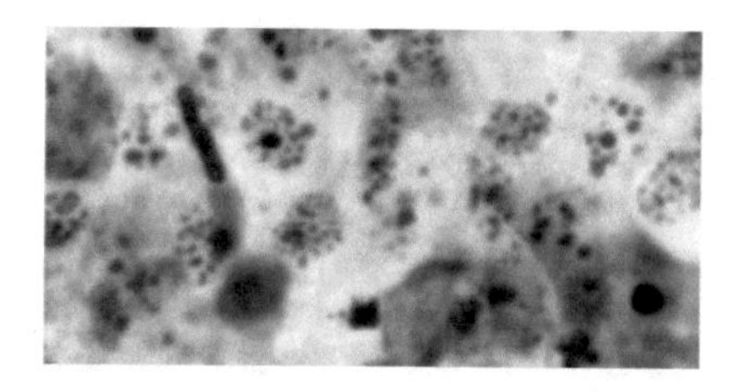

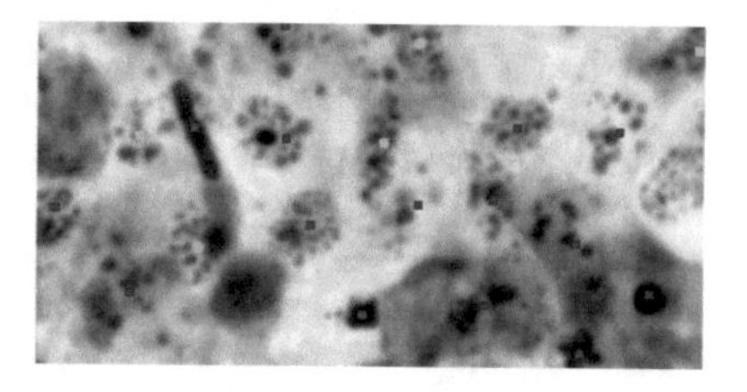

Fig. 9. Computers detection of the apoptosis, the necrosis and the nuclear morphology. Before and after the processing with the computer software.

7 Conclusion and Remarks

As the application of the image segmentation approach on the cell categorization problem is in its very beginning phase, a lot of the improvement still needs to be done. The score of the cell categories marked automatically correctly without any additional adjustment is between 70 to 85.

Nowadays, for simpler real-time data, for more complex data and global segmentation in tens of seconds and for local segmentation of more complex data, it is also real-time. Segmentation speed also depends on the quality of the input data, the dimensions of the image, the selection and the number of seed pixels. The program currently segments either the entire image (globally) or only part of it (locally), depending on what option the user chooses. In the future, it will also be possible to segment the color images and we would also like to program some other parts so that it is absolutely understandable and clear to the normal user. Of course, further optimization of the program is also one of the goals.

Acknowledgement. Mária Ždímalová acknowledges the Slovak Researcher and Development Agency, VEGA 1/0420/15 and APVV-14-0013. This work was supported as well by the project of the Slovak Researcher and Development Agency, APVV-15-0205.

References

1. Basavaprasad, B., Hegadi Ravindra, S.: A survey on traditional and graph theoretical techniques for image segmentation. Int. J. Comput. Appl. (0975-8887), 38–46 (2014). Recent Advances in Information Technology
2. Boykov, Y., Jolly, M.P.: Interactive graph cuts for optimal boundary and region segmentation of objects in N-D images. In: Proceedings of "International Conference on Computer Vision", Vancouer, Canada, vol. 1, pp. 105–112 (2001). ISBN 0-7695-1143-0
3. Drabiková, E., Fecková Škrabuľáková, E.: Decision trees-a powerful tool in mathematics and economic modelling. In: Proceedings of the 18-th International Carpathian Control Conference (ICCC), Palace Hotel, Sinaia, Romania, 28–31 May 2017, pp. 34–39 (2017)

4. Drabiková, E., Fecková Škrabuľáková, E.: Monitoring and controlling of economic tasks through tools of graph theory. Econ. Spectr. **7**(2), 1–8 (2017)
5. Ford Jr., L.R., Fulkerson, D.R.: Maximal flow through a network. Can. J. Math. **8**, 399–404 (1956)
6. Ford Jr., L.R., Fulkerson, D.R.: Flows in Networks. Princeton University Press, Princeton (1962)
7. Goldberg, A.V., Tarjan, R.E.: A new approach to the maximum flow problem. J. Assoc. Comput. Mach. **35**(4), 921–940 (1988). ISSN 0004-5411
8. Gómez, D., Yanez, J., Guada, C., Tinguaro Rodriguez, J., Montero, J., Zarrazola, E.: Fuzzy image segmentation based upon hierarchical clustering. Knowl. Based Syst. **87**, 25–37 (2015)
9. Kopani, M., Filon, B., Sevik, P., Krasnac, D., Misek, J., Polak, S., Kohan, M., Major, J., Ždímalova, M., Jakus, J.: Iron decomposition in rabbit cerebellem after exposure to generated and mobile GSM electromagnetic fields. Bratilslava Med. J. **118**(10), 575–579 (2017)
10. Loucký, J., Oberhuber, T.: Graph cuts in segmentation of a left ventricle from MRI data. Prague, Czech Technical University in Prague, COE Lecture Note, 2012, vol. 36, pp. 46–54 (2010)
11. Peng, B., Zhang, L., Zhang, D.: A survey of graph theoretical approaches to image segmentation. Pattern Recognit. **46**, 1020–1038 (2013)
12. Ždímalová, M., Bohumel, T., Plachá, Gregorovská, K., Weismann, P., El Falogy, H.: In: Kulczycki, P., Kowalski, P.A., Łukasik, S. (eds.) Contemporary Computational Science, p. 112. AGH-UST Press, Cracow (2018)
13. Ždímalová, M., Krivá, Z., Bohumel, T.: Graph cuts in image processing. In: 14th Conference on Applied Mathematics, APLIMAT 2015, Proceedings in Scopus, Institute of Mathematics and Physics, Faculty of Mechanical Engineering, STU in Bratislava, pp. 1–13 (2015)
14. XIn, J., Renje, Z., Shendong, N.: Image segmentation based on level set methods. Phys. Proc. **33**, 840–845 (2012)

A Methodology for Trabecular Bone Microstructure Modelling Agreed with Three-Dimensional Bone Properties

Jakub Kamiński, Adrian Wit, Krzysztof Janc, and Jacek Tarasiuk(✉)

Faculty of Physics and Applied Computer Science (WFiIS),
AGH – University of Science and Technology, al. Mickiewicza 30,
30-059 Kraków, Poland
tarasiuk@agh.edu.pl

Abstract. Bone tissue is a structure with a high level of geometrical complexity as a result of mutual distribution of a large number of pores and bone scaffolds. For the study of the mechanical properties of the bone, there is a demand to generated microstructures comparable to trabecular bone with similar characteristics. Internal structure of the trabecular and compact bone has a high impact of their mechanical and biological character. The novel methodology for the definition of three-dimensional geometries with the properties similar to natural bone is presented. An algorithm uses a set of parameters to characterize ellipsoids computed based on Finite Element Method (FEM). A comparative analysis of real trabecular bone samples and the corresponding generated models is presented. Additional validation schemas are proposed. It is concluded that computer-aided modelling appears to be an important tool in the study of the mechanical behavior of bone microstructure.

Keywords: Microstructure modelling · Trabecular bone · Porous structure

1 Introduction

The obtainment of precise information about the three-dimensional microstructure of a trabecular bone is a serious challenge in tissue engineering. Micro-computed tomography (μCT) is the most commonly used imaging technique with the great resolution and sensitivity. Despite the high cost of the measurement and the difficulties in obtaining sufficiently large and representative samples, it becoming more accessible and popular as the research tool. However, if it is necessary to characterize the large number of structures, with certain material and geometric properties, this technique may be insufficient. An approach of computer-generating microstructures, corresponding to samples obtained in μCT should me more profitable. Generation of the structure is developed by algorithm, which shall result in the structure corresponding to the real one, either parametrically or visually. It should be emphasized that there is no uniform approach necessary to obtain a geometric structure of any type, therefore, in every particular case, the individual characteristics of the material shall be taken into account.

P. Kulczycki et al. (Eds.): ITSRCP 2018, AISC 945, pp. 217–228, 2020.
https://doi.org/10.1007/978-3-030-18058-4_17

The idea of Digital Material Representation (DMR), successfully adapted in materials science [1–3], has also started to be used for the modelling of the microstructure of the trabecular bone, both in two-dimensional space using fractal paradigm, percolations and cellular automata [4], spheres modelling [5], an adaptation of non-periodic Voronoi diagrams [6], as well as in three dimensions, using identical, duplicated and parametrized cell-based polygons [7–9]. These types of structures, although they may ensure statistical correspondence, do not result in a models similar to the microstructure of the trabecular bone, nor do they accurately reflect the real micro-structure of the material in the explicit form. The approach proposed by the author of the last-mentioned works, shows the main benefit of this type of modelling-generating of artificial structures with an examination of the particular parameter change impact on the mechanical properties of bone tissue. However, these investigations covered the change of only some of the main parameters.

Generally, micro-structural modelling in combination with the Finite Element Method (FEM) may allow us to determine the material properties of the bone samples from different regions and topological variation. However, there are several combinations of parameters, that are worth FEM testing, but they do not necessarily occur in the measured bone samples. Works that utilize specific models defined directly with the usage of the finite elements and the adaptation of them to modelling of the bones structure are frequent [10–12]. Therefore, a novel model for the generation of microstructures in parametric agreement with real samples was considered. This allows us to enrich the traditional methods for the determination of the material properties based on experimental tests with an entirely virtual process, covering digital material definitions, numerical simulations and the digital analysis of the results. The models that are generated parametrically, do not require μCT imaging characterization, because the parameters can be chosen arbitrarily. Even though μCT scans can be the basis for the calculations, and in the presented study, real samples of trabecular bone tissue, with a cubic size of 5 mm, were used as a data for the Volumes of Interest (VOI), on which the calculations were based on. Samples were prepared from the distal end of the porcine femoral bone [13]. In this case, a voxel representation is used, that corresponds to an image filled with three-dimensional pixels. Cross-sections from μCT scans with a voxel size of 0.01 mm, used for the calculations, were segmented and binarized, following a previously developed semi-automated approach - see Janc et al. [14] for details.

2 Materials and Methods

Subsequent stages in the form of an algorithm were designed to generate three-dimensional models of trabecular bone. They can be formulated into four main parts:

1. characterization of the structure based on specific parameters,
2. selection of seed points distributions for the next step,
3. generation of geometries and locations of ellipsoids based on seed points,
4. transformation to the final microstructure in the voxel representation,

and are described in detail in the subsections below. The procedure that implements the defined objectives, can be consequently named by its parts as a: **C**(haracteristics) **D**(istributions) **E**(llipsoids) **V**(oxels) **M**odel.

2.1 Characteristics

In the first stage of the calculations, the structural parameters are defined. The Table 1 lists the main input parameters of the CDEV Model. The computations based on these parameters are limited to the dimensions of the given VOI (with the given cubic size). To select the best parameters defining the characteristics of the microstructure, there is a need to describe the methods for their measurement. Since there are ways to use three-dimensional methods developed for data from μCT, it is possible to organize them according to the similar characteristics of the microstructure. However, none of these methods are self-sufficient, and none of them can result in a full description of the material - striving for what requires a selection of the most reliable of them. Both the input parameters of the models, dedicated to creating virtual microstructures and reference parameters, that exist to describe samples from the trabecular bone, can be assigned to the corresponding groups. In the work of Lespessailles et al. [15], the authors divide all the different parameters into three main categories. The first of these contains classical morphological parameters, defined for the first time in the work of Parfitt et al. [16]. The second is constituted by parameters connected with the texture and the last refers to the topological analysis.

Table 1. Main input parameters of the CDEV Model.

Parameter	Description	Class
BV_FRAC	Percentage value of volume fraction	(0.0,1.0)
TB_MEAN	Averaged thickness of the foreground structure in pixels	—
BA_RATIO	Averaged ratio of the middle to the longest axis of ellipsoids	(0.0,1.0)
CA_RATIO	Averaged ratio of the shorter to the longest axis of ellipsoids	(0.0,1.0)
DA_VALUE	Degree of anisotropy, from isotropic to anisotropic	(0.0,1.0)
DA_MID	Middle axis of anisotropy, from transverse isotropic to orthotropic	(0.0,1.0)
PT_DIST	Averaged distance between seed points in pixels	—

Morphological Parameters

The bone volume fraction (BV/TV), is the ratio of the mineralized bone tissue volume (BV) to the total volume (TV). The microstructural mass distribution for a given sample is the linkage between volume density (ρ) with the actual density of the tissue (ρ_a) and porosity (n) in the following relation [17]:

$$n = 1 - \frac{\rho}{\rho_a} = \frac{TV - BV}{TV} \quad (1)$$

Measurement of the volume fraction can be performed either in the voxel representation or after surface reconstruction [18]. Two other ratios are connected with the second of these methods: the bone surface to the bone volume (BS/BV) and the bone surface to the total volume (BS/TV). Surface measurement is usually implemented using a Marching Cubes algorithm, where BS is the sum of the areas of the triangles from the surface mesh [19]. Classical parametrization also includes trabecular thickness (Tb.Th), trabecular separation (Tb.Sp) and trabecular number (Tb.N). A method for the model independent assessment of the thickness in three-dimensional images was presented by Hildebrand and Rüegsegger [20].

Textural Parameters

An analysis of the bone texture is generally made by measuring fractal dimension and anisotropy. Fractal analysis, utilizing the box-counting method [21], typically produces similar results for different samples, therefore it was not treated as a key parameter, but only as a reference for completeness. The orientation of trabeculae in the bone depends mainly on the mechanical load and can become anisotropic.

To determine the degree of anisotropy (DA), the mean intercept length (MIL) method is usually used, where the number of intersections of the structure in all directions are averaged [22, 23]. The ellipsoid, which describes the distribution of the MIL orientations most accurately, is adjusted statistically to the resulting cloud. It defines the fabric tensor, whose eigenvalues correspond to the lengths of the ellipsoid principal axes, and the eigenvectors determine the orientation of the principal axes [24]. The numerical value of DA is calculated as:

$$DA = 1 - \frac{e_{min}}{e_{max}} \tag{2}$$

where e_{min} is the length of the shortest axis, and e_{max} is the longest axis. As a result, 0 is the value for the isotropic structures, and 1 for the entirely anisotropic materials. DA is directly used as the main input parameter of the CDEV Model, from where the shortest axis and the longest axis can be determined.

Topological Parameters

Topological analysis focuses on studying and counting interconnections and cycles within the material, typically based on the skeletonization of the structure. In the historical approach, parameters for that type of characterization were defined on the basis of two opposite simplified models: rod-like for javelin-shaped and plate-like for discus-shaped structures.

The Structure Model Index (SMI) was proposed as the standard for the estimation of the rod-likeness and plate-likeness of a structure [25], measuring the change in the surface area as the volume increases infinitesimally, yielding 0 for plate-, 3 for rod- and 4 for sphere-like geometries. This parameter, however, can give values below 0 in the case of the trabecular bone, indicating ambiguous concave structure likeness. Therefore, other parameters are defined as a replacement for the SMI method, which in itself becomes only a reference.

A calculation of the Euler characteristic can give as a result the number of connected structures in a network. This method measures connectivity for different parts of the material. This is reflected in the number of connections in the structure that can be interrupted before the connection between two fragments will be lost [26, 27]. In the CDEV Model, connectivity varies indirectly by determining the distance between the seed points in the pixels, and Connectivity Density (Conn.D) parameter is associated with a number of seed points and can be interpreted as a trabecular number per unit volume.

2.2 Implementation

An original implementation of the described methodology was developed based on the proprietary console applications and scripts, as well as community tools dedicated to image processing and analysis. The CDEV Model was implemented using console applications. These correspond to three classes in C++ language: Distributor, Ellipsoider, Voxelizer. The cycle of data flow is complemented by using ImageJ tool, which is the public domain, Java-based image processing distribution that provides extensibility via plugins and recordable macros.

In Fig. 1 the data flow for the methodology, was illustrated, containing a processing pipeline for the ImageJ, starting from the cross-sections and ending with the 3D reconstruction after usage of the filters and structural analysis. The final microstructure can be exported to commonly used STL and U3D file formats, containing a triangular representation of a three-dimensional surface geometry.

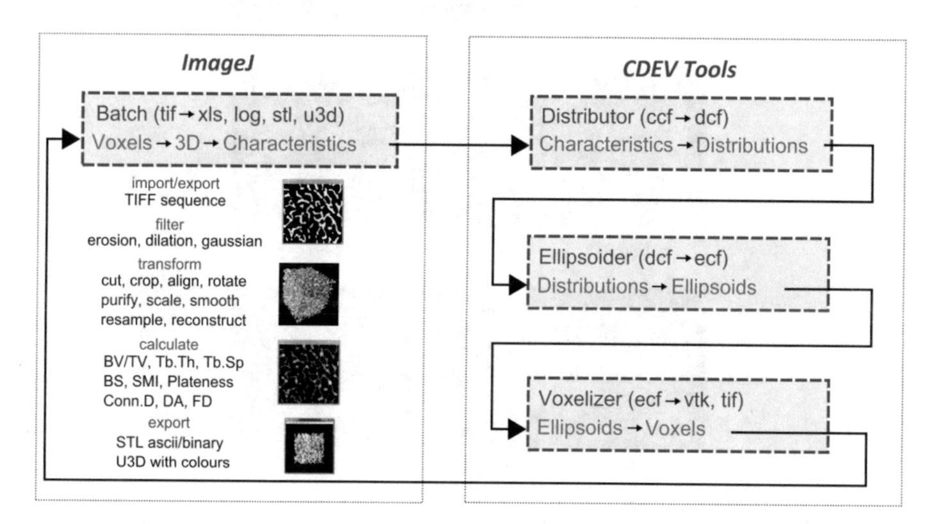

Fig. 1. Data flow diagram between ImageJ and CDEV tools: Distributor, Ellipsoider, Voxelizer.

3 Results and Observations

Properties of the CDEV Model have been tested and their effectiveness has been demonstrated using several attempts. The parameters of the CDEV Model were further tested in the case of the solid phase. Properties should be averaged over a representative volume, that for the trabecular bone should contain at least five middle distances between the trabeculae, which corresponds to the size of at least 3–5 mm [28]. Therefore, a comparative analysis of parameters and structures was carried for VOI cubic sizes equal to 5 mm. The parameters of the characteristics were noted for the corresponding samples and models:

1. morphological: BV/TV, Tb.Th [mm], Tb.Sp [mm], BS [mm^2],
2. topological: SMI, Plat.BA, Plat.CA, Conn.D [mm^{-3}],
3. textural: DA, FD.

3.1 Case Studies

In Fig. 2 a sequence of images for various case studies for the solid phase is presented. The algorithm can generate varied structures, with a view to: porosity (low, mid, and high), degree of anisotropy (isotropic, transverse isotropic, orthotropic) and model likeness (plate-like, plate/rod-like, rod-like and sphere-like).

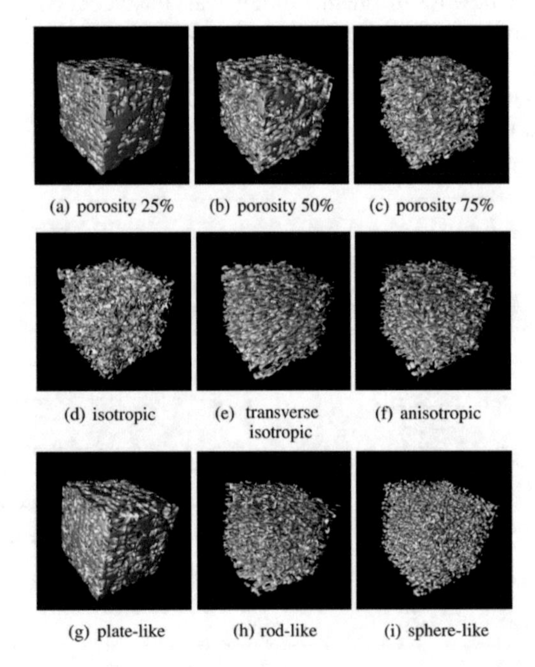

Fig. 2. Generated case studies: porosity (a–c), degree of anisotropy (d–f), model likeness (g–i).

An algorithm can be also utilized for the generation of space phase which is equivalent to the binary inverse of the structures obtained using the above approach. It can be used in the situations where the first method would not work as expected, probably in the cases of high percentage of plate-like fraction in structure, where pores are similar to ellipsoidal voids. In Fig. 3 a sequence of images for various sets of

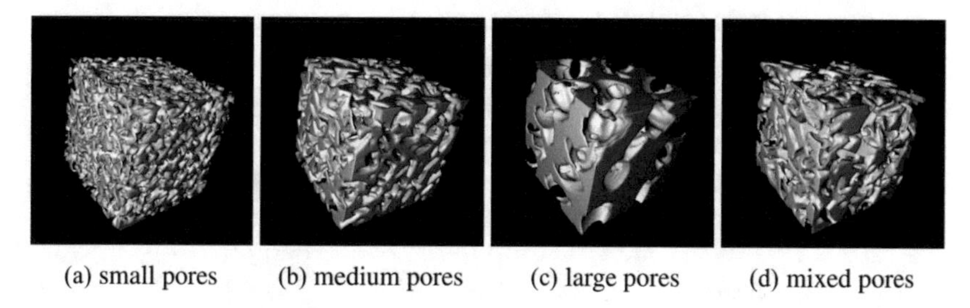

(a) small pores (b) medium pores (c) large pores (d) mixed pores

Fig. 3. Case studies for the generation of space phase with different size of the pores

parameters in the case of space phase is presented. Varied structures can be generated with different sizes of pores (small, medium, large, and mixed).

3.2 Parameters Tests

Although the interpretation of morphological parameters is similar for both the solid and space phase, ambiguities of interpretation in relation to topological and textural parameters for the latter case appear. Furthermore, the former case covers a wider range of structural combinations, that are possible to obtain, thus it is more flexible, and was subjected to further testing, and the comparative analysis part. Tables 2 and 3 show values for the parameters of the CDEV Model as an input and parameters of the characteristics as an output, for testing purposes. All tests were performed using 5 mm VOI cubic size and the same sequences of erosion and dilation filters.

Table 2. Input parameters for the CDEV Model testing purposes.

BV_FRAC	TB_MEAN	BA_RATIO	CA_RATIO	DA_VALUE	DA_MID	PT_DIST
0.250	12.500	0.500	0.259	0.800	0.100	29.412

Table 3. Output parameters for the CDEV Model testing purposes.

BV/TV	Tb.Th [mm]	Tb.Sp [mm]	BS [mm^2]	SMI	Plat. BA	Plat. CA	Conn.D [mm^{-3}]	DA	FD
0.213	0.121	0.324	595.043	2.923	0.616	0.361	15.724	0.541	2.707

The CDEV Model behaves as expected during the tests of the parameters, demonstrating the following behavior for input parameters and output parameters in Table 4.

Table 4. Relation between input and out parameters

Input parameter	Input value	Output parameter	Output value
BV_FRAC	0.300	BV/TV	0.254
TB_MEAN	13.333	Tb.Th	0.131
BA_RATIO	0.479	SMI	2.832
CA_RATIO	0.230	Plat.BA	0.602
		Plat.CA	0.341
BA_RATIO	0.543	SMI	2.832
CA_RATIO	0.296	Plat.BA	0.602
		Plat.CA	0.341
DA_VALUE	0.900	DA	0.666
DA_MID	0.500	DA	0.625
PT_DIST	27.273	Tb.Sp	0.295
		Conn.D	19.079

Table 5. CDEV Model parameters for generated models (5 mm VOI)—with sequences of filters.

Name	Filters	BV_FRAC	TB_MEAN	BA_RATIO	CA_RATIO	DA_VALUE	DA_MID	PT_DIST
$Model4_a$	5E 6D	0.287	12.600	0.500	0.222	0.910	0.100	31.250
$Model4_b$	6E 7D	0.245	10.000	0.500	0.205	0.900	0.100	27.778
$Model4_c$	5E 6D 100G	0.237	7.500	0.614	0.137	0.900	0.100	29.412
$Model5_a$	5E 6D	0.214	12.000	0.500	0.263	0.730	0.200	29.412
$Model5_b$	6E 7D	0.190	10.000	0.500	0.263	0.800	0.200	27.778
$Model5_c$	5E 6D 100G	0.195	7.500	0.528	0.140	0.800	0.100	29.412
$Model6_a$	5E 6D	0.287	12.800	0.502	0.275	0.914	0.100	27.778
$Model6_b$	6E 7D	0.287	12.800	0.502	0.275	0.914	0.100	27.770
$Model6_c$	5E 6D 100G	0.240	7.400	0.564	0.140	0.840	0.100	29.412
$Model7_a$	5E 6D	0.220	12.000	0.500	0.260	0.770	0.100	29.412
$Model7_b$	6E 7D	0.240	11.200	0.502	0.225	0.800	0.100	29.412
$Model7_c$	5E 6D 100G	0.180	7.400	0.601	0.154	0.770	0.100	29.412

3.3 Model Analysis

Additional usage of opening filtering (erosion and dilation sequences) for the models was performed, which increases connectivity and at the same time increases porosity. Therefore, in the next step additional dilation filtering was performed and input parameters were scaled, appropriately. For every structure, results for 3 corresponding models (a, b, c) and averages (abc) were noted. These models stand for:

- a - low values of standard deviations (for thickness, ratios, anisotropy and seed point coordinate randomization),
- b - high values of standard deviations,
- c - high values of standard deviations and additional usage of gaussian filtering, that smooths structures after the usage of erosion and dilation sequences.

Tables 4 and 5 show comparison of the input and output parameters for a 4 samples of the CDEV Model (Table 6).

Table 6. Parameters comparison for samples and generated models (5 mm VOI). Parameters of the characteristics were divided into three groups.

Name	BV/TV	Tb.Th [mm]	Tb.Sp [mm]	BS [mm^2]	SMI	Plat. BA	Plast. CA	Conn.D [mm^2]	DA	FD
Sample4	0.254	0.125	0.367	697.173	1.528	0.614	0.259	16.914	0.710	2.079
Model4$_a$	0.249	0.121	0.273	678.193	2.387	0.615	0.322	16.580	0.716	2.751
Model4$_b$	0.237	0.123	0.298	628.014	2.483	0.608	0.325	17.694	0.693	2.710
Model4$_c$	0.252	0.143	0.319	626.626	2.173	0.575	0.290	14.148	0.733	2.714
Model4$_{abc}$	0.246	0.129	0.279	644.277	2.348	0.599	0.312	16.141	0.714	2.725
Sample5	0.167	0.108	0.524	535.304	1.887	0.608	0.232	21.913	0.562	2.590
Model5$_a$	0.167	0.093	0.285	529.726	3.474	0.581	0.328	17.758	0.557	2.705
Model5$_b$	0.164	0.110	0.374	466.717	3.087	0.625	0.367	11.008	0.559	2.647
Model5$_c$	0.204	0.141	0.465	512.629	2.418	0.597	0.322	11.691	0.518	2.723
Model5$_{abc}$	0.178	0.115	0.375	503.024	2.993	0.601	0.339	13.486	0.545	2.692
Sample6	0.245	0.135	0.403	652.772	1.645	0.589	0.248	27.299	0.648	2.686
Model6$_a$	0.236	0.120	0.263	674.247	2.807	0.583	0.345	18.713	0.720	2.756
Model6$_b$	0.242	0.137	0.320	616.122	2.614	0.612	0.354	14.924	0.651	2.718
Model6$_c$	0.277	0.157	0.327	645.055	2.020	0.584	0.309	11.691	0.609	2.717
Model6$_{abc}$	0.252	0.138	0.303	645.138	2.480	0.593	0.336	13.486	0.660	2.730
Sample7	0.180	0.112	0.436	575.351	2.234	0.588	0.255	30.011	0.567	2.638
Model7$_a$	0.181	0.090	0.278	585.124	3.080	0.582	0.306	27.462	0.559	2.723
Model7$_b$	0.206	0.106	0.293	611.757	2.764	0.618	0.333	19.180	0.597	2.723
Model7$_c$	0.179	0.102	0.336	489.514	2.723	0.623	0.310	16.140	0.569	2.684
Model7$_{abc}$	0.189	0.099	0.302	562.132	2.856	0.608	0.316	20.827	0.575	2.710

Conclusions of the results are as follow:

1. Morphological parameters:
   - intentional focus on very good compatibility for the main parameters: BV/TV, Tb.Th, BS,
   - intentional reduction of the distance between the seed points for better connectivity values, resulting in too low a Tb.Sp value;
2. Topological parameters:
   - treatment of the first group as key parameters for overall compliance results in satisfactory comparability of Conn.D received for structures with relatively medium and good connectivity,
   - the problem of achieving high values of connectivity, while complying with the other parameters, appeared,
   - alongside with very good compatibility of Plat.BA, problems with Plat.CA, and consequently SMI, as a result of applying filters;

3. Textural parameters:
   - good compatibility of the DA value, also evidenced in the anisotropy of the three-dimensional structures,
   - natural good compatibility for FD, despite the fact that it was not directly taken into account in the model.

Incompatibilities in the case of SMI and Plat.CA, and also Tb.Sp values, can be reduced using greater values for deviations (randomizing structures) at first, and then use a gaussian filter and the associated reduction of the parameter Tb.Th, which refers to the definition of the models. This leads to a satisfactory convergence for the parameters and allows for the corresponding three-dimensional structures to be visually consistent (after the smoothing following applying a sequence of erosions and dilations). Finally, a high level of compliance for morphological and textural parameters, and relatively good compliance for topological parameters was achieved.

4 Discussion and Conclusion

The presented study is comparable to that of natural trabecular bone studies conducted by previous researchers. Although the authors had to take into account all the key parameters for trabecular bone microstructure from the literature, there may be other parameters to be considered, because only the appropriate combination of several indices will lead to success. Also, further validation is due to be implemented, focusing mainly on FEM and Additive Fabrication. It seems, however, that the presented approach can be used at this stage in several applications, also for other materials than bone.

4.1 Validation

Further validation can be done in a variety ways. A geometric comparative analysis, which was discussed above, can be applied to greater number of real samples, from various - other than from long bone - locations (for example a vertebral bone). Also, more diversified artificial structures can be defined to obtain. The study further suggests that the periodicity for obtained three-dimensional structures appears to be very useful. Such a repeatable model can be utilized as a Representative Volume Element (RVE), and greater VOI can be produced, than commonly could not be scanned in this case using μCT.

After FEM validation, enlarged replicas of the three-dimensional microstructures from the STL files can be sent directly to the Additive Fabrication tools, known as 3D printers. Continuously developed and accelerated techniques for making physical models are associated with an important aspect of the validation of computational models. In the case of the CDEV Model, various shapes for VOI (rectangular, cylindrical, and spherical) can be produced [29], and also mechanical testing, for example compression, can be performed [30]. To eliminate the effects of the surface and get well-reproduced models, the linear scaling of sizes is utilized to obtain the microstructure enlarged tens of times.

4.2 Summary

To summarize, it should be noted that the three-dimensional structures generated using the presented approach, reflect the expected characteristics of real bone samples at a satisfactory level, but they can also be used to produce structures beyond the scope of realistic parameters. The algorithm of the CDEV Model takes into account the most important parameters with the same weight, and additionally creates the possibility of a stronger compliance of certain parameters, at the expense of others. The auxiliary filters for the binarized cross-sections, discussed above, allow for the achievement of realistic three-dimensional reconstructions that can be used in many different ways, discussed above. A proposed solution in the form of the tools and scripts is freely available, and the use of parallel computations allow for obtainment of the representative VOI in a relatively fast way. The presented case studies demonstrate the flexibility of the CDEV Model, but the model itself cannot be treated as a universal algorithm. However, it is expected that the presented study can be applicable in a wide range of applications and may stimulate the interest of the investigators from various disciplines.

Acknowledgments. This work was partially financed by the Faculty of Physics and Applied Computer Science AGH. Adrian Wit has been partly supported by the EU Project POWR.03.02.00-00-I004/16.

References

1. Romanova, V., Balokhonov, R., Makarov, P., Schmauder, S., Soppa, E.: Simulation of elasto–plastic behaviour of an artificial 3D-structure under dynamic loading. Comput. Mater. Sci. **28**, 518–528 (2003)
2. Saylor, D., Fridy, J., El-Dasher, B., Jung, K.-Y., Rollett, A.: Statistically representative three-dimensional microstructures based on orthogonal observation sections. Metall. Mater. Trans. A **35**, 1969–1979 (2004)
3. Brahme, A., Alvi, M.H., Saylor, D., Fridy, J., Rollett, A.D.: 3D reconstruction of microstructure in a commercial purity aluminum. Scripta Mater. **55**, 75–80 (2006)
4. Haire, T., Ganney, P., Langton, C.: An investigation into the feasibility of implementing fractal paradigms to simulate cancellous bone structure. Comput. Methods Biomech. Biomed. Eng. **4**, 341–354 (2001)
5. Rajon, D.A., Jokisch, D.W., Patton, P.W., Shah, A.P., Watchman, C.J., Bolch, W.E.: Voxel effects within digital images of trabecular bone and their consequences on chord-length distribution measurements. Phys. Med. Biol. **47**, 1741–1759 (2002)
6. Ruiz, O., Schouwenaars, R., Ramírez, E.I., Jacobo, V.H., Ortiz, A.: Analysis of the architecture and mechanical properties of cancellous bone using 2D voronoi cell based models. In: Proceedings of the World Congress on Engineering 2010, WCE 2010, vol. I, pp. 1–6 (2010)
7. Kowalczyk, P.: Elastic properties of cancellous bone derived from finite element models of parameterized microstructure cells. J. Biomech. **36**, 961–972 (2003)
8. Kowalczyk, P.: Orthotropic properties of cancellous bone modelled as parameterized cellular material. Comput. Methods Biomech. Biomed. Eng. **9**, 135–147 (2006)
9. Kowalczyk, P.: Simulation of orthotropic microstructure remodelling of cancellous bone. J. Biomech. **43**, 563–569 (2010)

10. Lakatos, E., Bojtár, I.: Trabecular bone adaptation in a finite element frame model using load dependent fabric tensors. Mech. Mater. **43**, 1–9 (2011)
11. Donaldson, F.E., Pankaj, P., Law, A.H., Simpson, A.H.: Virtual trabecular bone models and their mechanical response. Inst. Mech. Eng. Proc. Part H J. Eng. Med. **222**, 1185–1195 (2008)
12. Wang, J.: Modelling young's modulus for porous bones with microstructural variation and anisotropy. J. Mater. Sci. Mater. Med. **21**, 463–472 (2010)
13. Mouton, E.: Caractérisation expérimentale des propriéteé élastiques apparentes de l'os spongieux porcin: essais mécaniques et imagerie, Master's thesis, École nationale d'ingénieurs de Metz (2011)
14. Janc, K., Tarasiuk, J., Bonnet, A.S., Lipinski, P.: Semi-automated algorithm for cortical and trabecular bone separation from CT scans. Comput. Methods Biomech. Biomed. Eng. **14**, 217–218 (2011)
15. Lespessailles, E., Chappard, C., Bonnet, N., Benhamou, C.L.: Imaging techniques for evaluating bone microarchitecture. Revue du Rhumatisme **73**, 435–443 (2006)
16. Parfitt, A.M., Drezner, M.K., Glorieux, F.H., Kanis, J.A., Malluche, H., Meunier, P.J., Ott, S.M., Recker, R.R.: Bone histomorphometry: standardization of nomenclature, symbols, and units. report of the ASBMR histomorphometry nomenclature committee. J. Bone Miner. Res. **2**, 595–610 (1987)
17. Doblaré, M.: Mechanical behaviour of bone tissue. computational models of bone fracture. In: 15th CISM-IUTAM Summer School on "Bone Cell and Tissue Mechanics", International Centre for Mechanical Sciences, pp. 1–25 (2007)
18. Hildebrand, T., Laib, A., Müller, R., Dequeker, J., Rüegsegger, P.: Direct three-dimensional morphometric analysis of human cancellous bone: microstructural data from spine, femur, iliac crest, and calcaneus. J. Bone Miner. Res. **14**, 1167–1174 (1999)
19. Lorensen, W.E., Cline, H.E.: Marching cubes: a high resolution 3D surface construction algorithm. SIGGRAPH Comput. Graph. **21**, 163–169 (1987)
20. Hildebrand, T., Rüegsegger, P.: A new method for the model-independent assessment of thickness in three-dimensional images. J. Microsc. **185**, 67–75 (1997)
21. Fazzalari, N.L., Parkinson, I.H.: Fractal dimension and architecture of trabecular bone. J. Pathol. **178**, 100–105 (1996)
22. Harrigan, T.P., Mann, R.W.: Characterization of microstructural anisotropy in orthotropic materials using a second rank tensor. J. Mater. Sci. **19**, 761–767 (1984)
23. Odgaard, A.: Three-dimensional methods for quantification of cancellous bone architecture. Bone **20**, 315–328 (1997)
24. Cowin, S., Doty, S.: Tissue Mechanics. Springer, New York (2007)
25. Hildebrand, T., Rüegsegger, P.: Quantification of bone microarchitecture with the structure model index. Comput. Methods Biomech. Biomed. Eng. **1**, 15–23 (1997)
26. Odgaard, A., Gundersen, H.: Quantification of connectivity in cancellous bone, with special emphasis on 3-D reconstructions. Bone **14**, 173–182 (1993)
27. Toriwaki, J., Yonekura, T.: Euler number and connectivity indexes of a three-dimensional digital picture. FORMATOKYO **17**, 183–209 (2002)
28. Harrigan, T.P., Jasty, M., Mann, R.W., Harris, W.H.: Limitations of the continuum assumption in cancellous bone. J. Biomech. **21**, 269–275 (1988)
29. Woo, D., Kim, C., Kim, H., Lim, D.: An experimental–numerical methodology for a rapid prototyped application combined with finite element models in vertebral trabecular bone. Exp. Mech. **48**, 657–664 (2008)
30. Woo, D., Kim, C., Lim, D., Kim, H.: Experimental and simulated studies on the plastic mechanical characteristics of osteoporotic vertebral trabecular bone. Curr. Appl. Phys. **10**, 729–733 (2010)

RMID: A Novel and Efficient Image Descriptor for Mammogram Mass Classification

Sk Md Obaidullah(✉), Sajib Ahmed, Teresa Gonçalves, and Luís Rato

Department of Informatics, University of Évora, Évora, Portugal
sk.obaidullah@gmail.com, jack6148@gmail.com, {tcg,lmr}@uevora.pt

Abstract. For mammogram image analysis, feature extraction is the most crucial step when machine learning techniques are applied. In this paper, we propose RMID (Radon-based Multi-resolution Image Descriptor), a *novel* image descriptor for mammogram mass classification, which perform *efficiently* without any clinical information. For the present experimental framework, we found that, in terms of area under the ROC curve (AUC), the proposed RMID outperforms, upto some extent, previous reported experiments using histogram based hand-crafted methods, namely Histogram of Oriented Gradient (HOG) and Histogram of Gradient Divergence (HGD) and also Convolution Neural Network (CNN). We also found that the highest AUC value (0.986) is obtained when using only the carniocaudal (CC) view compared to when using only the mediolateral oblique (MLO) (0.738) or combining both views (0.838). These results thus proves the effectiveness of CC view over MLO for better mammogram mass classification.

Keywords: Image descriptor · Mammogram image · Breast cancer · Classification

1 Introduction

Breast Cancer is the most frequent cancer among women, impacting over 1.5 million women each year and is also the cause of the highest number of cancer related death among women. In 2015, 570,000 women died from breast cancer – which is approximately 15% of all cancer deaths among women [1]. If breast cancer is early detected, it is one of the most treatable types of cancer. The primary imaging modality for breast cancer is done by a low cost X-Ray based technique which is known as mammography. Based on the processing modalities, mammography can be classified into two types: (i) Screen Film Mammography (SFM) and (ii) Full Field Digital Mammography (FFDM). For SFM, images are captured on the film, whereas for FFDM, images are directly stored in the digital computer. As per the reported study in literature [2,3], both type of mammography, have almost equal ability to detect suspicious lesions in the breast. The present work deals with the SFM images which are available through the

P. Kulczycki et al. (Eds.): ITSRCP 2018, AISC 945, pp. 229–240, 2020.
https://doi.org/10.1007/978-3-030-18058-4_18

BCDR-F03 dataset [27,29], one of the latest breast imaging film mammography benchmark datasets.

In recent years many computational approaches have been proposed for computer assisted diagnostic of breast cancer; these methods are known as Computer Aided Diagnostic methods or, in short, CAD [4]. A double checking procedure by radiologists is normally used to reduce the number of false-negative cases, but this has an obvious cost associated as the number of radiologists are not in general adequate in our health centres. Alternatively, a CAD system can help one radiologist to verify her/his observations with the result of the automated system without requiring another radiologist in the same place. That is why the importance of developing CAD system is in demand.

Presently various CAD systems have been proposed in the literature. The general framework for a traditional CAD system consists of three parts: (i) preprocessing the mammogram images for ROI extraction, (ii) feature extraction and finally (iii) classification. Recently, deep learning based approaches are also reported in literature which replaces the extraction of hand-crafted features by combining step (ii) and (iii) in a single stage. In literature works are reported where image descriptors are combined with clinical information for better classification accuracy [4]. The present work focuses on an image descriptor based classification of masses from mammogram images. We have not considered any clinical information.

Among the reported image descriptors, Constantinidis et al. [5] and Belkasim et al. [6] considered the Zernike moment based descriptor to classify masses; texture based classification of calcification and masses using Haralick features [7] was reported by different authors [8–12]. Haralick texture features were employed in other areas of medical imaging also [13] and a comparison of texture features and a deep learning approach is reported in [14]. Wavelet [15,16] and curvelet [17] analysis based feature descriptors are also used by different authors and combination of intensity and texture descriptors was explored by Ramos et al. [18]. Histogram of oriented gradient (HOG) based features was employed along with the clinical information for mammogram image classification by Moura and Guevara [4] and Arevalo et al. [29] used a convolution neural network to separate malignant and benign masses without using any clinical information reporting the effectiveness of a deep learning based approach over the traditional hand-crafted one.

In this paper, we propose a novel image descriptor based on radon transform over multi-resolution images. The block diagram of the proposed method is shown in Fig. 1. First film mammogram dataset is considered and images are categorized based on CC and MLO views; ROI are then extracted and their contrast is enhanced; next step computes a feature vector, followed by classification and performance comparison of multiple classifiers.

The rest of the paper is organized as follows: the contributions are reported in Sect. 2 where we discuss about the proposed RMID and the design of classifiers; experimental details are reported in Sect. 3, which include dataset description, experimental setup and results with a comparative study. Finally we conclude the paper in Sect. 4.

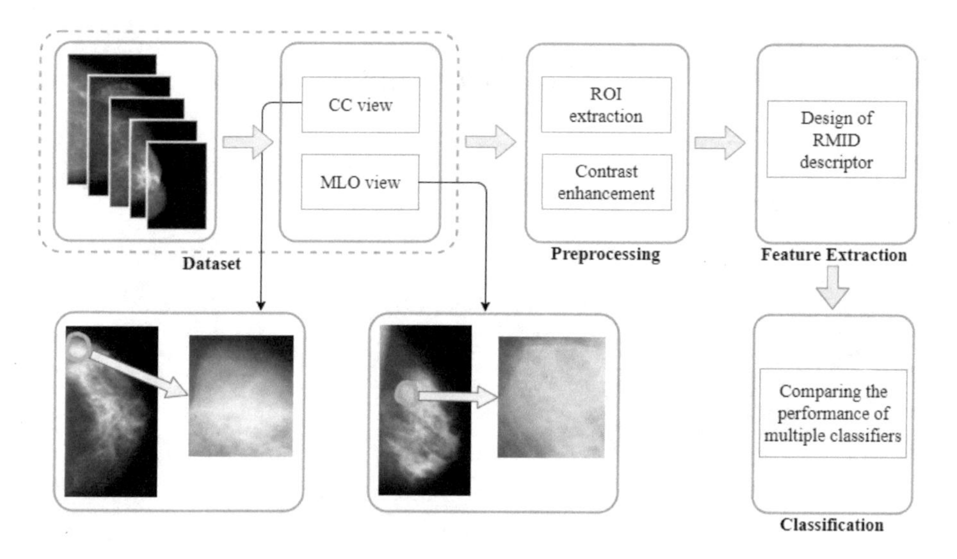

Fig. 1. Block diagram of the proposed method

2 Contribution Outline

As mentioned earlier, we propose RMID, a novel feature descriptor for classification of mammogram masses. In particular, our target is to classify malignant and benign masses from mammogram images using this novel image descriptor. We describe RMID in detail in the following sub section.

2.1 RMID Principals and Design

RMID uses the concepts of radon transform and multi-resolution analysis. The basic principal of these techniques are described below.

Radon Transform. Radon transform is an integral transform that consists of a set of projections of a pattern at different angles [19], as illustrated in Fig. 2 where, the part (a) shows the projection principal and part (b) shows different angles considered for the present work. It is a mapping of a function $f(x,y)$ to another function $fR(x,y)$ defined on the 2D space of lines in the plane, whose value at a particular line is equal to the line integral of the function over that line for the given set of angles. In other words, the radon transform of a pattern $f(x,y)$ and for a given set of angles may be assumed as the projection of all non-zero points. The projection output is the sum of the non-zero points for the image pattern in each direction (angle between 0 to π). Finally it results forming a matrix. The matrix elements are related to the integral of $f(x,y)$ over a line $Lin(\rho,\theta)$ defined by $\rho = x\cos\theta + y\sin\theta$ and can formally be expressed as

$$fR(\rho,\theta) = \int_{-\infty}^{\infty}\int_{-\infty}^{\infty} f(x,y)\delta(x\cos\theta + y\sin\theta - \rho)dxdy$$

where $\delta(.)$ is the Dirac delta function, $\delta(x) = 1$, if $x = 0$ and 0 otherwise. Also, $\theta \in [0, \pi]$ and $\rho \in]-\infty, \infty[$. For the radon transform, Lin_i be in normal form (ρ_i, θ_i).

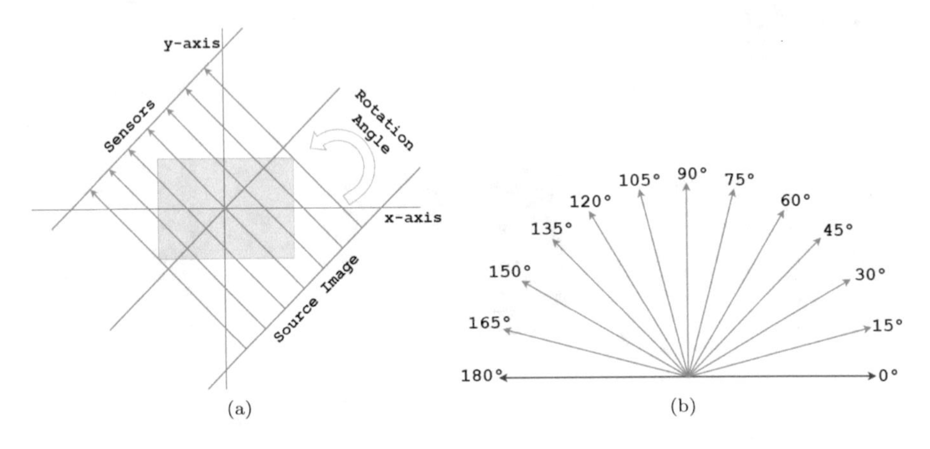

Fig. 2. Illustrating the radon transform theory: (a) generation of radon spectrum, (b) different angular directions considered for the present work to compute the line integral.

Multi-resolution Analysis. The time-frequency response of a signal (for present work it is an image) is represented through wavelet transform. Daubechies wavelets [20], which belongs to the family of discrete wavelet techniques, are used for the present work. Wavelets are used for multi-resolution analysis and their advantage includes: computational ease with minimum resource and time requirements. These orthogonal wavelets are characterized by maximum number of vanishing moments for some given support. We decompose an image into different frequencies with different resolutions for further analysis. The family of Daubechies wavelet is denoted as '*dbN*', where the wavelet family is denoted by the term '*db*' and the number of vanishing moments is represented by 'N'. An image can be represented by the combination of different components of different coefficients. For the present work, the wavelet decomposition has been done at level 1 for *db1*, *db2* and *db3* which capture the constant, linear and quadratic coefficients of an image component. Four sub-band images namely, approximation coefficients (cA), horizontal coefficients (cH), vertical coefficients (cV), and diagonal coefficients (cD) are generated by this process for each of the *db1*, *db2* and *db3* part, resulting a total of 12 sub-band images.

Design of RMID. RMID is developed by combining the idea of radon transform over multi-resolution analysis. The texture pattern of benign and malignant masses are different. The malignant mass region has more irregular in comparison to benign masses whose boundary regions are more regular in shape. Radon transform computes the line integral of a set of pixels over a specified direction.

So, if we compute radon transform on benign and malignant masses the line integral value will be different for each case. Figure 3 shows the radon spectrum of benign and malignant masses: Fig. 3(a) is a benign mass and its radon spectrum is shown in Fig. 3(c); Fig. 3(b) is a malignant mass and its radon spectrum is shown in Fig. 3(d).

There are several methods available for image decomposition like, divide by "n" and quad-tree decomposition among others. Here, we chose wavelet, as different directional approximation can be done through wavelet decomposition. From each of the sub-band images i.e. on *cA*, *cH*, *cV* and *cD*, we compute the radon spectrum. Finally, statistical values are computed from those radon-wavelet spectrum which are used to construct the feature vector.

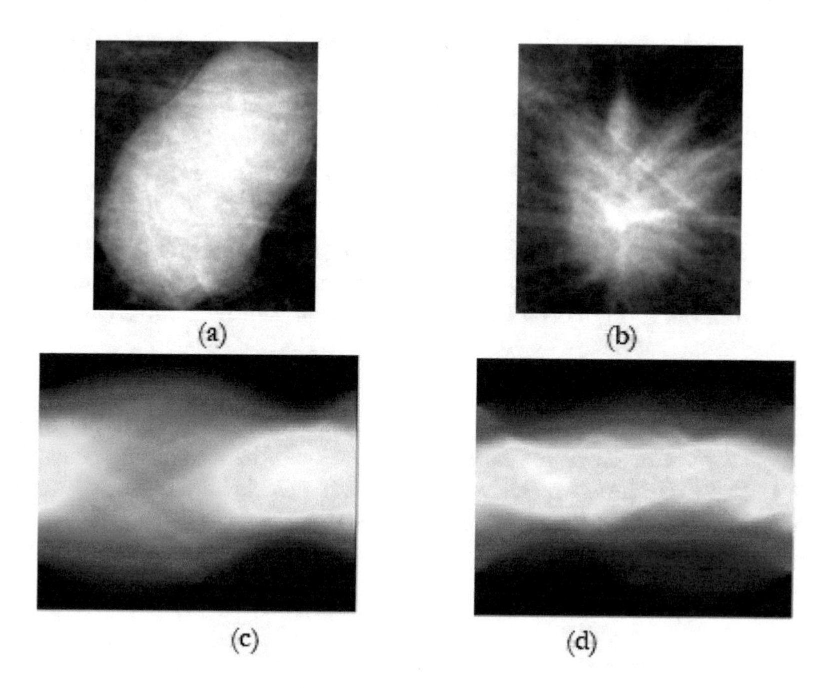

Fig. 3. ROI extracted from mammogram images and their radon spectrum (a) benign mass, (b) malignant mass, (c) radon spectrum of figure a, (d) radon spectrum of figure b

Feature Vector Generation. In what follows, we summarize the generation of feature vector using RMID:

- First, ROIs were extracted from the original mammogram images and contrast enhancement was done on each ROI. The enhanced ROI was then stored as a gray-scale image.
- Wavelet decomposition at level 1 was done using Daubechies method for *db1*, *db2* and *db3*. This step generates 04 sub band images for each coefficients resulting a total of 12 sub-band images.

- Radon transform is applied on the original ROI image and each of the 12 sub-band images generated on the previous step. At this step we generate a total of 13 radon spectrum.
- From each of the 13 radon spectrum we compute one energy and three statistical features value. Altogether this step generates 52 features (04×13). Then, we compute 04 statistical features namely: entropy, mean, standard deviation and maximum coefficient of radon spectrum from the original gray-scale image too, so overall we get a feature vector of dimension 56 ($52 + 04$).

2.2 Classifiers

In our study, six different classifiers were used to train and classify the masses. They are Bayesian network (BN), Linear discriminant analysis (LDA), Logistic, Support vector machine (SVM), Multilayer perceptron (MLP) and Random Forest (RF). We compare the performance of these classifiers to find the best one. These classifiers are briefly explained below.

Bayesian Network. For the Bayesian network (BN) we used K2, a hill-climbing technique which is a famous score-based algorithm that recovers the underlying distribution in the form of directed acyclic graph efficiently. Details can be found in [23].

Linear Discriminant Analysis. In linear discriminant analysis [24], we model the data as a set of multivariate normal distributions where a common covariance matrix exist with different mean vectors for different classes. LDA partition the feature-space by using a hyper-plane (HP), where two sides of the HP represent two classes. The class pattern is determined from the test dataset based on which side of the plane the classes lie.

Logistic. Logistic regression is used as a classification algorithm to assign observations to a discrete set of classes. The logistic classifier transforms its output using a logistic function (sigmoid) and returns a probability value. This probability value is then mapped into two or more classes [25].

Support Vector Machine. SVM classifies the data by constructing a hyperplane (HP) on the high dimensional feature space. Different linear and non linear kernels can be used. For the present work, we used SVM tuned by a linear kernel since it's fast and presented promising results.

Multilayer Perceptron. MLP is one of the most widely used classifiers. Here, the chosen configuration was *56-hl-2*, where 56 is the number of feature values, *hl* is the number of nodes in the hidden layer and 2 is the number of classes.

hl can be determined empirically, by considering it as a function of the feature dimension (fifty six) and output classes (two). In our experiment the value of hl is considered as 29, i.e $(56 + 2)/2$.

Random Forest. Theoretically, a random forest is defined as a collection of unpruned decision trees which are trained on bootstrap samples using random feature selection in the tree generation process. Among a large number of generated trees, each tree votes for a popular class and, by combining all, a decision is taken [26].

3 Experiments

3.1 Dataset and Pre-processing

A benchmark film mammography dataset known as BCDR-F03 [28,29] from the Breast Cancer Digital Repository, a wide-ranging public repository composed of Breast Cancer patients' cases from Portugal [4], was used in our experiment. The BCDR-F03 is one of the latest benchmarked film mammography dataset which consists of 668 film mammogram images. Out of these 668 image there are 736 biopsy proven masses containing 426 benign masses and 310 malignant masses from 344 patients. Thus, in many cases a single image contains more than one masses.

For present work, we have considered one mass per image having a total of 668 masses. Out of these 668 masses, 662 images are considered for classification after removal of few extremely low resolution images. The samples provided are available in two different views namely carniocaudal (CC) and mediolateral oblique (MLO) view. In our data we have 328 CC views and 334 MLO views (almost equal ratio for fair comparison). Figure 4 shows different mammogram views with the lesions marked.

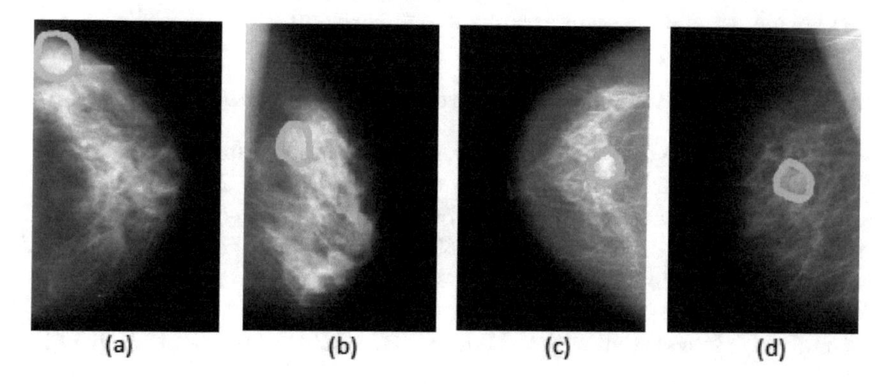

Fig. 4. Different mammogram views, (a) LCC, (b) LO, (c) RCC, (d) RO. The green boundary is the ROI.

The pre-processing step includes (i) ROI extraction and (ii) contrast enhancement. ROI was extracted based on the information provided by the radiologist; an annotated file with the ROI coordinate information is provided along with the BCDR-F03 dataset. Using automated techniques ROIs were extracted and stored in separate folders based on different view types. Further, they were categorized into two class folders namely benign and malignant. Next, ROIs contrast enhancement was performed since original film mammograms are of very low contrast due to several factors (poor lighting condition, orientation, etc.); contrast is enhanced by subtracting the mean of the intensities in the image to each pixel. Figure 5 shows one original ROI and its contrast enhanced version.

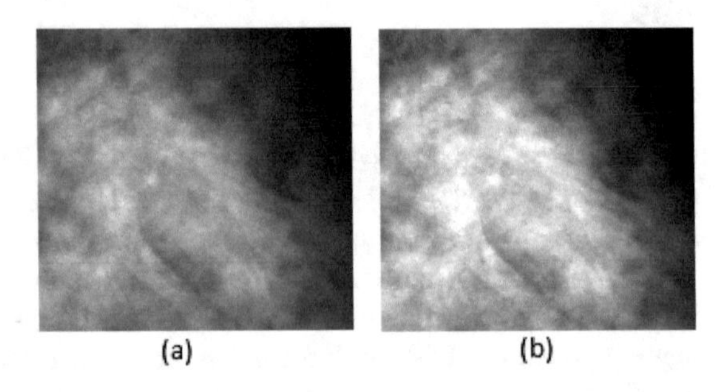

Fig. 5. Contrast enhancement, (a) original low contrast ROI, (b) contrast enhanced image

3.2 Evaluation Metrics

To measure the performance of the system, we use the Area Under the Curve of the Receiver Operator Characteristic (AUC). The ROC curve is created by plotting the true positive rate against the false positive rate. The AUC is a measure of discrimination also used in previous works on this dataset [4,29], allowing then to make a comparison with our method.

3.3 Evaluation Strategy and System Configuration

We carried out three different type of tests: (i) mammogram mass classification from CC view, (ii) mammogram mass classification from MLO view and (iii) mammogram mass classification with both views combined. In each case, the dataset was divided into 60:40 ratio (60% data for training and rest 40% for testing), following the split of previous works for fair comparison [4,27,29].

Regarding the resources, all experiments were carried out using MATLAB 2017b software in a system with 2.8 GHz CPU, 8 GB RAM, 4 GB NVIDIA GPU.

3.4 Results and Analysis

In the present work, we not only propose a novel image descriptor for mammogram mass classification but also study the performance of different classifiers. In addition, the analysis of which image view is better for mass classification is also done. Table 1 shows the performance of six different classifiers (BN, LDA, Logistic, SVM, MLP and RF) for three different mammogram views: CC, MLO and both combined (best values are presented in bold face). From the first column of the table, which shows the output of the CC view, LDA and RF perform best with 0.986 AUC among the six classifiers. For the MLO view, RF shows the highest AUC of 0.738 and when both the views are combined (i.e. all images of CC and MLO are considered together) RF also shows highest AUC of 0.838. Given these results, it is possible conclude that RF is the best performer irrespective of image view.

Table 1. Mammogram mass classification results for CC, MLO and combined (CC + MLO) views on test dataset measured in AUC

Classifier	CC	MLO	CC + MLO
Bayesian Network (BN)	0.934	0.690	0.816
Linear Discriminant Analaysis (LDA)	**0.986**	0.672	0.807
Logistic	0.958	0.674	0.811
Support Vector Machine (SVM)	0.977	0.682	0.783
Multilayer Perceptron (MLP)	0.985	0.618	0.813
Random Forest (RF)	**0.986**	**0.738**	**0.838**

Previous Relevant Work – Analogy. Prior to this study, Moura et al. [4] proposed one histogram based image descriptor known as HGD and tested the method on the BCDR-F01 dataset, which is a subset of the BCDR-F03. They reported the highest AUC of 0.787 by HDG and 0.770 by traditional HOG. Nonetheless, these results are not solely based on image descriptor as clinical information was also used. Recently, Arevalo et al. [27,29] proposed a deep learning based approach on BCDR-F03 dataset. Applying a convolution neural network (CNN) they obtained a AUC of 0.822; then, using CNN combined with the hand-crafted HGD features, the overall AUC was boosted upto 0.826, showing an improvement of 0.40%. Nonetheless, these two works are not comparable in true sense as different experimental framework were considered, i.e. for [4], a 80:20 data split for training and testing was considered while for [27,29] the split was 60:40.

Table 2 shows the results reported and allows a qualitative comparison to our method (based on the present experimental consideration and framework). The proposed RMID performs significantly well without any clinical information as compared to the traditional approaches, thus proving the effectiveness of RMID as image descriptor for mass classification.

Table 2. Performance comparison of proposed technique with baseline results (CC + MLO views combined)

Methods		AUC
Hand-crafted techniques [4]	HOG	0.770
	HGD	0.787
Deep-learning based approach [27,29]	CNN	0.822
Deep learning + Hand-crafted techniques [29]	CNN + HDG	0.826
Proposed method	RMID	**0.838**

4 Conclusions and Future Work

No doubt, automated diagnostics of breast cancer from mammogram images will support the radiologists by double checking their observations. Several methods are proposed in the literature for mammogram mass classifications from film mammogram images, but most of the time these descriptors show promising performance if combined with clinical information. In the present work, we propose a novel image descriptor which performs well without clinical information. The proposed RMID along with the random forest classifier shows an AUC of 0.986 for CC view, 0.738 for MLO view and 0.838 for the combined view. For mammogram mass classification the CC view is more effective than MLO one. In this experiment, we found that, CC view shows a 33.60% improvement over MLO.

To support the conclusions drawn on this work, namely the superiority of RMID descriptor and the different discriminating powers of the CC and MLO views, our plan for future work includes evaluating the performance of the proposed descriptor on other publicly available film mammogram datasets and performing a statistical analysis over the results. Further, we intend to compare the performance of digital mammography and film mammography and observe the performance of RMID along with the clinical information.

Acknowledgement. The preliminary version of this paper was presented at the 3rd Conference on Information Technology, Systems Research and Computational Physics, 2–5 July 2018, Cracow, Poland [30]. The authors are thankful for considering the paper in the proceedings.

References

1. http://www.who.int/cancer/prevention/diagnosis-screening/breast-cancer/en/. Accessed 15 Feb 2018
2. Skaane, P., Hofvind, S., Skjennald, A.: Randomized trial of screen-film versus full-field digital mammography with soft-copy reading in population-based screening program: follow-up and final results of Oslo II study. Radiology **244**(3), 708–17 (2007)

3. Pisano, E.D., Hendrick, R.E., Yaffe, M.J.: for the Digital Mammographic Imaging Screening Trial (DMIST) Investigators Group: Diagnostic accuracy of digital versus film mammography: exploratory analysis of selected population subgroups in DMIST. Radiology **246**(2), 376–83 (2008)
4. Moura, D.C., López, M.A.G.: An evaluation of image descriptors combined with clinical data for breast cancer diagnosis. Int. J. Comput. Assist. Radiol. Surg. **8**, 561–574 (2013)
5. Constantinidis, A.S., Fairhurst, M.C., Rahman, A.F.R.: A new multi-expert decision combination algorithm and its application to the detection of circumscribed masses in digital mammograms. Pattern Recognit. **34**(8), 1527–1537 (2001)
6. Belkasim, S.O., Shridhar, M., Ahmadi, M.: Pattern-recognition with moment invariants-a comparative-study and new results. Pattern Recognit. **24**(12), 1117–1138 (1991)
7. Haralick, R.M., Shanmuga, K., Dinstein, I.: Textural features for image classification. IEEE Trans. Syst. Man Cybern. **3**(6), 610–621 (1973)
8. Yu, S.Y., Guan, L.: A CAD system for the automatic detection of clustered microcalcifications in digitized mammogram films. IEEE Trans. Med. Imaging **19**(2), 115–126 (2000)
9. Dhawan, A.P., Chitre, Y., Kaiser, B.C., Moskowitz, M.: Analysis of mammographic microcalcifications using gray-level image structure features. IEEE Trans. Med. Imaging **15**(3), 246–259 (1996)
10. Wang, D., Shi, L., Ann, H.P.: Automatic detection of breast cancers in mammograms using structured support vector machines. Neurocomputing **72**(13–15), 3296–3302 (2009)
11. Dua, S., Singh, H., Thompson, H.W.: Associative classification of mammograms using weighted rules. Expert Syst. Appl. **36**(5), 9250–9259 (2009)
12. Sahiner, B., Chan, H.P., Petrick, N., Helvie, M.A., Hadjiiski, L.M.: Improvement of mammographic mass characterization using spiculation measures and morphological features. Med. Phys. **28**(7), 1455–1465 (2001)
13. Claudia, M., Enrique, A., Maria T., Víctor G.C.: Tissues classification of the cardiovascular system using texture descriptors. In: Medical Image Understanding and Analysis, MIUA 2017, pp. 123–132 (2017)
14. Alison, O.N., Matthew, S., Erin, B., Keith, G.: A comparison of texture features versus deep learning for image classification in interstitial lung disease. In: Medical Image Understanding and Analysis, MIUA 2017, pp. 743–753 (2017)
15. Ferreira, C.B.R., Borges, D.B.L.: Analysis of mammogram classification using a wavelet transform decomposition. Pattern Recogn. Lett. **24**(7), 973–982 (2003)
16. Rashed, E.A., Ismail, I.A., Zaki, S.I.: Multiresolution mammogram analysis in multilevel decomposition. Pattern Recogn. Lett. **28**(2), 286–292 (2007)
17. Meselhy, E.M., Faye, I., Belhaouari, S.B.: A comparison of wavelet and curvelet for breast cancer diagnosis in digital mammogram. Comput. Biol. Med. **40**(4), 384–391 (2010). https://doi.org/10.1016/j.compbiomed.2010.02.002
18. Ramos-Pollán, R., Guevara-López, M., Suárez-Ortega, C., Díaz-Herrero, G., Franco-Valiente, J., Rubio-del-Solar, M., de Posada González, N., Vaz, M., Loureiro, J., Ramos, I.: Discovering mammography-based machine learning classifiers for breast cancer diagnosis. J. Med. Syst. **36**(4), 2259–2269 (2011)
19. Deans, S.R.: Applications of the Radon Transform. Wiley Interscience Publications, New York (1983)
20. Mallat, S.G.: A theory for multiresolution signal decomposition: the wavelet representation. IEEE Trans. Pattern Anal. Mach. Intell. **11**(7), 674–693 (1989)

21. Huhn, J., Hullermeier, E.: FURIA: an algorithm for unordered fuzzy rule induction. Data Min. Knowl. Discov. **19**(3), 293–319 (2009)
22. Breiman, L.: Random forests. Mach. Learn. **45**(1), 5–32 (2001)
23. Bielza, C., Li, G., Larrañaga, P.: Multi-dimensional classification with Bayesian networks. Int. J. Approx. Reason. **52**, 705–727 (2011)
24. Mika, S., Ratsch, G., Weston, J.: Fisher discriminant analysis with kernels. In: Conference on Neural Networks for Signal Processing IX, pp. 41–48 (1999)
25. http://mlcheatsheet.readthedocs.io/en/latest/logistic_regression.html. Accessed 15 Feb 2018
26. Santosh, K.C., Antani, S.: Automated chest X-ray screening: can lung region symmetry help detect pulmonary abnormalities. IEEE Trans. Med. Imaging (2017). https://doi.org/10.1109/TMI.2017.2775636
27. Arevalo, J., González, F.A., Ramos-Pollán, R., Oliveira, J.L., Lopez, M.A.G.: Convolutional neural networks for mammography mass lesion classification. In: International Conference of the Engineering in Medicine and Biology Society (2015)
28. http://bcdr.inegi.up.pt. Accessed 15 Feb 2018
29. Arevalo, J., González, F.A., Ramos-Pollán, R., Oliveira, J.L., Lopez, M.A.G.: Representation learning for mammography mass lesion classification with convolutional neural networks. Comput. Methods Programs Biomed. **127**, 248–257 (2016)
30. Obaidullah, Sk. Md., Sajib, A., Teresa, G., Luis, R.: RMID: a novel and efficient image descriptor for mammogram mass classification. In: Kulczycki, P., Kowalski, P.A., Łukasik, S. (eds.) Contemporary Computational Science, p. 203. AGH-UST Press, Cracow (2018)

Instrumentals/Songs Separation for Background Music Removal

Himadri Mukherjee[1(✉)], Sk Md Obaidullah[2], K. C. Santosh[3], Teresa Gonçalves[2], Santanu Phadikar[4], and Kaushik Roy[1]

[1] Department of Computer Science, West Bengal State University, Kolkata, India
himadrim027@gmail.com, kaushik.mrg@gmail.com
[2] Department of Informatics, University of Evora, Évora, Portugal
sk.obaidullah@gmail.com, tcg@uevora.pt
[3] Department of Computer Science, The University of South Dakota, Vermillion, SD, USA
santosh.kc@usd.edu
[4] Department of Computer Science and Engineering, Maulana Abul Kalam Azad University of Technology, Kolkata, India
sphadikar@yahoo.com

Abstract. The music industry has come a long way since its inception. Music producers have also adhered to modern technology to infuse life into their creations. Systems capable of separating sounds based on sources especially vocals from songs have always been a necessity which has gained attention from researchers as well. The challenge of vocal separation elevates even more in the case of the multi-instrument environment. It is essential for a system to be first able to detect that whether a piece of music contains vocals or not prior to attempting source separation. In this paper, such a system is proposed being tested on a database of more than 99 h of instrumentals and songs. Using line spectral frequency-based features, we have obtained the highest accuracy of 99.78% from among six different classifiers, viz. BayesNet, Support Vector Machine, Multi Layer Perceptron, LibLinear, Simple Logistic and Decision Table.

Keywords: Background track · Vocals · Line spectral frequency · Framing

1 Introduction

Technology has had a profound impact in every sphere and the music industry has not been an exception to this. Audio engineers now have various advanced solutions to help them with music production. One of the primary requirements of musicians has always been for such a technology that can enable them to separate background tracks from vocals. This can be extremely helpful for acapella extraction for rearrangements. It can also help musicians in understanding minute technicalities of background tracks who have little audio engineering

P. Kulczycki et al. (Eds.): ITSRCP 2018, AISC 945, pp. 241–251, 2020.
https://doi.org/10.1007/978-3-030-18058-4_19

knowledge. The separation of vocals from music is itself a difficult task which elevates even more in the case songs due to presence of multiple instruments. A system of this sort can also help towards voice activity detection in songs as well and aid the separation of individual instruments in songs for further analysis. It is essential to be able to distinguish instrumentals from songs prior to extracting instrumental portions from the songs and perform any kind of analysis.

In this paper, such a system is proposed which tries to segregate instrumentals and songs using line spectral frequency (LSF)-based features. The system has been pictorially illustrated in Fig. 1. It has been tested with multiple feature dimensions and various classifiers whose details are presented in the subsequent paragraphs.

In the rest of the paper, Sects. 2, 3 and 4 describe the related works, datasets and proposed methodology, respectively. Section 5 highlights the details of the results while conclusion and future work are presented in Sect. 6.

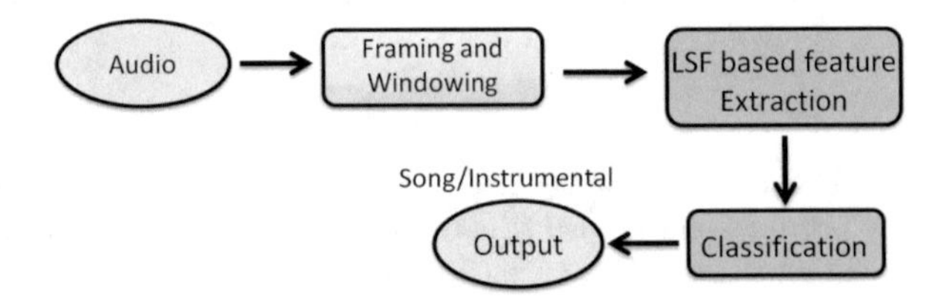

Fig. 1. Pictorial view of the system.

2 Related Work

Leung *et al.* [1] used a supervised variant of independent component analysis namely ICA-FX for the task of segregating instrumentals and voices. They had also used general likelihood ratio based distance and SVM based classification; using 5 and 25 pop songs for training and testing respectively, they obtained a highest individual accuracy of 80.04%. Chanrungutai *et al.* presented a system for separating vocals from music with the aid of a non negative matrix factorization based technique. They performed pitch extraction on the separated voices; their data consisted of both real backing tracks as well as MIDI ones. A detailed account of the results is presented in [2].

Rocamora *et al.* [3] studied various audio descriptors for the task of music and voice segregation and concluded the fact that mel frequency cepstral coefficient (MFCC)-based approach is the most appropriate. They also presented a statistical classification technique with the help of a reduced descriptor set for detecting voice in songs and obtained a highest accuracy of 78.5%. Hsu *et al.* performed separation of music accompaniments and unvoiced singing voice on the MIR-1K dataset. They followed the computational auditory scene analysis framework in their experiments whose details are presented in [4].

Rafii *et al.* [5] adopted a repetitive musical structure identification based approach for segregating voice and music; they experimented with the MIR-1K dataset and obtained a highest global normalized signal to distortion ratio of 1.11. On another work Rafii *et al.* [6] presented a system named REPET for the task of speech and music separation; they experimented with 1000 song clips and 14 songs and extended the system to aid in the pre-processing stage for detecting pitch to help in melody extraction. Liutkus *et al.* [7] further extended REPET to handle background variations as well as long excerpts in order to process full songs.

Ghosal *et al.* [8] adopted a random sample and consensus based approach for the purpose of separating songs and instrumentals; they experimented on a dataset of 300 instrumentals and songs each and obtained an accuracy of 95%. Mauch *et al.* [9] obtained an accuracy of 89.8% for the task of instrumental solo detection using a combination of four features in the thick of MFCC, pitch fluctuation, MFCC of re-synthesised predominant voice and normalised amplitude of harmonic partials.

3 Dataset Development

Data is an essential entity of any experiment. It is always important to ensure that the dataset contains real life characteristics as far as possible. In our experiment, audio clips of songs and instrumentals were extracted from various websites like YouTube [10]. The top three languages of India namely English, Hindi and Bangla [11] were considered in the case of songs. Songs of different genres and timelines were chosen in order to ensure that the dataset covered various qualities of songs in terms of rendering and audio engineering. The song clips had either background music or sections of instrument only parts. Different artists were chosen for collecting instrumental clips in order to get data of various types like genre, playing style, tonality and technicality. Instrumental covers of various songs, as well as original compositions using different instruments like guitar, violin and piano, along with background music constituted the instrumental part of the dataset.

This data was used to generate 4 datasets (D_1–D_4) having clips of lengths 5, 10, 15 and 20 s, respectively. The details of the generated datasets along with that of the original data is presented in Table 1.

Table 1. Number of instrumental and song clips for the different datasets.

Datatset (clip length)	Song (49:19:48)	Instrumental (49:50:15)
D_1 (5 s)	35116	35718
D_2 (10 s)	17362	17771
D_3 (15 s)	11431	11798
D_4 (20 s)	8500	8805

4 Proposed Method

4.1 Pre Processing

The audio clips were split into smaller frames as the spectral characteristics tend to show a lot of deviation for longer clips. The clips were partitioned into 256 sample point wide frames with a 100 point overlap as presented in [12]. Next, the frames were subjected to a windowing function (Hamming Window as presented in [13]) in order to get rid of the jitters which might lead to spectral leakage at the time of frequency based analysis. The mathematical representation of hamming window $B(n)$ is given by Eq. (1) where the value of r ranges between the frame boundary of size R

$$B(n) = 0.54 - 0.46 \cos\left(\frac{2\pi r}{R-1}\right). \tag{1}$$

4.2 Feature Extraction

Line spectral frequency [14] is a method for representing linear predictive coefficients with better interpolation properties. Here, a signal is considered as the output of an all pole filter ($H(z)$). The inverse filter of $H(z)$ is represented by Eq. 2, where r_1, $r_2 \ldots r_T$ designate the predictive coefficients up to the order T

$$R(z) = 1 + r_1 z^{-1} + r_2 z^{-2} + r_3 z^{-3} + \ldots\ldots + r_T z^{-T}. \tag{2}$$

$R(z)$ is decomposed into polynomials $F(z)$ and $G(z)$ as shown in Eq. 3 and Eq. 4, respectively, whose zeros lie on the unit circle. They are also interlaced with each other, thus helping in computation

$$F(z) = R(z) + z^{-(T+1)} R(z^{-1}) \ and \tag{3}$$

$$G(z) = R(z) - z^{-(T+1)} R(z^{-1}). \tag{4}$$

Each of the datasets were used for extraction of 5, 10, 15 and 20 dimensional standard line spectral frequency features. Since the clips were of disparate lengths, a disparate number of frames were produced for the clips, producing features of disparate dimension. In order to tackle this problem, the band wise sums for the energy values were computed which were then used to compute the ratio of distribution of energy across the bands. Along with this, the band numbers were also added to the feature set graded in descending order based on total energy content.

It was experimentally found that a clip of just 2 s produced 880 frames; if a 5 dimensional LSF was extracted for the clip then a feature set of 4400 dimension was obtained. However with the help of the proposed technique, this dimension was brought down to just 10 (5 ratio distribution values and 5 band numbers). Thus, LSF values of 5, 10, 15 and 20 dimensions produced even dimensional feature sets of 10, 20, 30 and 40 dimensions which were independent of the length of the clips. Trends of the feature values for the songs and instrumental clips for the 40 dimensional features (best result) is presented in Fig. 2.

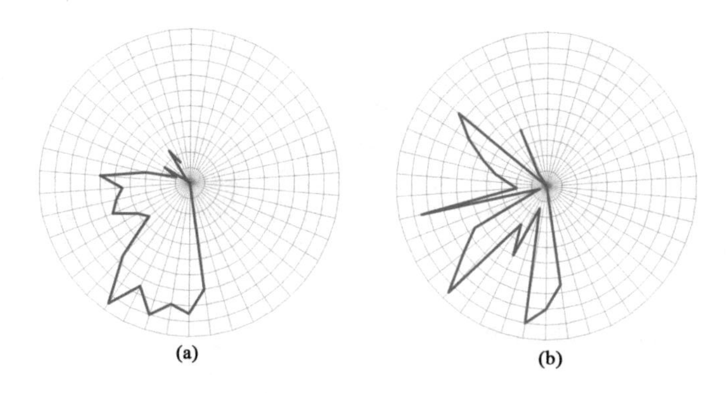

Fig. 2. 40 dimensional feature values for (a) Song, (b) Instrumental

4.3 Classification

Each feature set for each datasets was fed into different classifiers popularly used in pattern recognition problems in the thick of BayesNet (BN) [15], Support Vector Machine (SVM) [16], LibLinear (Lib) [17], Multi Layer Perceptron (MLP) [21], Simple Logistic (SL) [18] and Decision Table (DT) [19].

BayesNet: is a bayesian classifier that makes use of a Bayes Network for learning with the aid of different quality parameters and search algorithms. The base class provides data structures like conditional probability distributions, structure of network, etc. and different facilities which are similar to that of the Bayesian Network learning algorithm like K2.

Support Vector Machine: is a supervised learning algorithm that can be used for classification as well as regression analysis. SVM builds a bi-class model from a set of training instances and then associates each instance to either class.

LibLinear: is a functional and linear type of classifier which is suitable for either large number of instances or large feature sets. It is also suitable for regression problems.

Decision Table: is one of the simplest supervised learning algorithms; it consists of 2 parts namely, schema which defines the features to be included in the table and body which embodies the set of instances along with their feature values and class labels.

Multi Layer Perceptron: is a feed forward variant of an artificial neural network which maps an input and output set; it consists of nodes which are connected by links having weights associated to it. It is one of the most popular classifiers in pattern recognition problems.

Simple Logistic: is a classifier used to build linear logistic regression models. The classifier has a base learner associated with it along with number of iterations which aids to automatically select attributes.

The extremely popular open source classification tool named Weka [20] was used in the present experiment. We used 5 fold cross validation for all the classifiers with default parameters. The details of the obtained results are presented in the subsequent section.

5 Result and Discussion

The accuracies obtained for datasets D_1–D_4 using different classifiers are presented in Tables 2, 3, 4 and 5.

Table 2. Obtained accuracies for D_1 using different classifiers and feature dimensions.

Dimension	BN	MLP	DT	SL	SVM	Lib
10	83.38	86.22	86.51	84.86	84.86	84.86
20	67.01	72.83	72.92	61.71	88.15	57.37
30	69.47	72.36	71.82	61.53	72.21	50.69
40	95.30	**99.78**	94.96	62.36	73.01	54.55

Table 3. Obtained accuracies for D_2 using different classifiers and feature dimensions.

Dimension	BN	MLP	DT	SL	SVM	Lib
10	67.38	69.08	72.06	60.24	64.57	60.24
20	75.351	92.51	94.42	64.64	96.89	62.47
30	89.62	94.96	96.62	62.16	94.98	57.03
40	98.27	98.18	96.76	73.97	61.21	52.27

Table 4. Obtained accuracies for D_3 using different classifiers and feature dimensions.

Dimension	BN	MLP	DT	SL	SVM	Lib
10	67.53	70.26	71.73	60.61	61.89	60.62
20	78.56	99.30	99.04	61.97	99.31	59.56
30	93.27	97.44	97.36	61.49	92.55	52.84
40	99.57	99.52	94.13	89.91	59.67	72.56

From Table 2 it is observed that the highest and lowest accuracies of 99.78% and 50.69% were obtained using MLP and LibLinear, respectively, which are the

Table 5. Obtained accuracies for D_4 using different classifiers and feature dimensions.

Dimension	BN	MLP	DT	SL	SVM	Lib
10	85.25	85.26	85.42	60.77	84.19	60.63
20	83.55	96.32	96.51	69.11	96.16	66.51
30	90.97	99.39	97.47	62.62	95.53	51.77
40	90.97	92.58	90.93	70.43	64.85	61.84

overall best and worst results among all the classifiers with default parameters. The same behaviour is found for D_2 dataset as can be seen in Table 3; the highest and lowest accuracies being 98.27% and 52.27%. On the other hand, for D_3 (Table 4), the highest and lowest accuracies were 99.57% and 52.84%, obtained using BayesNet and LibLinear, respectively. Highest and lowest accuracies of 99.39% and 51.77% were obtained using MLP and LibLinear, respectively for D_4 (Table 5).

Table 6 shows the highest and lowest accuracies obtained for all experiments based on the feature dimension. It can be seen from the Table that LibLinear produced the lowest accuracy for every feature dimension. It can also be observed that the top 2 results were obtained using MLP on D_1 (shortest clips in experiment) and D_4 (longest clips in experiment) which shows the effectiveness of MLP.

Table 6. Highest and lowest accuracies obtained for all experiments based on feature dimension along with the classifier and dataset.

Dimension	Highest	Lowest
10	86.51 (D_1, DT)	60.24 (D_2, SL; D_2, Lib)
20	99.31 (D_3, SVM)	57.37 (D_1, Lib)
30	99.39 (D_4, MLP)	50.69 (D_1, Lib)
40	99.78 (D_1, MLP)	52.27 (D_2, Lib)

It is also observed from Tables 2, 3, 4 and 5 that highest accuracies of 99.57%, 99.31%, 84.86% were obtained for BayesNet, SVM and LibLinear respectively. The lowest accuracies for the same classifiers were found to be 67.01%, 59.67% and 50.69% respectively. In the case of MLP, Simple Logistic and Decision Table, highest accuracies of 99.78%, 89.91% and 97.47% respectively were obtained. The lowest accuracies for the same were found to be 69.08%, 60.24%, 71.73% respectively.

Concluding, the classifiers can be organized in the following manner based on their best performance: MLP, BayesNet, SVM, Decision Table, Simple Logistic and LibLinear. MLP is very suitable for audio signal based applications as demonstrated in [12,13,21] which is depicted in the obtained results as well.

The confusion matrix for the best result (D_1, MLP with 40 dimensional features) is presented in Table 7 where it can be observed that 0.19% of the song clips and 0.25% of the instrumental clips were confused with each other. One possible reason for this is that during the generation of the dataset (splitting of clips into shorter clips), some of the instrumental parts from the songs might have got entirely isolated for the 5 s clips (it was observed that various songs had instrumental sections of more than 5 s at a stretch) which interfered with the trained model.

Table 7. Confusion matrix for D_1 with 40 dimensional features using MLP showing the number of correctly and misclassified instances.

	Song	Instrumental
Song	35051	65
Instrumental	89	35629

Since the best result was obtained for MLP, we further experimented with it by varying the number of training iterations (ephocs); the obtained accuracies are depicted in Table 8.

Table 8. Accuracies using MLP for D_1 with 40 dimensional features for different training iterations.

Iterations	100	200	300	400	500	600	700	800
Accuracy (%)	99.66	99.72	99.72	99.79	99.78	99.78	99.78	99.78
Iterations	900	1000	1100	1200	1300	1400	1500	1600
Accuracy (%)	99.79	99.80	99.79	99.80	99.80	99.80	99.79	99.80
Iterations	1700	**1800**	1900	2000	2100	2200	2300	2400
Accuracy (%)	99.80	**99.81**	99.81	99.81	99.80	99.81	99.80	99.81

From the Table it can be observed that the highest accuracy (99.81%) was obtained for 1800 iterations and that value maintained for further iterations. The confusion matrix for this experiment is presented in Table 9 where it can be seen that the number of misclassified samples for both the classes decreased with respect to the default configuration of MLP as shown in Table 7.

Table 9. Confusion matrix for D_1 with 40 dimensional features using MLP at 1800 learning iterations.

	Song	Instrumental
Song	35052	64
Instrumental	74	35644

We had further experimented by varying the number of folds in cross validation for the same setup (best accuracy with lowest number of training iterations); the details are presented in Table 10. We had varied the number of folds for cross validation from 2–10 to observe the performance of MLP for test and training sets of different sizes. It can be seen from the Table that the best accuracy was obtained for 5 and 7 folds which further decreased on increasing the number of folds.

Table 10. Accuracies for different number of cross validation folds of cross using MLP for D_1 with 40 dimensional features and 1800 training iterations.

# Folds	2	3	4	5	6	7	8	9	10
Accur. (%)	99.36	99.75	99.69	99.81	99.71	99.81	99.78	99.75	99.73

5.1 Statistical Significance Tests

Friedman test [22] on the 40 dimensional feature set of D_1 was carried out to check for statistical significance. The dataset was divided into 5 parts (N) and all the 6 classifiers (k) were involved in the test. The distribution of ranks and accuracies are presented in Table 11.

Table 11. Rank and accuracies for parts of D_1.

Classifiers		Parts of D_1					Mean rank
		1	2	3	4	5	
MLP	Acc	99.64	99.99	99.98	100.0	99.92	1.8
	Rank	(3.0)	(1.0)	(2.0)	(2.0)	(1.0)	
BN	Acc	99.9	99.72	100.0	100.0	99.46	1.7
	Rank	(1.5)	(2.0)	(1.0)	(2.0)	(2.0)	
SL	Acc	99.9	61.8	78.45	99.97	75.81	3.9
	Rank	(1.5)	(4.0)	(5.0)	(4.0)	(5.0)	
DT	Acc	95.4	92.98	99.18	97.72	96.01	3.6
	Rank	(4.0)	(3.0)	(3.0)	(5.0)	(3.0)	
SVM	Acc	76.33	55.12	96.05	89.99	76.97	4.8
	Rank	(5.0)	(5.0)	(4.0)	(6.0)	(4.0)	
Lib	Acc	55.01	50.66	50.76	100.0	62.36	5.2
	Rank	(6.0)	(6.0)	(6.0)	(2.0)	(6.0)	

The Friedman statistic (χ_F^2) [22] was calculated with the help of Table 11 in accordance with Eq. 5 where R_j corresponds to the mean rank of the jth classifier.

$$\chi_F^2 = \frac{12N}{k(k+1)} \left[\sum_j R_j^2 - \frac{k(k+1)^2}{4} \right]. \quad (5)$$

The critical values of χ_F^2 for the above setup was found to be 11.070 and 9.236 at significance levels of 0.05 and 0.10 respectively; we got a value of 15.54 for χ_F^2 thereby rejecting the null hypothesis.

6 Conclusion and Future Work

In this paper, a system for segregating instrumentals and songs is presented using line spectral frequency based features. We have applied different popular classifiers on the feature sets and obtained the highest result using MLP algorithm. It was also observed that LibLinear produced most of the accuracies in the lower side.

As future work we will experiment with a larger and more robust dataset and observe the performance of various other classifiers. We will also experiment with other machine learning techniques including deep learning based approaches and use different features to further minimize the errors. We also plan to pre-process the data for noise removal to make the system more robust which is critical for live audio. The system will be further extended to detect instrument sections from songs in real time to separate the vocals and instrument tracks.

Note. The preliminary version of this paper was presented at the 3rd Conference on Information Technology, Systems Research and Computational Physics, 2–5 July 2018, Cracow, Poland [23].

References

1. Leung, T.W., Ngo, C.W., Lau, R.W.: ICA-FX features for classification of singing voice and instrumental sound. In: Proceedings of the 17th International Conference on Pattern Recognition, ICPR 2004, vol. 2, pp. 367–370. IEEE (2004)
2. Chanrungutai, A., Ratanamahatana, C.A.: Singing voice separation for mono-channel music using non-negative matrix factorization. In: International Conference on Advanced Technologies for Communications, ATC 2008, pp. 243–246. IEEE (2008)
3. Rocamora, M., Herrera, P.: Comparing audio descriptors for singing voice detection in music audio files. In: 11th Brazilian Symposium on Computer Music, São Paulo, Brazil, vol. 26, p. 27 (2007)
4. Hsu, C.L., Jang, J.S.R.: On the improvement of singing voice separation for monaural recordings using the MIR-1K dataset. IEEE Trans. Audio Speech. Lang. Process. **18**(2), 310–319 (2010)
5. Rafii, Z., Pardo, B.: A simple music/voice separation method based on the extraction of the repeating musical structure. In: 2011 IEEE International Conference on Acoustics, Speech and Signal Processing (ICASSP), pp. 221–224. IEEE (2011)

6. Rafii, Z., Pardo, B.: Repeating pattern extraction technique (REPET): a simple method for music/voice separation. IEEE Trans. Audio Speech Lang. Process. **21**(1), 73–84 (2013)
7. Liutkus, A., Rafii, Z., Badeau, R., Pardo, B., Richard, G.: Adaptive filtering for music/voice separation exploiting the repeating musical structure. In: 2012 IEEE International Conference on Acoustics, Speech and Signal Processing (ICASSP), pp. 53–56. IEEE (2012)
8. Ghosal, A., Chakraborty, R., Dhara, B.C., Saha, S.K.: Song/instrumental classification using spectrogram based contextual features. In: Proceedings of the CUBE International Information Technology Conference, pp. 21–25. ACM (2012)
9. Mauch, M., Fujihara, H., Yoshii, K., Goto, M.: Timbre and melody features for the recognition of vocal activity and instrumental solos in polyphonic music. In: ISMIR, pp. 233–238 (2011)
10. https://www.youtube.com/. Accessed 1 Mar 2018
11. https://www.ethnologue.com/. Accessed 1 Mar 2018
12. Mukherjee, H., Obaidullah, S.M., Phadikar, S., Roy, K.: SMIL-a musical instrument identification system. In: International Conference on Computational Intelligence, Communications, and Business Analytics, pp. 129–140. Springer, Singapore (2017)
13. Mukherjee, H., Phadikar, S., Rakshit, P., Roy, K.: REARC-a Bangla Phoneme recognizer. In: 2016 International Conference on Accessibility to Digital World (ICADW), pp. 177–180. IEEE (2016)
14. Paliwal, K.K.: On the use of line spectral frequency parameters for speech recognition. Digit. Sig. Process. **2**(2), 80–87 (1992)
15. Friedman, N., Geiger, D., Goldszmidt, M.: Bayesian network classifiers. Mach. Learn. **29**(2–3), 131–163 (1997)
16. Cristianini, N., Shawe-Taylor, J.: An Introduction to Support Vector Machines and Other Kernel-Based Learning Methods. Cambridge University Press, Cambridge (2000)
17. Fan, R.E., Chang, K.W., Hsieh, C.J., Wang, X.R., Lin, C.J.: LIBLINEAR: a library for large linear classification. J. Mach. Learn. Res. **9**, 1871–1874 (2008)
18. Sumner, M., Frank, E., Hall, M.: Speeding up logistic model tree induction. In: European Conference on Principles of Data Mining and Knowledge Discovery, pp. 675–683. Springer, Heidelberg (2005)
19. Kohavi, R.: The power of decision tables. In: European Conference on Machine Learning, pp. 174–189. Springer, Heidelberg (1995)
20. Hall, M., Frank, E., Holmes, G., Pfahringer, B., Reutemann, P., Witten, I.H.: The WEKA data mining software: an update. ACM SIGKDD Explor. Newsl. **11**(1), 10–18 (2009)
21. Mukherjee, H., Halder, C., Phadikar, S., Roy, K.: READ-a Bangla Phoneme recognition system. In: Proceedings of the 5th International Conference on Frontiers in Intelligent Computing: Theory and Applications, pp. 599–607. Springer, Singapore (2017)
22. Demšar, J.: Statistical comparisons of classifiers over multiple data sets. J. Mach. Learn. Res. **7**, 1–30 (2006)
23. Mukherjee, H., Obaidullah, Sk.Md., Santosh, K.C., Gonçalves, T., Phadikar, S., Roy, K.: Instrumentals/songs separation for background music removal. In: Kulczycki, P., Kowalski, P.A., Lukasik, S. (eds.) Contemporary Computational Science, p. 204. AGH-UST Press, Cracow (2018)

Content-Based Recommendations in an E-Commerce Platform

Łukasz Dragan and Anna Wróblewska(✉)

Faculty of Mathematics and Information Science, Warsaw University of Technology, ul. Koszykowa 75, 00-662 Warsaw, Poland
a.wroblewska@mini.pw.edu.pl

Abstract. Recommendation systems play an important role in modern e-commerce services. The more relevant items are presented to the user, the more likely s/he is to stay on a website and eventually make a transaction. In this paper, we adapt some state-of-the-art methods for determining similarities between text documents to content-based recommendations problem. The goal is to investigate variety of recommendation methods using semantic text analysis techniques and compare them to querying search engine index of documents. As a conclusion we show, that there is no significant difference between examined methods. However using query based recommendations we need more precise meta-data prepared by content creators. We compare these algorithms on a database of product articles of the biggest e-commerce marketplace platform in Eastern Europe - Allegro. (The primary version of this paper was presented at the 3rd Conference on Information Technology, Systems Research and Computational Physics, 2–5 July 2018, Cracow, Poland [4].)

Keywords: Content-based recommendations · Natural language processing · Distributional semantics · Word embeddings

1 Introduction

Recommendation systems are commonly used in e-commerce services. Such systems give profit to both a user as well as a website owner. It allows the user to get information s/he could not find in other case, or even know that such information exists. Recommendations also attract users to a service, making them more likely to buy products, increasing website company's profit thereby. The key issue of a recommendation generation process is how suggested items are relevant to these which the user is interested in at the moment of browsing the service.

Generally we can divide recommendation systems into two groups: collaborative and content-based filtering. The first one assumes that a user is likely to be interested in items which also other users similar to her or him were interested in.

P. Kulczycki et al. (Eds.): ITSRCP 2018, AISC 945, pp. 252–263, 2020.
https://doi.org/10.1007/978-3-030-18058-4_20

In this paper we are focusing on the second group—content-based filtering—in which recommended items are similar to those the user liked so far.

A detailed problem comes from the on-line buying and auction site Allegro. Allegro—the biggest marketplace platform in Eastern Europe—contains a section with text articles concerning products available via the platform. Next to an article displayed by a user, there is a list of references to articles somehow similar to the given one. Common resolution to such a problem is to use search engine to find similar texts based on defined queries, keywords or a title manually attached to the given article. These short descriptions of texts should define the overall content thus summarizing it in a very concise way.

In this paper we strive to check if some distributional semantic text analysis methods, including newly proposed word embeddings, are able to supplement query-based methods using search engine common algorithms. Examined methods are topic modeling methods: latent semantic analysis [2], latent Dirichlet allocation [1] and some word embeddings methods: word2vec [6], fastText [3] and GloVe [8] together with similarity metrics.

In Sect. 2 we make an analysis of an article dataset achieved from a e-commerce platform and also we describe data preprocessing techniques we performed. Section 3 is about methods used for building the model of text similarity, how to use and evaluate this model. In Sect. 4 we present our results and try to interpret them. The last section concludes the paper and proposes a direction of future work.

2 Dataset

Our dataset was achieved form a widely-used e-commerce platform. It consists of 20 thousands text articles describing several categories of products available via the platform. A single record consists of article contents and meta-data attached by an author of the article. As significant meta-data we consider "category" and "keywords". Keywords are a set of words aiming to describe contents and help the search engine in its search task. Categories are arranged in a tree hierarchy: each article has assigned a list of categories ordered according to relations from more general to the more precise ones (homonymy and hyperonymy).

All the articles are written in Polish. The vocabulary based on the articles set contains many industry-specific words like brands and models names, books titles and technical words. Moreover raw articles contain some tags responsible for images and hyper-links present on the website.

A standard activity before building a model is to pre-process of a raw dataset (Fig. 1).

The steps we performed as a preprocessing phase are as follows:

1. Cleansing the text from the redundant, previously mentioned tags. From the viewpoint of semantic analysis they are useless or even noxious. That is why we remove them using properly constructed regular expressions. An example of such a tag is `[2_new.jpg] (http: // (...)'2_new.jpg')` placing the

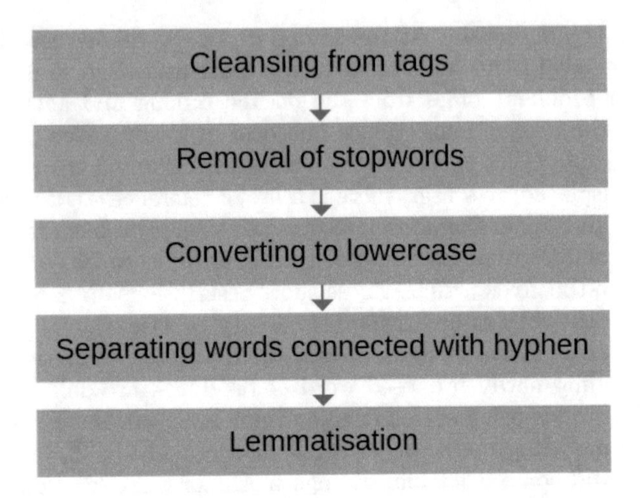

Fig. 1. Preprocessing pipeline.

picture in the middle of the text (the content of the URL removed due to confidentiality reasons).

2. Removal of stopwords—usually short words carrying very little information about the actual document contents, e.g. "in", "from", "because" etc. Removing them reduces the number of words in a document and the processing time thereby.
3. Converting all words of a document to lowercase letters. It helps to unify words with the same meaning, but written using both uppercase and lowercase letters, e.g. in the middle or at the beginning of a sentence.
4. Separating words connected with a hyphen. Experience at a later lemmatisation stage shows that the tool used for this purpose (Morfologik [7]) does not cope with these types of words and leaves them in the unchanged grammatical form. This made it necessary to build our own tool that breaks such words into sub-words compatible with the lemmatiser.
5. Tokenization and lemmatisation. This is the most important element of the process. It performs separating text into individual words and converting the words with the same meaning, and different grammatical form to the same base form. Complexity of the Polish language and the number of exceptions that this language has are making this problem harder than in many other languages. To carry out this operation, we used the Morfologik tool [7].

The above steps lead the data to the state in which we can apply techniques of text analysis. The dictionary built on the preprocessed corpus contains 98.174 unique words and 7.409.145 all words (with repetitions). The overwhelming majority of the words of the dictionary built on the corpus are the words appearing rarely (Fig. 2).

Most common words are: "sam" (alone), "uwaga" (attention), "ważny" (important), "należeć" (to belong), "wybrać" (to choose). The rarest words

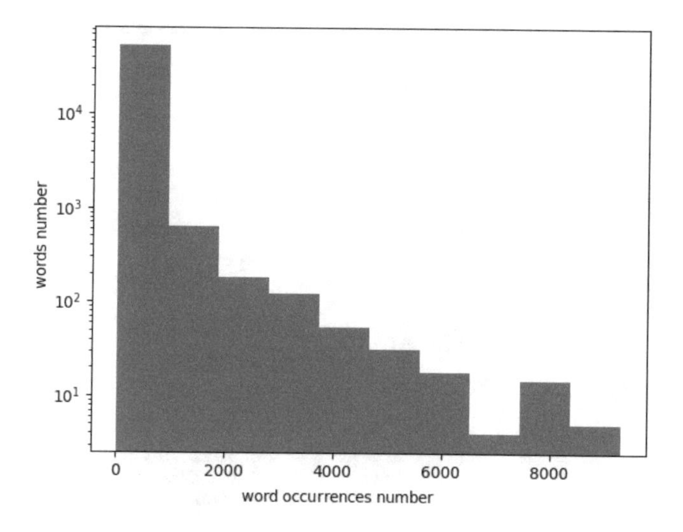

Fig. 2. Histogram of number of words occurrences in logarithmic scale.

include typos, very atypical words and brands or models. Most articles are of similar length. An average article length is about 370 words (Fig. 3).

3 Methods

In our research we focused on word embeddings methods which we compared with some well known topic modeling methods (lsi, lda) and the common method of meta-keywords querying of a search engine (search engine used was *Elastic-Search*). Word embeddings are techniques aimed at representing words as dense vectors, usually of dimensionality between 50 up to 1000. Sudden increase of popularity of word embedding methods started with introducing *Word2Vec* by Mikolov et al. in 2013 [6]. This method converts words to vectors using a shallow neural network and outperforms earlier approaches in terms of time complexity. Besides basic Word2Vec we also examined a derivative method *FastText* and similar one—*GloVe*.

3.1 Model Building

All models are built on preprocessed dataset received from Allegro. Word2Vec authors offer a pre-trained model but only for English language. There exists a pre-trained models for Polish language [9]. However we decided to train models independently using our entire articles corpus. Such a procedure allows to have model of language more adequate to our particular domain of e-commerce products and their characteristics descriptions. To be honest it introduces a little bias into our tests because we train our language models on the dataset we used to compare statistics. Yet analyzing literature we came to know that it is

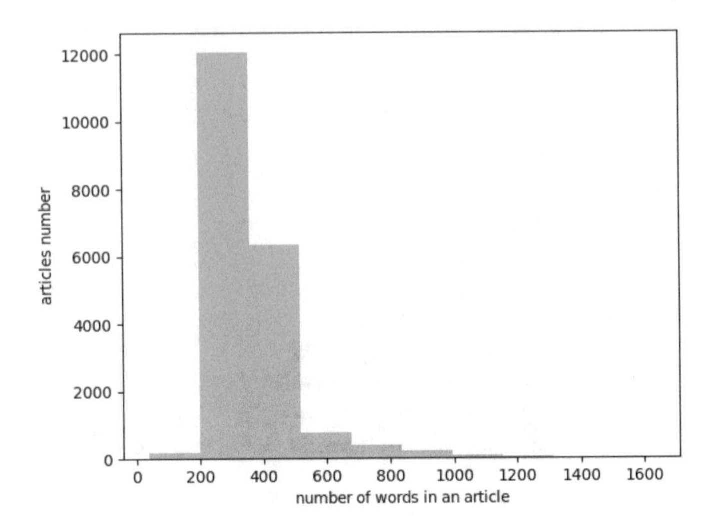

Fig. 3. Articles length histogram.

very important to have domain-specific language models. Thus using pre-trained models from general sources like National Polish Corpus [12] or Wikipedia is not sufficient to our task, because of possible lack of quite important words like brand and model names etc.

Mostly we used default hyper-parameters that came with gensim [10] implementation of examined methods. The hyper-parameters we considered as key ones were: number of topics (in case of topic modeling methods) and vectors dimensionality (word embedding methods). The key hyper-parameters were chosen based on preliminary automatic evaluation. Next, the best versions of every method were compared using user evaluation schema.

Word embedding methods do conversion from words to vectors. In order to compare whole documents we needed to represent documents as vectors. For this purpose we examined two approaches. First one was a simple calculation of an average of vectors being representation of words included in an article. The second one was Word Mover's Distance [5]. Unfortunately it turned out to be too slow to be used in production system so we abandoned this technique.

3.2 Evaluation

Evaluation of recommendations generation methods is a nontrivial task. A degree of similarity of two articles can be perceived differently by different persons. In order to compare the results of the methods used to determine a similarity between articles, a formalization of certain measures of this similarity was necessary. To decide which recommendation method performs best with respect to a given particular product article, we sum relevancies of n best articles recommended by the method to that given article. Evaluation techniques we used differ in the way of that relevancy determination.

We measure relevancy of recommended articles in two ways as follows:

1. First one was based on meta-tags similarity: the more mutual categories or keywords two articles had the more similar we consider them to be. E.g. categories distance between two articles is a length of a common part of paths from the root of the category tree to nodes representing the articles (Fig. 4). Due to the fact that this technique is automatic and every article has meta-tags attached we had no limits in use of this technique.
2. The second method we used is based on similarity ratings between pairs of articles made by real users. For each tested method we choose 5 most similar articles for 50 randomly chosen base articles. Next we split them into pairs: base article—similar article; it gave us $5 * 50 = 250$ pairs for a method. After that 5 persons individually evaluated a similarity of pairs of articles giving scores in scale 1-10. Finally we took an average score for each pair and it allowed us to calculate an average score for each method. Together with tested methods we evaluated also simple search keywords query based method (our baseline) and randomly generated pairs of articles just for a comparison purpose.

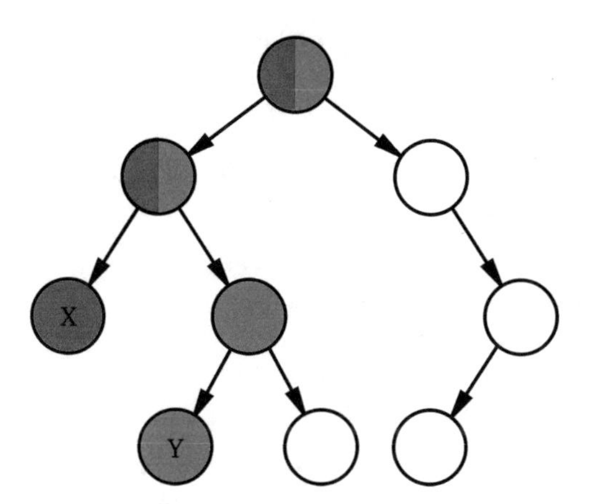

Fig. 4. Sample category tree. In this example articles X and Y have two mutual categories.

4 Results and Discussion

In this section we present results of performed tests. As mentioned earlier, first of all we chose some hyper-parameters values based on automatic evaluation methods. Results featured in Figs. 5, 6, 7, 8 and 9 show that both categories and keywords based evaluation methods give similar results. In every case both methods indicate the same hyper-parameter value as the one giving the best score.

Categories-based method usually gives very similar results, however differences of keywords-based score between different hyper-parameters values reaches up to 30%.

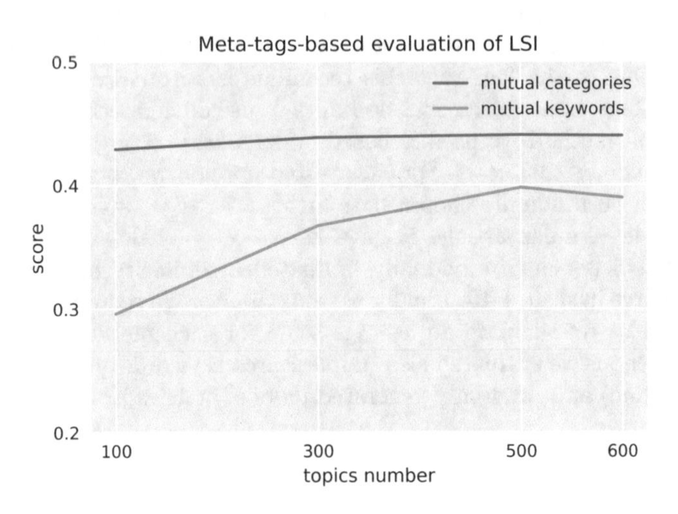

Fig. 5. Comparison of meta-tags based evaluation methods scores for LSI depending on topics number.

Eventual hyperparameters values for particular methods we chose are contained in below enumeration.

1. Latent semantic indexing [2] with 500 topics.
2. Latent Dirichlet allocation [1] with 900 topics.
3. Word2Vec [6] with 300 length vectors.
4. FastText [3] with 1000 length vectors.
5. GloVe [8] with 1000 length vectors.

In order to compare aforementioned methods we performed an evaluation with users. The evaluated methods were: LSI, LDA, word2vec, GloVe, FastText, our baseline—ElasticSearch-based method and randomly chosen pairs of articles (Table 1).

User evaluation suggests that none of the tested methods is significantly different from others (Fig. 10). The difference between the best method (FastText) and the worst one (LDA) is around 9%. Each of the tested methods also gave a significantly better result than the random method.

To answer the question posed at the beginning of this work, i.e. whether the examined text analysis methods adapted to content-based recommendation task are able to match the previous method used in Allegro, we perform a statistical test which goal is to check whether there are grounds to believe that the results of any of the methods is statistically significantly different from the rest of the examined methods.

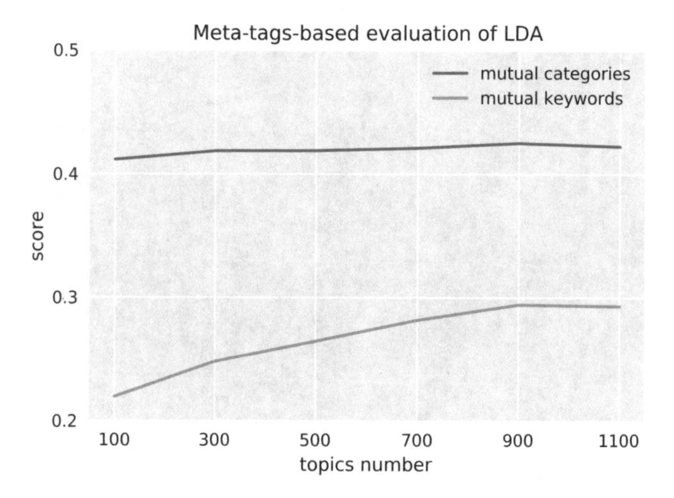

Fig. 6. Comparison of meta-tags based evaluation methods scores for LDA depending on topics number.

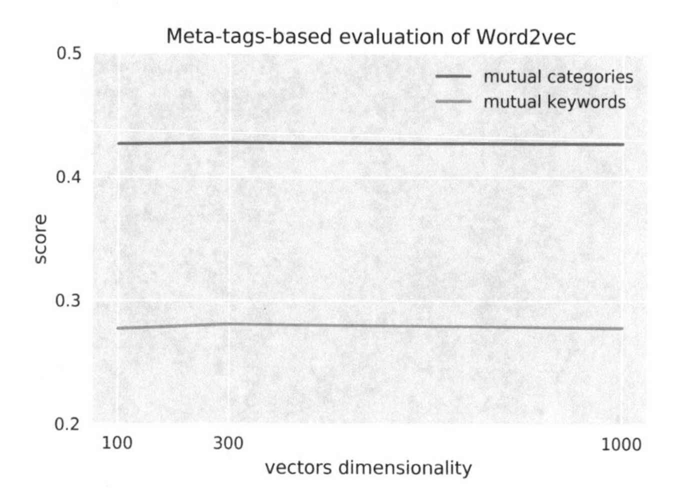

Fig. 7. Comparison of meta-tags based evaluation methods scores for word2vec depending on vectors dimensionality.

Table 1. Comparison of user evaluation scores for best hyper-parameters configurations of selected methods.

Method	lsi	lda	word2vec	GloVe	fastText	baseline	random
Score	8.415	7.895	8.185	8.610	**8.680**	8.470	1.345

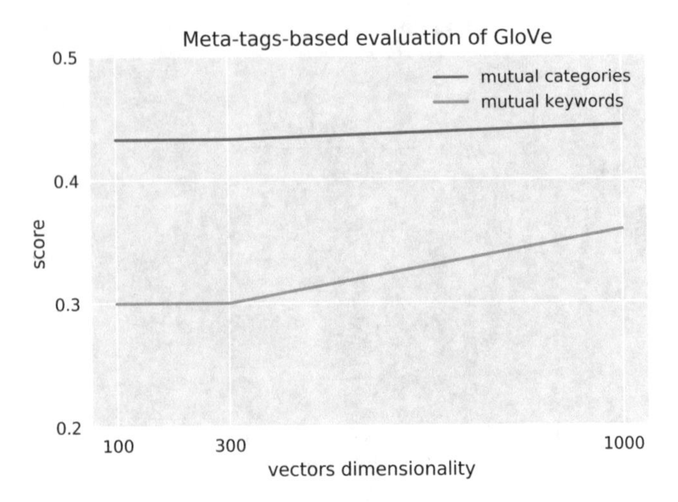

Fig. 8. Comparison of meta-tags based evaluation methods scores for GloVe depending on vectors dimensionality.

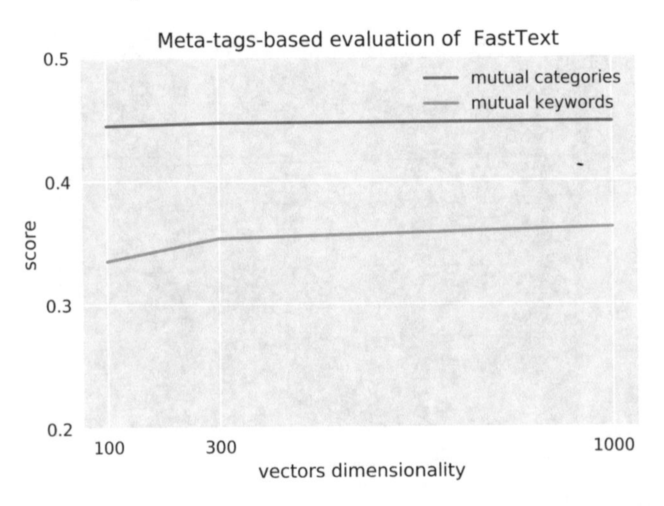

Fig. 9. Comparison of meta-tags based evaluation methods scores for FastText depending on vectors dimensionality.

The statistical test that we carry out is the Kruskal-Wallis test. In the test the null hypothesis H_0 assumes an equality of distributions in populations from which samples originated. As input data the Kruskal-Wallis test takes samples corresponding to expert evaluation of each method consisting of an average score made by users for the most relevant recommendation for each of the examined base articles.

In the test, the level of significance is $\alpha = 0.05$. Finally, as the result of the Kruskal-Wallis test we received $p = 0.0571 > \alpha$, on the basis of which we state

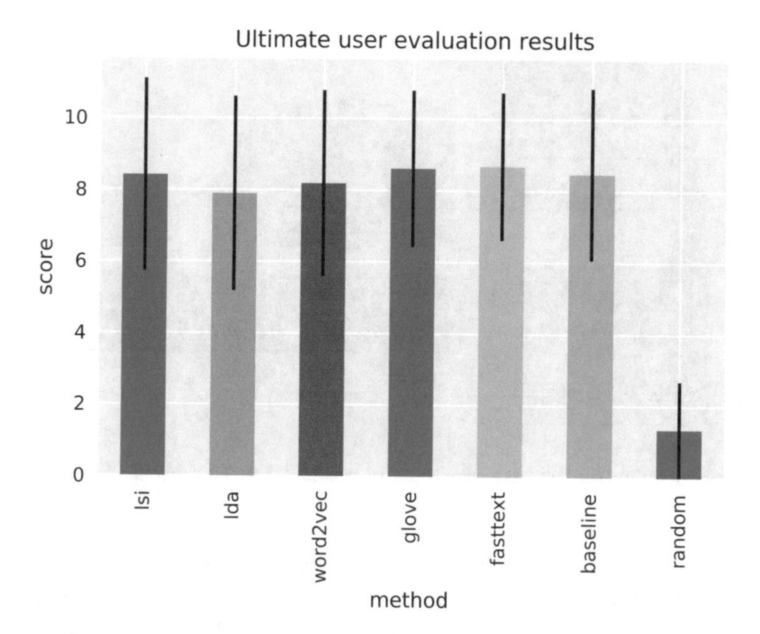

Fig. 10. Comparison of user-evaluated methods results. Vertical lines symbolize standard deviation.

that there are no reasons to reject the null hypothesis—the quality difference between recommendations generated by the tested methods is not statistically significant. It means that neither of the tested semantic methods sticks out from others, nor the previous Allegro method is significantly better/worse than other tested methods.

5 Conclusion

After analyzing the test results we can conclude that each of the examined methods of natural language semantic analysis in their best configuration can be adapted to generate article content-based recommendations. This means we can replace our baseline (querying with manually generated keywords) with e.g. FastText distributional similarity and then any user should not feel any differences in recommendations quality.

Using methods of semantic text analysis allows to capture the hidden similarities between documents, where the documents combine not the same words or synonyms, but some abstract concepts related to each other. An important advantage of semantic methods compared to the baseline is the fact that they are based only on article contents. This frees the authors of the articles from providing additional meta-data, which the baseline method uses in large extent.

Tests carried out in this paper can be the basis for full design of a recommendation system utilizing word embeddings that could be used in a real application (Fig. 11).

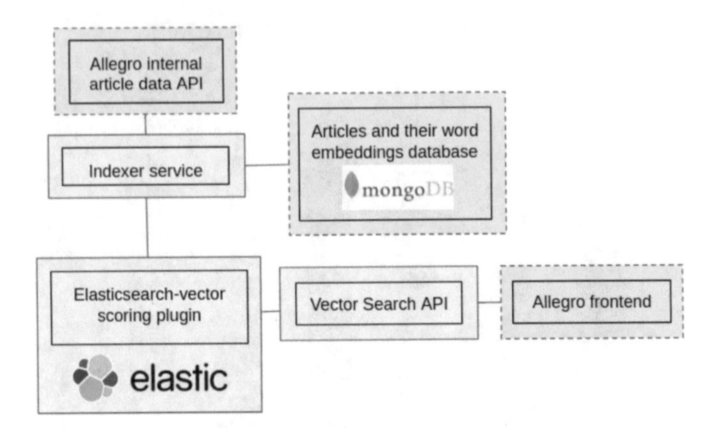

Fig. 11. Sample architecture of a system based on word embeddings similarity.

In this application a user makes request via web browser in order to display the content of an article. Each article displayed by the user is accompanied by recommended articles. These recommendations come from "Vector Search API", which is queried with a vector representation of the requested article. The API uses Elasticsearch engine with special plug-in installed [11], which enables searching with vector representations of documents. Elasticsearch stores vector representations of articles made by "Indexer Service" from two sources: an original text articles database and vector representations (e.g. FastText) of vocabulary words stored in a database optimized to read.

Acknowledgments. This paper provides description of graduate work by Łukasz Dragan, that was conducted and supervised by Anna Wróblewska. This work was made with cooperation of Allegro team, that provided business case and the valuable dataset of 20 thousands product articles available through the platform https://allegro.pl/artykuly.

The work was conducted as Anna Wróblewska was an employee of Allegro and after that during cooperation as a research advisor from Warsaw University of Technology.

The work was also partially supported as the RENOIR Project by the European Union's Horizon 2020 Research and Innovation Programme under the Marie Skłodowska-Curie grant agreement No. 691152 and by Ministry of Science and Higher Education (Poland), grant Nos. W34/H2020/2016.

References

1. Blei, D.M., Ng, A.Y., Jordan, M.I.: Latent dirichlet allocation. J. Mach. Learn. Res. **3**, 4–5 (2003)
2. Deerwester, S., Dumais, S.T., Furnas, G.W., Landauer, T.K., Harshman, R.: Indexing by latent semantic analysis. J. Am. Soc. Inf. Sci. **41**(6), 391–407 (1990)
3. Joulin, A., Grave, E., Bojanowski, P., Mikolov, T.: Bag of Tricks for Efficient Text Classification. Facebook AI Research (2016)

4. Dragan, Ł., Wróblewska, A.: Contemporary computational science. In: Kulczycki, P., Kowalski, P.A., Łukasik, S. (eds.), p. 22. AGH-UST Press, Cracow (2018)
5. Kusner, M.J., Sun, Y., Kolkin, N.I., Weinberger, K.Q.: From Word Embeddings to Document Distances. In: International Conference on Machine Learning (ICML) (2015)
6. Mikolov, T., Chen, K., Corrado, G., Dean, J.: Efficient estimation of word representations in vector space. In: International Conference on Machine Learning (ICML) (2013)
7. http://morfologik.blogspot.com/
8. Pennington, J., Socher, R., Manning, C.D.: GloVe: Global Vectors for Word Representation. Computer Science Department, Stanford University, Stanford, CA 94305 (2014)
9. Kedzia, P., Czachor, G., Piasecki, M., Kocoń, J.: Vector representations of polish words (Word2Vec method)., Wrocław University of Technology (2016). https://clarin-pl.eu/dspace/handle/11321/327
10. https://github.com/RaRe-Technologies/gensim
11. https://github.com/MLnick/elasticsearch-vector-scoring
12. Przepiórkowski, A., Bańko, M., Górski, R., Lewandowska-Tomaszczyk, B. (eds.) Narodowy Korpus Jezyka Polskiego/National Corpus of Polish Language. Publisher PWN, Warsaw (2012). (in Polish). http://nkjp.pl

Computational Physics and Applied Mathematics

Generative Models for Fast Cluster Simulations in the TPC for the ALICE Experiment

Kamil Deja[1(✉)], Tomasz Trzciński[1], and Łukasz Graczykowski[2]
for the ALICE Collaboration

[1] Institute of Computer Science, Warsaw University of Technology, Warsaw, Poland
kdeja@stud.elka.pw.edu.pl, t.trzcinski@ii.pw.edu.pl
[2] Faculty of Physics, Warsaw University of Technology, Warsaw, Poland
lukasz.graczykowski@pw.edu.pl

Abstract. Simulating the possible detector response is a key component of every high-energy physics experiment. The methods used currently for this purpose provide high-fidelity results. However, this precision comes at a price of a high computational cost, which renders those methods infeasible to be used in other applications, e.g. data quality assurance. In this work, we present a proof-of-concept solution for generating the possible responses of detector clusters to particle collisions, using the real-life example of the Time Projection Chamber (TPC) in the ALICE experiment at CERN. We introduce this solution as a first step towards a semi-real-time anomaly detection tool. It's essential component is a generative model that allows to simulate synthetic data points that bear high similarity to the real data. Leveraging recent advancements in machine learning, we propose to use state-of-the-art generative models, namely Variational Autoencoders (VAE) and Generative Adversarial Networks (GAN), that prove their usefulness and efficiency in the context of computer vision and image processing. The main advantage offered by those methods is a significant speedup in the execution time, reaching up to the factor of 10^3 with respect to the GEANT3, a currently used cluster simulation tool. Nevertheless, this computational speedup comes at a price of a lower simulation quality. In this work we show quantitative and qualitative limitations of currently available generative models. We also propose several further steps that will allow to improve the accuracy of the models and lead to the deployment of anomaly detection mechanism based on generative models in a production environment of the TPC detector.

1 Introduction

High-energy physics (HEP) experiments, including those conducted at the Large Hadron Collider (LHC) [1], rely heavily on detailed simulations of the detector response in order to accurately compare recorded data with theoretical Monte Carlo models. This simulated data is used to understand the physics behind

P. Kulczycki et al. (Eds.): ITSRCP 2018, AISC 945, pp. 267–280, 2020.
https://doi.org/10.1007/978-3-030-18058-4_21

the real collisions registered in the experiment. The simulations are also used to adjust the experimental data for detector's inefficiencies and various conditions observed in the detector during the data taking procedure.

Traditional approaches to the simulation of the detector response use Monte Carlo methods to generate collisions. To properly simulate events, those methods need to synthesize propagation of every simulated particle through the detector's medium and model every interaction with the experimental apparatus using a transport package, usually GEANT3 [2], GEANT4 [3] or FLUKA [4]. This approach comes with disadvantages. First of all, those packages require specific implementations of the interactions of highly energetic particles with matter that occur inside the detector. In practice, the physics regulating those interactions is complex and often difficult to simulate. Secondly, the simulation methodology that relies on Monte Carlo methods is computationally expensive and scales linearly with the amount of simulated collisions.

To address the above shortcomings of a Monte Carlo-based simulation, we propose a fundamentally different approach to simulations in high-energy physics experiments that relies on generative models. The proof-of-concept solution, that implements this approach is based on the recently proposed models, such as Variational Autoencoders [5] and Generative Adversarial Networks [6]. The main idea behind using those models for simulations is to learn how to synthesize the situation of the detector after a collision from the data, instead of modeling every interaction of the particles.

Although this approach implies several limitations, especially linked to the inability of modeling the entire collision data, we believe that the proposed method can be successfully applied in other high-energy physics applications, *e.g.* anomaly detection. This is mainly thanks to the computation efficiency of the proposed approach. We will address the problem of modeling the entire collision data in the future work.

As a first step towards implementing our solution in production, we perform a set of experiments that aim at simulating clusters of the Time Projection Chamber (TPC) [7,8] detector of the ALICE (A Large Ion Collider Experiment) [9] experiment at the LHC. We present quantitative, qualitative and computational cost evaluation of our method for several generative models and their variations, including standard Variational Autoencoder, Deep Convolutional Generative Adversarial Networks [10] and Long Short Term Memory [11] based networks. We also show how to improve the results obtained from those models using a recently proposed training method called a Progressive GAN [12].

To our knowledge, this work represents the first attempt to use machine-learning generative models for clusters simulation in any high-energy physics experiment currently operating at CERN. The main contribution of this paper is a prototype of a proof-of-concept method for event simulation in the TPC detector that relies on generative models. We provide here an extensive overview of the recently proposed generative models that can be used in this context, as well as a detailed evaluation of their current performances when applied to cluster simulation. Evaluation results show a promising speedup over currently used Monte Carlo based simulation methods reaching up to 10^3, however, at the expense of simulation quality. We address this problem in the future work and show how to improve the generative models in the context of cluster simulation.

The remainder of this paper is organized as follows. In the next section, we describe related works. We then overview generative models that can be used in the proposed solutions and present the core idea behind its adaptation. Finally, we describe the dataset and present the results of our evaluation. In the last section, we conclude this work and outline the next steps.

2 Related Work

Currently used methods start from a collision simulation using a Monte Carlo generator, than different transport packages such as GEANT3 [2], GEANT4 [3] or FLUKA [4] are used to synthesize propagation of particles. At first, they provide space points where a given energetic particle has deposited some or its energy, as well as simulate the secondary particle showers in the detectors which leave their signal. Those space points have to be further translated to a simulated electronic signal from the detector sensors, as it would appear in a real collision. In the tracking detectors, those electronic signals, called *digits*, are then used to calculate *clusters*, a set of space-time digits from the adjacent sensitive elements of the detector that were presumably generated by the same particle crossing these elements.

Machine learning generative models, which we intend to use in the context of high-energy physics, have already proven their potential in numerous research areas related to text-to-image generation [13], conditional image generation for criminal identification purposes [14] or even drug design [15]. Their main research directions, however, are still restricted to applications related to images and videos. The task we aim to address in this paper, differs significantly from those applications. Simulating precise responses of a detector require high-fidelity of the obtained results, measured both qualitatively and quantitatively. This is not the case for artificially generated images or videos where subjective assessments are often the only way to evaluate generated output.

Despite those problems, there are several attempts to use Generative Adversarial Networks for simulating parts of the physics processes, like the very first one in the field: calorimeters response simulation [16]. In that work, the authors generate particle showers in electromagnetic calorimeters. Following the LAGAN processing [17], they adapt a DCGAN architecture to generate a number of 2D images, which they merge, to produce a final 3D Calorimeter response. Although our work draws inspiration from [16], we tackle a different problem related to detector's response simulation, while [16] focuses on generating particle showers.

3 Generative Models

In this section we overview two generative machine learning models that we use in our solution and their architectures. First, we describe a Variational Autoencoder

(VAE), and propose two different models of a VAE that fit our task. We use them as an evaluation baseline for the second method which is based on the Generative Adversarial Network (GAN) algorithm. We present those methods in the following sections. We also outline an adaptation of a recently proposed progressive GAN training algorithm [12] for cluster simulation, which according to our studies performs the best out of all evaluated approaches.

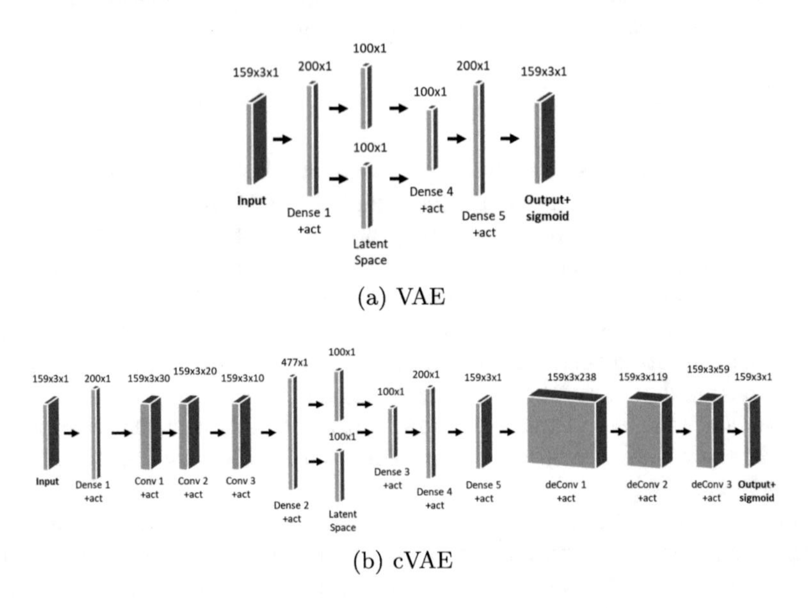

Fig. 1. Variational Autoencoders architectures evaluated in the context of detector's response simulation. Each block represent a network layer with it's size above.

3.1 Variational Autoencoder

Variational Autoencoder is a machine learning method that relies on an autoencoder idea proposed in [18]. It is implemented through an artificial neural network that can be used for dimensionality reduction and data compression. The simplest architecture of an autoencoder consists of an input layer with N neurons, where N is a dimensionality of the data, a hidden layer with $K < N$ neurons and the output layer with N neurons, exactly the same size as the input. The goal of an autoencoder is to re-generate the same output as received on the input, while internally reducing the latent representation to a lower number of dimensions.

The Variational Autoencoder (VAE) [5] extends this idea by focusing on the generalisation part of the encoder architecture. VAE re-generates new samples straight from the latent space by modifying the distribution of noise populated in the hidden layer. To ensure that it generates meaningful results, additional normalization constraint is added to the training loss function. It forces a neural network to ensure a normal distribution of the values retrieved as the output of the hidden layer neurons.

In this work, we focus on two architectures of Variational Autoencoders that can be applied in the context of detector's response simulation. First of them, dubbed simply VAE, is a fully connected network with one hidden layer. Second one, which we call cVAE, is a convolutional model with three convolutional and deconvolutional layers in encoder and decoder (generator) layers. Figure 1 shows both architectures.

3.2 Generative Adversarial Networks

A second family of generative machine learning methods discussed in this work is based on the Generative Adversarial Network [6]. The main concept behind this unsupervised generative model is to train two neural networks to play a *min-max game* between each other. The first one – a *generator* – is trained to generate new data points whose distribution resembles the distribution of real data. The second one – trained as a *discriminator* – aims at predicting whether a given example comes from a real or synthetic distribution. During the training process, both networks are updated iteratively, one at a time, which leads to a simultaneous optimization of loss functions of both generator and discriminator. The *min-max* loss function introduced in [6], encompassing both networks can be written in the following manner presented in the Eq. 1:

$$\underset{G}{min}\,\underset{D}{max}\,V(D,G) = \mathbb{E}_{x \sim p_{data}(x)}[\log D(x)] + \mathbb{E}_{z \sim p_z(z)}[\log(1 - D(G(z)))], \quad (1)$$

where $p_{data}(x)$ is the distribution of real data, x is a sample from p_{data}, $p(z)$ is a distribution of a noise vector z input into a generator and G and D represent a discriminator.

Architectures. The general framework of Generative Adversarial Networks is rather flexible in terms of the networks architecture used as a generator and discriminator. Hence, we propose several architectures that fit to the context of simulating detector's response in a high-energy physics experiment.

More precisely, we evaluate the following models:

- **Multi-layered Perceptron (MLP):** a baseline Generative Adversarial Network that consists of 4 fully connected layers used as a discriminator network with a generator built exactly in the reverse symmetrical way,
- **Recurrent GAN based on Long-Short Term Memory (LSTM):** a recurrent network that comprises LSTM units proposed in [11],
- **Deep Convolutional Generative Adversarial Network (DCGAN):** inspired by [13] multi layered network, with two dimensional convolutional/ de-convolutional layers, as shown in Fig. 2.

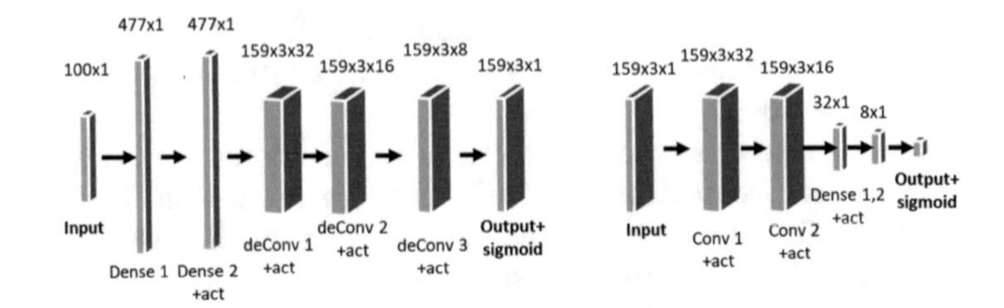

Fig. 2. Architecture of the DCGAN model. Each block represent a network layer with it's size above.

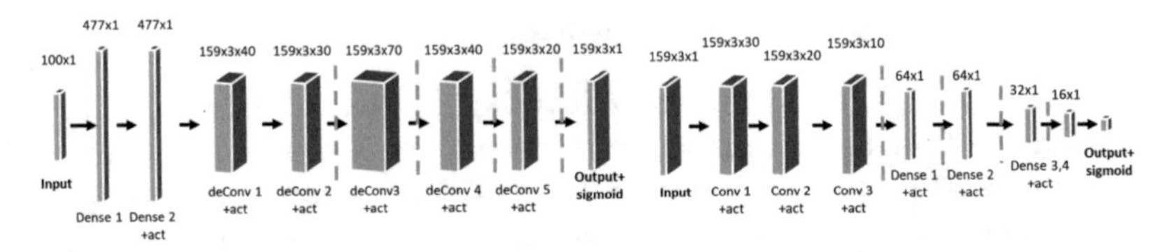

Fig. 3. Architecture for the Progressive GAN model. Dotted lines represent consecutive grow of the network with progressive steps. Each block represent a network layer with it's size above.

Progressive GAN. A typical approach to increase the performance of neural networks is to add more layers to the model, hence increasing the depth of a resulting neural network. This, however, comes at a price of more complex training procedure. To evaluate how much performance gain we can obtain by growing our network, we rely on a recently proposed progressive GANs training method introduced in [12].

As presented in Fig. 2, the DCGAN model we evaluate is rather shallow, *i.e.* it consists of only 3 convolutional/deconvolutional layers. We therefore employ the progressive training method for our DCGAN and grow our network by gradually increasing the number of stacked layers. This way, we are able to enhance the resolution of the resulting simulation by gradually increasing the number of possible locations of simulated cluster responses. This is of our particular interest, since in the context of high-energy physics the resolution of the simulated cluster's response has to be relatively high. Our DCGAN is therefore grown in 5 progressive steps, ending with 6 convolutional/deconvolutional layers. Figure 3 shows the final architecture obtained by the training method.

Implementation Details. To train all networks, we use dropout [19], Leaky ReLu activation [20], and batch normalization [21]. As our loss function, we take a binary cross-entropy loss. We optimize our loss function with ADAM, a stochastic optimization method [22]. We initialize network weights with the glorot algorithm [23]. The neural networks are implemented in Python using Keras library [24] with TensorFlow backend [25]. Our implementations are based on those enlisted in the large scale study [26].

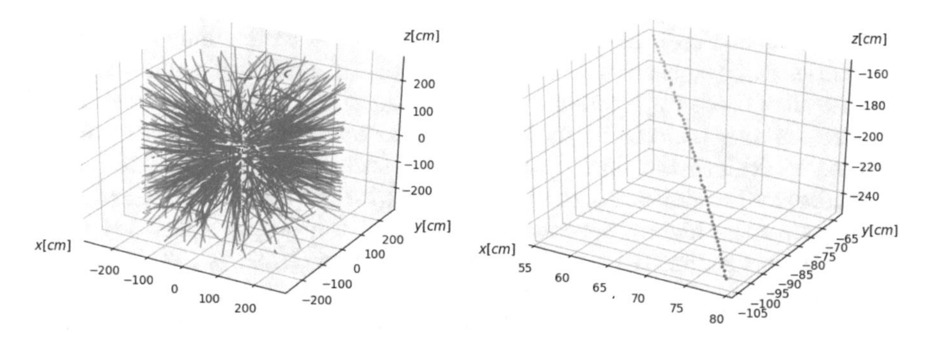

Fig. 4. Input data examples: **(left)** a full simulation of space points occurring after particles collision in the TPC detector of the ALICE Experiment; **(right)** clusters generated by a single particle.

4 Dataset

To evaluate our generative models, we use a dataset of 3D trajectories of particles after collision generated with a Monte Carlo simulation method of PYTHIA 6.4 [27] Perugia-0 [28] model. Our dataset therefore contains a sample of proton–proton collisions at the center of mass energy of $\sqrt{s} = 7$ TeV. After the simulated collision, the generated particles were transported, using GEANT3 [2] package, through the detector medium so that the full simulation of the detector response as well as the track reconstruction were performed. The trajectories correspond to the real traces observed in the TPC detector of the ALICE Experiment at LHC, CERN and the experimental conditions corresponded to 2010 data-taking period of the experiment. Figure 4 shows sample visualisation of the particle trajectories from the collected dataset.

Since the size of a dataset generated this way results in petabytes of data, we randomly sample it to contain around 100 events corresponding to approximately 60.000 data samples.

Each data sample contains a series of 3D coordinates (x, y, z) corresponding to the trajectory of a given particle after the collision. The minimal and maximum values of the coordinates depend on the detector size, which is a cylinder-shaped structure of approximately 5 m × 5 m × 5 m. Since the collision does not have to occurs in the central point of the detector, the minimal distance from the [0, 0, 0] point, where the trajectories are collected, is 848 mm. The precision of the recorded coordinates is limited to the resolution of the TPC read-out pads, which is between 0.8 to 1.2 mm, depending on the size of the pad [7]. For normalization purposes, the input data coordinates are scaled to fit the $[0, 1]$ interval in each dimension. We also apply other normalization procedures, such as zero-padding the samples for particle trajectories consisting less than 159 points (maximal value in our dataset). Although additional characteristics of the particles can be registered, *e.g.* its energy and speed, for the purpose of the evaluation presented in this work we restrict our data samples to 3D coordinate values.

As shown in Fig. 4, after the collision occurs, up to few hundreds particle trajectories can be observed inside the detector. Although we could attempt to simulate all of those trajectories at once using generative models, we postulate to generate separate trajectories for individual particles first and then merge them together to achieve the final goal. This approach allows us to better control the studies and evaluate the results iteratively, as more trajectories are added. Furthermore, the highest reported resolution of the output generated by the state-of-the-art generative models, such as VAEs and GANs, is 1024×1024 [12]. This resolution is relatively low, when compared to the results obtained in the high-energy physics experiments. To circumvent this limitation of the generative models, we transform our simulation problem into a so-called transactional form and simulate individual points with their (x, y, z) coordinates, which we can then link together to form a full trajectory.

5 Results

In this section, we evaluate the proposed methods based on generative models in terms of both quality of the generated results and the computational cost. As our baseline model, we use the Monte Carlo-based method currently used in the ALICE Experiment to generate the full simulation of the TPC response of the generated particle showers, propagated through the detectors medium by the GEANT3 package. We use the most recent implementation provided as part of the O2 software [8, 29].

To evaluate the quality of the trajectories generated by our method, we refer to the physical property of the particle tracks observed after the collision. When particles move through the detector medium, they form a track whose shape can be defined as a helix of particular parameters. We leverage this phenomenon and calculate for each method an evaluation metric that measures the distance of generated trajectory from a theoretically ideal shape of helix, approximated with an arc. We report this metric as a mean squared error (MSE) between the ideal helix shape and the simulated one and average it across all simulated particles. We give the result both in the coordinate unit system and metric system, scaled using detector's dimensions. Although this approach does not take into account the physical properties of the particles resulting in different parameters of the helix created by each particle, it allows us to compare several unsupervised simulation methods that are not conditioned on particle characteristics. The proposed metric validates the quality of the shape generated by the evaluated simulation methods and serves as a first approximation of a method performance. We intend to extend the proposed metric to handle more complex evaluation procedure in future work.

We also evaluate the computation cost of generating the results for various methods. To that end, we execute the code 10 times on a standalone machine with a Intel Core i7-6850K (3.60 GHz) CPU (using single core, no GPU acceleration) and record the average execution time. We then normalize it with respect to the baseline method of simulation, namely GEANT3, which is currently used to simulate the detector's response.

Table 1. Quality of the Generative models, and their performance comparing to the GEANT3 based simulation solution.

Method	MSE	MSE (mm)	speedup
GEANT3 (*current simulation*)	0.00017	0.085	1
Random (*estimated*)	0.33000	166.155	N/A
GAN-MLP	0.11077	55.385	10^4
GAN-LSTM	0.10879	54.395	10^4
VAE	0.07483	37.415	10^4
DCGAN	0.05236	26.18	10^2
cVAE	0.02666	13.33	10
proGAN	**0.00176**	**0.88**	**30**

The results of our evaluation are presented in Table 1. For the baseline simulation method GEANT3, the average mean squared error (MSE) is equal to 0.00017 or 0.085 mm. When using a random generator to generate a set of 3D points, the estimated expected MSE is 0.33. Compared to those baselines, the best performing progressive DCGAN model (proGAN) achieves a satisfactory MSE of 0.00176. This result corresponds to a sub-centimeter accuracy of our method, while providing over 30-fold speedup with respect to the baseline method of simulation, a Monte Carlo-based GEANT3. We believe that this result is a promising sign and should inspire further research. Finally, one needs to remember that the goal of our simulation is to provide a benchmark set of positive results that will then be used to assess the health of a detector through a complimentary anomaly detection system. Therefore, the accuracy achieved by proGAN is a good step in this direction.

Regarding the other evaluated models, the Variational Autoencoder (VAE) achieves the MSE value of 0.075, while providing a speedup of over four orders of magnitude. The sample results generated using this approach are presented in Fig. 5(a). In fact, this is the least complex modal of all evaluated in this paper and therefore both qualitative and quantitative results it achieves are relatively good. Furthermore, we are able to improve the performance of the VAE by substituting the fully connected layers with the convolutional ones, as it is done in the convolutional Variational Autoencoder (cVAE). With this modification, we can reduce the MSE error to approximately 0.0266. However, because of a significant increase in model complexity, it is also considerably slower than the VAE. The observed speedup with respect to the GEANT3 solution reaches only one order of magnitude. As shown in Fig. 5(b), the qualitative results generated by the convolutional VAE appear to be more random, although their resolution is higher than in the case of VAE.

As far as the models based on the Generative Adversarial Networks are concerned, the results obtained by the Multilayer Perceptron (MLP) method indicate that this approach, as well as the recurrent network method based on the

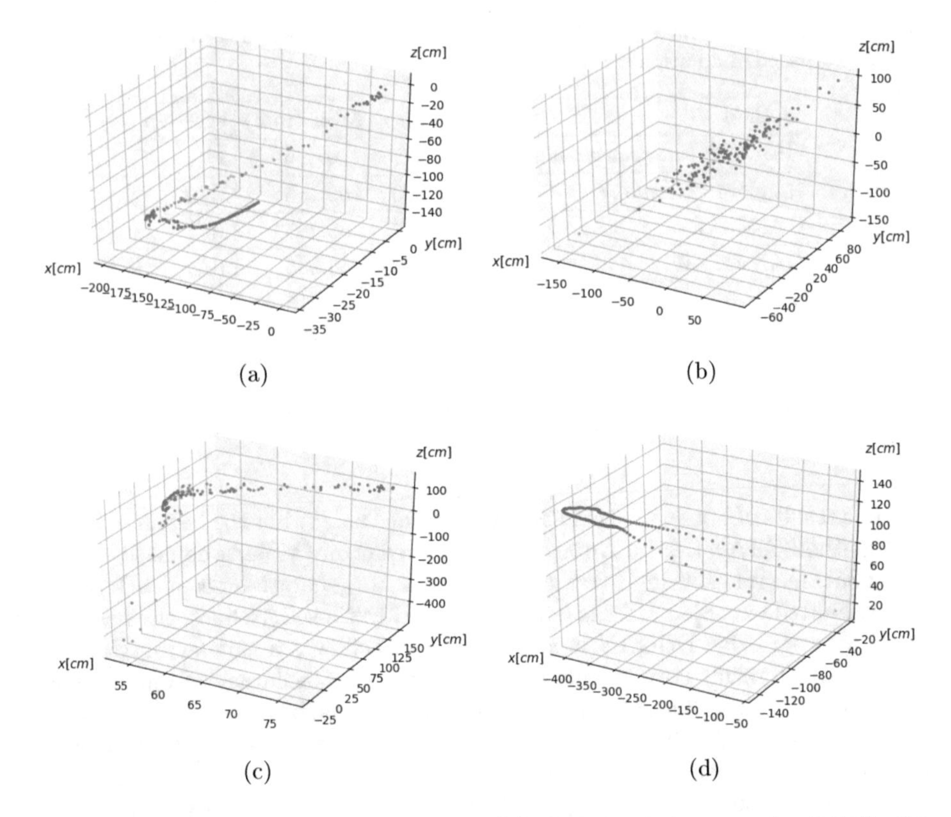

Fig. 5. Exemplar results generated by the **(a)** Variational Autoencoder (VAE), **(b)** convolutional Variational Autoencoder (cVAE), **(c)** the Deep Convolutional GAN (DCGAN) and **(d)** the progressive DCGAN (proGAN).

LSTM units, do not achieve sufficient accuracy. Although their speedup over the baseline simulation method reaches four orders of magnitude, the quantitative and qualitative results are only slightly better than those obtained by a random generator.

However, when analysing the results obtained by the Deep Convolutional GAN (DCGAN) model, we can see that extending neural network architectures with convolutional layers significantly reduces the MSE value, while still providing a reasonable speedup over the baseline that reaches two orders of magnitude. We explain that performance gain through the fact that convolutional layers can better simulate the working conditions of 3D simulation, as well as allow for deeper and more powerful models. The MSE value that is achieved by the evaluated DCGAN architecture is approximately 0.052 and is comparable to the results produced with VAE and cVAE architectures. The qualitative results can be seen in Fig. 5(c).

Extending the DCGAN model using a progressive training method (proGAN) allows us to improve the quality of the results and reach an unprecedented MSE value of 0.00176. The qualitative results generated by the proGAN also indicate that this method is a promising path of research, especially when additional conditioning on the particle type is added to the model. Although the computational complexity of the proGAN model is higher than the competing approaches, it still offers a massive 30-fold speedup over the baseline GEANT3 simulation method.

To emphasize the true potential of the simulation methods based on the generative models, we computed the execution times for different methods while increasing the number of simulated clusters, i.e. model outputs. Figure 6 shows the results of this experiment. Although the computational cost of all the methods increases linearly with the number of simulated cluster responses, the improvement achieved by the generative models is massive and grows with increasing number of clusters. It is worth mentioning that the presented results are obtained for a single-core CPU computation, while additional hardware-based speedup of another order of magnitude was observed when using the GPU-based implementation for the neural network methods. Although Monte Carlo simulations can also benefit from such an acceleration, its iterative character does not allow the same improvement as in the case of deep neural networks.

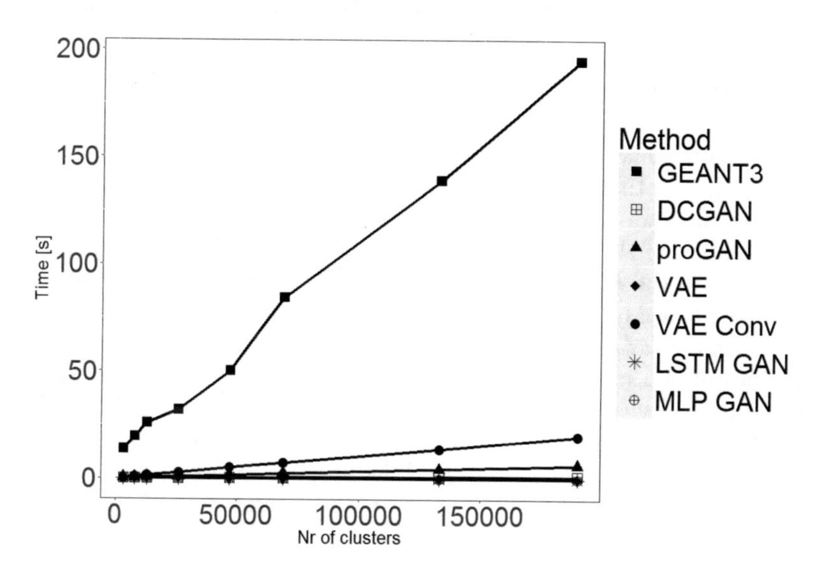

Fig. 6. Comparison of performance between GEANT 3 simulation and machine learning generative models.

6 Future Work

In this work, we give an overview and preliminary evaluation results of possible machine learning generative methods that can be used for cluster simulation.

We intend to investigate these approaches further and implement a fully working solution which can be deployed in production, *i.e.* as a part of the ALICE Experiment at LHC.

To that end, we plan to extend the described models by conditioning our model on the initial characteristics of the collision. More precisely, we envision extending the input to the model with initial particles momenta generated by PYTHIA or other currently used simulation tool. This shall reduce the variance of the model and increase the quality of the observed results.

On top of replacing the complex methods for clusters simulation that are currently in use, we want to employ the generative models as a part of the quality assurance tool. We draw our inspiration from the work of [30] where the Generative Adversarial Networks are used to discover anomalies in medical imaging data. We plan to design a similar solution for the high-energy physics experiments to discover faulty outputs of the particle detectors in a semi-real-time manner. To that end, we plan to use the efficient and highly effective generative models to simulate multiple healthy responses of the detector and compare the registered parameters of the detector with the simulated ones. If the synthetically generated responses do not agree with the real conditions, the model shall label the condition as an anomaly and alert the appropriate services.

7 Conclusion

In this work, we demonstrated the potential of machine learning generative methods namely Generative Adversarial Networks and Variational Autoencoders, for cluster simulation in the high-energy physics experiments. We proved the possible application of those methods using the example dataset generated for the TPC detector in the ALICE Experiment at LHC. We evaluated several architectures that are based on generative models, such as Variational Autoencoders and Generative Adversarial Networks, and compared their results in terms of quality and computational cost. We also adapted a progressive training technique to enhance the quality of generated samples. Although the quality of the best performing method is not yet comparable to the quality of currently used simulation methods, the computational speedup of the proposed method is unprecedented and reaches up to 10^4 when compared to the currently employed GEANT 3 simulation technique.

Acknowledgements. The authors acknowledge the support from the Polish National Science Centre grant no. UMO-2016/21/D/ST6/01946. The GPUs used in this work were funded by the grant of the Dean of the Faculty of Electronics and Information Technology at Warsaw University of Technology (project II/2017/GD/1).

The preliminary version of this paper was presented at the 3rd Conference on Information Technology, Systems Research and Computational Physics, 2–5 July 2018, Cracow, Poland [31].

References

1. Evans, L., Bryant, P.: LHC machine. JINST **3**, S08001 (2008)
2. Brun, R., Bruyant, F., Carminati, F., Giani, S., Maire, M., McPherson, A., Patrick, G., Urban, L.: GEANT Detector Description and Simulation Tool. CERN-W5013, CERN-W-5013, W5013, W-5013 (1994)
3. Agostinelli, S., et al.: GEANT4: a simulation toolkit. Nucl. Instrum. Meth. **A506**, 250–303 (2003)
4. Ferrari, A., Sala, P.R., Fasso, A., Ranft, J.: FLUKA: a multi-particle transport code. CERN-2005-010, SLAC-R-773, INFN-TC-05-11 (2005)
5. Kingma, D.P., Welling, M.: Auto-encoding variational bayes, CoRR, vol. abs/1312.6114 (2013)
6. Goodfellow, I., Pouget-Abadie, J., Mirza, M., Xu, B., Warde-Farley, D., Ozair, S., Courville, A., Bengio, Y.: Generative adversarial nets. In: NIPS (2014)
7. Dellacasa, G., et al.: ALICE: Technical Design Report of the Time Projection Chamber. CERN-OPEN-2000-183, CERN-LHCC-2000-001 (2000)
8. Abelev, B., et al.: Upgrade of the ALICE experiment: letter of intent. J. Phys. **G41**, 087001 (2014)
9. Aamodt, K., et al.: The ALICE experiment at the CERN LHC. JINST **3**, S08002 (2008)
10. Radford, A., Metz, L., Chintala, S.: Unsupervised representation learning with deep convolutional generative adversarial networks, CoRR, vol. abs/1511.06434 (2015)
11. Hochreiter, S., Schmidhuber, J.: Long short-term memory. Neural Comput. **9**(8), 1735–1780 (1997)
12. Karras, T., Aila, T., Laine, S., Lehtinen, J.: Progressive growing of gans for improved quality, stability, and variation, CoRR, vol. abs/1710.10196 (2017)
13. Reed, S., Akata, Z., Yan, X., Logeswaran, L., Schiele, B., Lee, H.: Generative adversarial text to image synthesis, CoRR, vol. abs/1605.05396 (2016)
14. Yan, X., Yang, J., Sohn, K., Lee, H.: Attribute2image: conditional image generation from visual attributes. In: ECCV (2016)
15. Kadurin, A., Aliper, A., Kazennov, A., Mamoshina, P., Vanhaelen, Q., Khrabrov, K., Zhavoronkov, A.: The cornucopia of meaningful leads: applying deep adversarial autoencoders for new molecule development in oncology. Oncotarget **8**(7), 10883 (2017)
16. Paganini, M., de Oliveira, L., Nachman, B.: CaloGAN: simulating 3d high energy particle showers in multi-layer electromagnetic calorimeters with generative adversarial networks, CoRR, vol. abs/1705.02355 (2017)
17. de Oliveira, L., Paganini, M., Nachman, B.: Learning particle physics by example: location-aware generative adversarial networks for physics synthesis. Comput. Softw. Big Sci. **1**, 4 (2017)
18. Hinton, G.E., Salakhutdinov, R.R.: Reducing the dimensionality of data with neural networks. Science **313**(5786), 504–507 (2006)
19. Srivastava, N., Hinton, G., Krizhevsky, A., Sutskever, I., Salakhutdinov, R.: Dropout: a simple way to prevent neural networks from overfitting. JMLR **15**(1), 1929–1958 (2014)
20. Glorot, X., Bordes, A., Bengio, Y.: Deep sparse rectifier neural networks. In: International Conference on Artificial Intelligence and Statistics, pp. 315–323 (2011)
21. Ioffe, S., Szegedy, C.: Batch normalization: accelerating deep network training by reducing internal covariate shift. In: ICML (2015)

22. Kingma, D.P., Ba, J.: Adam: a method for stochastic optimization, CoRR, vol. abs/1412.6980 (2014)
23. Glorot, X., Bengio, Y.: Understanding the difficulty of training deep feedforward neural networks. In: International Conference on Artificial Intelligence and Statistics, pp. 249–256 (2010)
24. Chollet, F., et al.: Keras (2015). https://github.com/fchollet/keras
25. Abadi, M., et al.: Tensorflow: a system for large-scale machine learning. In: 12th USENIX Symposium on Operating Systems Design and Implementation (OSDI 2016), pp. 265–283 (2016)
26. Lucic, M., Kurach, K., Michalski, M., Gelly, S., Bousquet, O.: Are gans created equal? A large-scale study, CoRR, vol. abs/1711.10337 (2017)
27. Sjostrand, T., Mrenna, S., Skands, P.Z.: PYTHIA 6.4 physics and manual. JHEP **05**, 026 (2006)
28. Skands, P.Z.: Tuning Monte Carlo generators: the perugia tunes. Phys. Rev. D **82**, 074018 (2010)
29. Suaide, A.A.D.P., Prado, C.A.G., Alt, T., Aphecetche, L., Agrawal, N., Avasthi, A., Bach, M., Bala, Barnafoldi, R., Bhasin, A., et al.: O2: a novel combined online and offline computing system for the alice experiment after 2018. J. Phys. Conf. Ser. **513**, 012037 (2014). IOP Publishing
30. Schlegl, T., Seeböck, P., Waldstein, S.M., Schmidt-Erfurth, U., Langs, G.: Unsupervised anomaly detection with generative adversarial networks to guide marker discovery. In: International Conference on Information Processing in Medical Imaging, pp. 146–157. Springer (2017)
31. Deja, K., Trzciński, T., Graczykowski, Ł.: Generative models for fast cluster simulations in the TPC for the ALICE experiment. In: Kulczycki, P., Kowalski, P.A., Łukasik, S. (eds.) Contemporary Computational Science, p. 2. AGH-UST Press Cracow (2018)

2D-Raman Correlation Spectroscopy Recognizes the Interaction at the Carbon Coating and Albumin Interface

Anna Kołodziej[1], Aleksandra Wesełucha-Birczyńska[1](✉), Paulina Moskal[1], Ewa Stodolak-Zych[2], Maria Dużyja[3], Elżbieta Długoń[2], Julia Sacharz[1], and Marta Błażewicz[2]

[1] Faculty of Chemistry, Jagiellonian University, Kraków, Poland
birczyns@chemia.uj.edu.pl
[2] Faculty of Materials Science and Ceramics, AGH - University of Science and Technology, Kraków, Poland
[3] Technolutions, Łowicz, Poland

Abstract. Carbon materials open new perspectives in biomedical research, due to their inert nature and interesting properties. For biomaterials the essential attribute is their biocompatibility, which refers to the interaction with host cells and body fluids, respectively. The aim of our work was to analyze two types of carbon layers differing primarily in topography, and modeling their interactions with blood plasma proteins. The first coating was a layer formed of pyrolytic carbon C (CVD) and the second was constructed of multi-walled carbon nanotubes obtained by electrophoretic deposition (EPD), both set on a Ti support. The results of the performed complex studies of the two types of model carbon layers exhibit significant dissimilarities regarding their interaction with chosen blood proteins, and the difference is related to the origin of a protein: whether it is animal or human. Wettability data, nano scratch tests were not sufficient to explain the material properties. In contrast, Raman microspectroscopy thoroughly decodes the phenomena occurring at the carbon structures in contact with the selected blood proteins interface. The 2D correlation method selects the most intense interaction and points out the different mechanism of interactions of proteins with the nanocarbon surfaces and differentiation due to the nature of the protein and its source: animal or human. The 2D-correlation of the Raman spectra of the MWCNT layer + HSA interphase confirms an increase in albumin β-conformation. The presented results explain the unique properties of the C-layers (CVD) in contact with human albumin.

Keywords: Multi-walled carbon nanotubes · Pyrolytic carbon · Carbon coatings · Raman microspectroscopy · Plasma blood protein

1 Introduction

With the growing needs of modern societies, as well as their ability to cope with, nanotechnology methods represent a unique position [2]. The presented research topic is devoted to certain aspects in the field of materials for the needs of nanomedicine,

P. Kulczycki et al. (Eds.): ITSRCP 2018, AISC 945, pp. 281–295, 2020.
https://doi.org/10.1007/978-3-030-18058-4_22

which is a rapidly growing area. Size in nano-scale results in nano-additives direct involvement in the processes and even biological structures [2–8]. Innovative biomaterials have contributed to the field of controlled drug delivery applications [9–11], cardiovascular diseases [12–14] and orthopedics [4, 15], while novel materials with carbon nanotubes coatings are desirable for application as sensors and neural electrodes or as a platform for Central Nervous System diseases [16–18]. The nano-sized dimension causes that by selecting nanoparticle that reinforcing material one can affect structural, mechanical and electronic properties of nanocomposite material, which induce as a consequence its characteristics in the interaction with the biological environment [19]. In this way the nanoparticle by modern nanotechnology may be adjusted to modify, as intended, the polymer matrix [20–22], or to alter the surface of the synthetic material that comes into contact with the tissue of the living organism [12, 23–27].

Carbon materials are attractive due to their unique properties and large variety of carbon structures, so they are often used as a modifying particle [28]. Two types of carbon structures have been selected to this research: pyrolytic carbon and multiwall carbon nanotubes (MWCNTs). Pyrolytic carbon is classified as a group of turbostratic carbons, which have a similar structure to graphite. Graphite consists of carbon atoms that are covalently bonded in hexagonal arrays. These arrays are stacked and held together by weak interlayer binding. Pyrolytic carbon and other turbostratic carbons differ in that the neighboring graphitic layers are disordered, resulting in wrinkles or distortions within layers. This provides pyrolytic carbon improved durability compared to graphite [29]. Although this type of structure has been the most popular material available for artificial mechanical heart valves for about half a century there are some requirements to consider. MWCNTs are fibrous nanostructures [30], although recently used ones are synthesized [28, 30]. They can be also found naturally or as a byproduct of industrial processes [31, 32]. The Ijima discovery aroused a lot of interest for CNTs because of their features: small size and mass, high strength, and high electrical conductivity [33].

For biomaterials the most important attribute is their biofunctionality and biocompatibility, which refers to mechanical characteristic and to the interaction with host cells and fluids, respectively [34]. The biological response to the synthetic material is determined at the interface of the surface of a biomaterial and cell. One must be aware that the interface is complex, as biomolecules and synthetic material are composed of three-dimensional entities [35]. Therefore, the number of surface parameters such as: surface morphology, roughness (micro and nano), wettability, and the degree of crystallinity are parameters that result from the surface properties and significantly affect the biological properties of the material [26, 27, 36–38].

Studies on protein adsorption are regarded as method used to assess biomaterials, not only that in blood-contacting applications. It is considered that the adsorption of selected proteins, may indicate the specific biological properties of the examined nanomaterials. There is usually analyzed the interaction of the surface with albumin, the blood plasma protein of the next highest abundance after hemoglobin, which is involved in transport and storage and regarded as an inhibitor of blood clotting [39].

Key blood proteins are generally selected to model carbon layers interaction with a liquid tissue. The commonly used are proteins of animal origin, that represent adequately human proteins. However, our earlier studies have shown that animal equivalents differently affect the surface of a carbon material than human proteins [26, 27]. Thus, modeling interactions with proteins of animal origin can be questioned in some cases. As target proteins, albumin from chicken egg white (Alb), bovine serum albumin (BSA) and human serum albumin (HSA), were used to check their influence on the surface.

We employed conventional methods of material engineering such as wetting angle measurement and electron microscopic techniques (SEM) to obtain the important characteristics of the materials surface. To check the mechanical properties of the layers on the nano-level the Nano Scratch Tester was applied to study the adhesion and scratch resistance of representative coatings incubated with selected proteins. Then, materials were tested using Raman spectroscopy. The Raman spectroscopy was chosen, as an important spectroscopic technique to test short-range ordering. It has been used to describe the structure of two reference carbon layers and then adsorption on these layers of selected plasma proteins. Finally, a 2D Raman correlation spectroscopy was applied to the collected Raman spectra allowed for the resolution of the phenomena occurring in the carbon layer surface in contact with the blood proteins. In this mathematical analysis as an external perturbation the spatial position in which the spectrum was measured was taken into account.

The preliminary version of this paper was presented at the 3rd Conference on Information Technology, Systems Research and Computational Physics, 2–5 July 2018, Cracow, Poland [1].

2 Materials and Methods

2.1 Preparation of Carbon Layers on Titanium

Titanium plates (Grade 2 according to ASTM B265) in the form of discs with a diameter of 12 mm and a thickness of 0.5 mm were chosen as suitable substrates for the deposition of carbon layers, which were: pyrolytic carbon (C (CVD)) and carbon nanotubes (MWCNTs) layer.

The first, pyrolitic C-layer was obtained in the Plasma-enhanced chemical vapor deposition process (PECVD; Elettrorava, Italy) on titanium Ti surfaces. All depositions were performed by the RF PECVD method, wherein the plasma was generated by radio frequency waves of 13.56 MHz and of power 60 W. The formation of the layers was preceded by ion-etching in argon plasma during 10 min. In the case of the titanium substrate this stage caused surface cleaning and the elimination of the TiOx surface layer. The C-layer layer was deposited at room temperature (25 °C) during 30 min from a methane (gas flow 10 cm^3/min) and argon (gas flow 75 cm^3/min) mixture. Argon was used as inert gas. During the deposition process, the chamber pressure was kept constant (53 Pa).

The second layer of MWCNTs (#1213NMGS, Nanostructured & Amorphous Materials, Inc., USA; outside diameter 10–30 nm; core diameter 5–10 nm, length of 1–2 µm and purity >95%.) was produced in the electrophoretic deposition (EPD).

Full details on the oxidizing procedure, preparation of the suspension, titanium handling and EPD set-up are discussed in our previous studies [25–27].

The albumin from chicken egg white (Alb), bovine serum albumin (BSA), human serum albumin (HSA), were purchased from Sigma-Aldrich (Poland). Both carbon coatings were incubated in 1% albumin solution for 15 min.

2.2 SEM Images Analysis

The morphology and chemical composition of the obtained coatings were examined using a scanning electron microscope Nova Nanosem (FEI) equipped with an adapter for EDS X-ray microanalysis (EDAX). The system was operated with 10–15 kV accelerating voltage, high vacuum mode.

2.3 Wettability Measurements

The contact angle measurements were taken using a direct method (DSA 10 Kruss goniometer). The tests were performed at room temperature applying the sitting drop technique (the drop of deionized water of 0.15–0.25 μl in volume). Wettability [θ] is the contact angle for a liquid droplet on a solid surface, in the point where three phases meet: solid, liquid and gas. The measurements were taken five times in order to establish variability and the standard deviation (SD) that is equal to $\pm 2.5\%$. All experiments were performed for reference and both tested surfaces and also for the selected blood proteins conditioned with tested surfaces. All conducted tests were under ambient conditions.

2.4 The Nano Scratch Test

All tests were carried on a platform with the NST (Nano Scratch Tester) head made by CSM Instruments SA (currently Anton Paar TriTec) (Switzerland). The test parameters were set as follows: the load Fn increased linearly from 0, 1 to 5 mN on the 3 mm distance, the speed of loading was set for 10 mm/min and the Rockwell certified indenter radius was equal to 2 μm.

Two nanocomposite layers were analyzed: the C-layer grown in the CVD process and the MWCNTs deposited in the Ti support, after incubation with white chicken egg albumin and human serum albumin (HSA).

2.5 Raman Microspectroscopy

A Renishaw inVia spectrometer, connected to a Leica microscope, was used for the measurements of the Raman spectra. The beam from a 514.5 nm Ar^+ ion Modu Laser was focused on the samples by 100x magnifying, a high numerical aperture (NA = 0.9) top-class Leica objective for standard applications. Laser power was kept low, c.a. 1–3 mW at the sample, to ensure minimum disturbances to the samples.

2.6 2D Raman Correlation

The generalized 2D correlation analysis based on the Noda method [40–42] was performed using Raman spectra as an input data for generating the correlation maps. The spatial position was regarded as an external perturbation [43]. The five points, morphologically similar were measured for each sample, and they were regarded as dynamical spectra in the 2D correlation. 2Dshige, v.1.3 software was employed [44].

3 Results and Discussion

3.1 SEM Images of Carbon Layers

SEM investigations indicate dissimilarities in the topography of both materials. Two types of carbon structures differing primarily in the topography. The first coating was a layer formed of pyrolytic carbon (CVD) (Fig. 1A) and the second was constructed of multi-walled carbon nanotubes obtained by electrophoretic deposition (EPD) (Fig. 1B), both set on a Ti support.

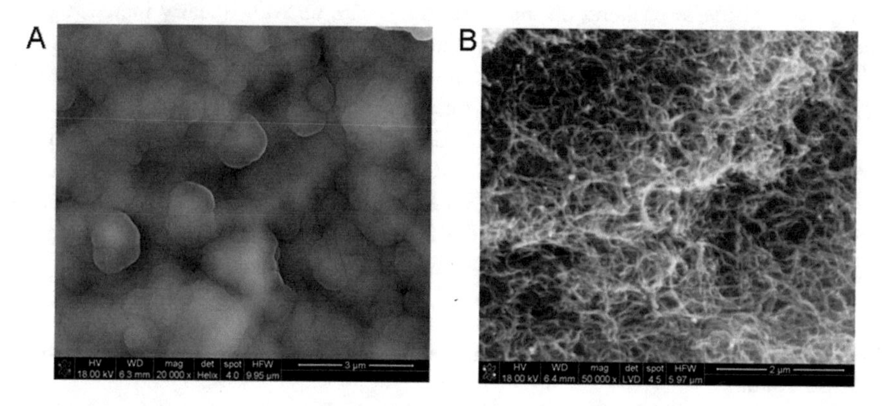

Fig. 1. SEM image of: (A) C (CVD) layer 20000x magnification, (B) MWCNTs (EPD) layer 50000x magnification.

3.2 Wettability of Carbon Layers

The surfaces wettability was analyzed by the static sessile drop method. The contact angles of water droplets on the top face of the reference C (CVD) layer are $82.2 \pm 2.8°$, respectively (Fig. 2A). The C (CVD) layer does not change the very nature of the surface with respect to the titanium substrate [26]. The difference in the contact angles for C (CVD) incubated with Alb ($78.4 \pm 2.3°$) and HSA ($84.5 \pm 1.9°$) with comparison to a reference the C (CVD) layer is not so significant, it fits within the limit of 5% (Fig. 2A). However, the C (CVD) incubation in BSA leads to the creation of a film with hydrophobic characteristics and contact angle of $112.0 \pm 6.7°$.

The contact angles on the top face of the MWCNTs nanocomposite layer is $25.0 \pm 0.9°$ (Fig. 2B). This value implies a hydrophilic character of the surface of the

MWCNTs coating. The Alb and HSA form a layer having a wettability 62.6 ± 1.3° and 57.3 ± 0.3°, respectively. The BSA conditioned MWCNTs nanocarbon layer reaches the highest contact angle of 77.1 ± 1.3°.

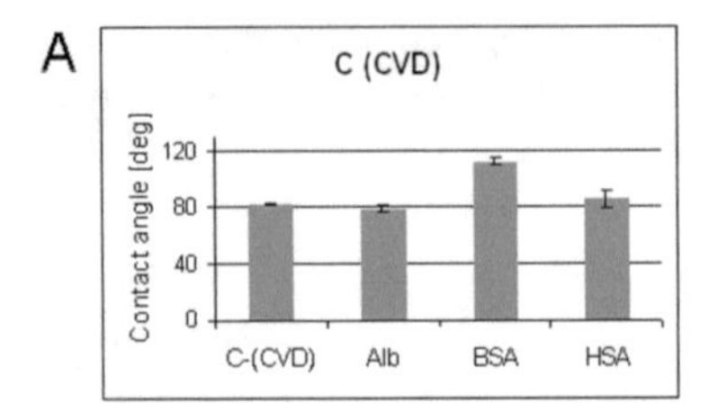

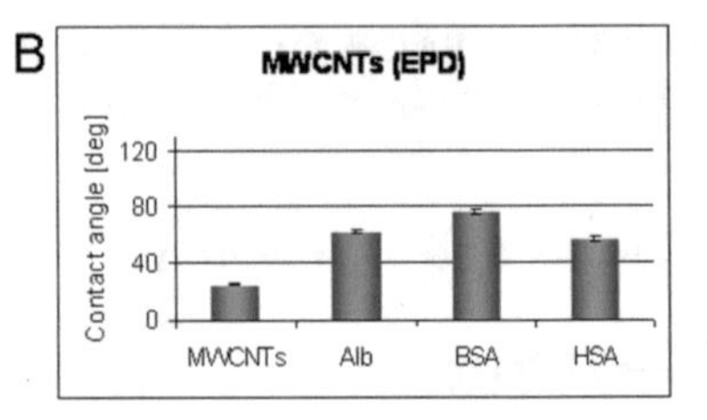

Fig. 2. Wettability of studied reference carbon samples (A) C (CVD) layer; (B) MWCNTs (EPD) layer, and after incubation with selected proteins.

3.3 Nano Scratch Test

The test consists of three phases: first phase - collecting profile of the sample (Pre Scan), second phase pressing the indenter in the sample with linearly increasing load and moving the sample at a defined length to scratch off the coating (test phase) and the third phase collecting profile inside scratch (Post Scan). After the performed scratch the panoramic photograph is taken and the obtained features are analyzed.

HSA has the best adhesion to the C (CVD) layer (Table 1). This coating did not break during the performed test and the Ti substrate is not visible. It can be observed that the C (CVD) layer with a thin HSA sheet looks like a "pressed and smeared" during the test and therefore presents the best adhesion to the Ti base. In addition, violation of coating, chipping or cracking up to the 5 mN load is not observed.

Table 1. Summary of the Scratch Test parameters (linear scrach; load range 0.1–5 mN; loading rate 10 mN/min).

Sample	Critical load [mN]
C (CVD) + Alb	coating failed at 2.16 ± 0.20
C (CVD) + HSA	coating is not destroyed
MWCNTs + Alb	coating failed at 1.63 ± 0.10
MWCNTs + HSA	coating failed at 2.17 ± 0.16

For the MWCNTs nanolayer, the HSA film is de-laminated with a load of about 2.0–2.5 mN. It is observed that the entire carbon nanolayer along with the thin film of protein is detached from the Ti substrate.

Adhesion of the Alb to the carbon C (CVD) layer is better, but the difference is small, the surface is scratched with a load of about 2.16 mN. The surface of the Alb on the MWCNTs nanolayer is scratched off easily, the visible surface of the Ti substrate is noticeable from the start of the test which means that the critical load for this layer is below 1.63 mN.

3.4 Raman Microspectroscopy Characterization of Nanocarbon Layers and Nanocarbon Interaction with Selected Blood Proteins

The Raman band positions and assignments of the reference carbon layers excited by the 514.5 nm laser line are collected in Table 2. The first, the C (CVD) layer is formed of pyrolytic carbon anisotropic materials so only the main G- and D-bands are seen, what confirms the graphite-type arrangement in this coating (Fig. 3A). For the second, MWCNTs nano-layer, in addition to G- and D- also characteristic 2D, D′ and D + D′ bands are visible confirming more complex organization in this coating (Fig. 3B).

Table 2. Observed Raman bands [cm^{-1}] and their assignments for carbon coatings excited with 514.5 nm laser line.

Sample	Peak position [cm^{-1}]	Assignment [21–29, 45–48]
C (CVD)	1349	D-mode; breathing mode that requires a defect for its activation
MWCNTs/(EPD)	1364	
C (CVD)	1539	G-mode; E_{2g} mode at the Brillouin zone center
MWCNTs (EPD)	1595	
MWCNTs (EPD)	1633	D′-mode; effect of double resonance as an intravalley process
MWCNTs (EPD)	2713	2D (G′); second order of the D peak
MWCNTs (EPD)	2956	D + D′; combination of phonons with different momenta, requires a defect for theirs activation

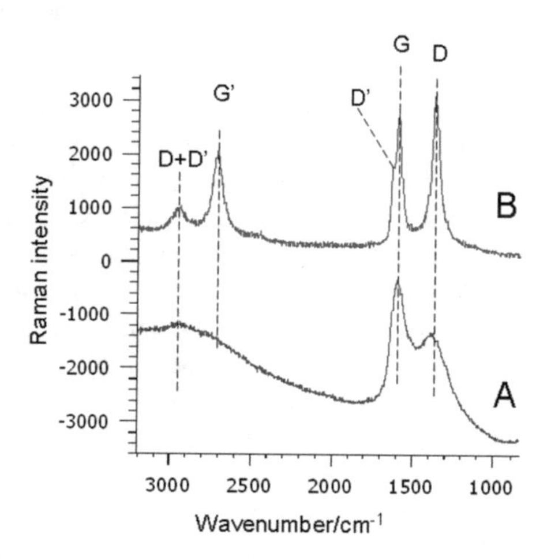

Fig. 3. Raman spectra for reference carbon layers (A) C (CVD), and (B) MWCNTs (EPD); 514.5 nm excitation line.

The G-band at ca.1580 cm^{-1} is described to the sp^2 carbon materials and is assigned to the high frequency E_{2g} optical phonon [45]. The G-band position for the C (CVD) differs from that for the MWCNTs (EPD) layer, which is more heterogeneous [46, 47].

Taking advantage of the carbon particle sensing properties, the interaction with proteins can be observed and determined. This interaction occurs at the border between the two phases and depends on both of them, on the type of carbon material and protein (Fig. 4). The shift of position of the G-band show the changes in relation to the reference coating. Furthermore, the positions of the G-band for the animal albumins, Alb and BSA, differ from that of HSA, what confirms the interaction occurring between the thin protein layer and the coating, and their specificity for different types of albumin.

Therefore, you need to consider what properties of the carbon layers are crucial in the application you are working on.

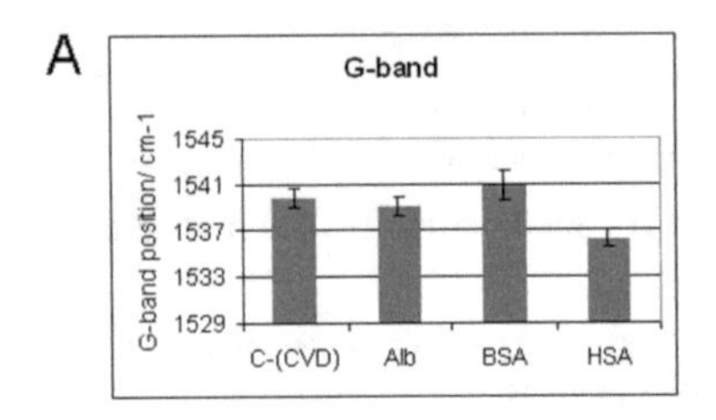

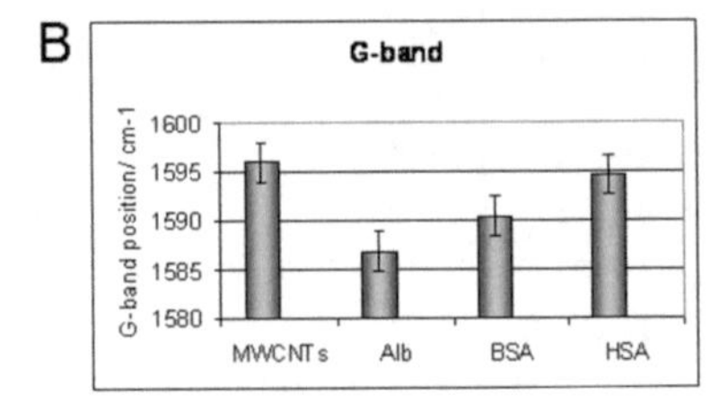

Fig. 4. Position of the G-band for the reference carbon layers (A) C (CVD), and (B) MWCNTs (EPD), and after incubation with the selected proteins; 514.5 nm excitation line.

3.5 2D Raman Correlation Spectroscopy Characterization of Carbon Layers Interaction with Selected Blood Proteins

Carbon materials have very high polarizability, therefore Raman spectroscopy is the method of choice for their analysis. However, the studied relations are primarily in the interphase space, so in addition to carbon material, the complex biological structures of albumin are involved in the interaction. The ideal solution would be to analyze the spectra of both components in their impact, therefore, the vibration range of amide I is interesting. Unfortunately, the Raman signal from the carbon component on the phase boundary might cover the spectrum of the protein. So, one can treat this issue as an analysis of the response of the system to the applied perturbation which is displayed as characteristic variations in the spectrum. Hence, a two-dimensional correlation analysis was used to control changes in the structure of the proteins on the interaction with the carbon layers in order to decode the relations hidden in the Raman spectra [41, 43]. Summarizing, in the correlation spectroscopy variable intensities were linked with a location on a sample characterizing the respective protein film - carbon coating interactions. The 2D spectrum indicates the clear differentiation between the origins of the Raman spectral signals [41].

Synchronous signal fluctuations indicate a common chemical constituent at ca. 1588 and 1345 cm^{-1} and also at ca. 1613 and 1313 cm^{-1} originating from the graphite G- and D- band components of the studied coating for the C (CVD) and MWCNTs layer, respectively.

The nonsynchronous signal fluctuations indicate chemically dissimilar components, thus the map pattern differs for the two carbon coatings and additionally animal albumins are clearly distinguishable from that of human (Figs. 5 and 6). The asynchronous correlation map for the C (CVD) + Alb layer in the 1530–1720 cm^{-1} range shows intensive negative cross-peaks originating from the amide I component bands at 1650 cm^{-1} (α-helix conf.), 1665 cm^{-1} (β-sheet conf.) and 1682 cm^{-1} (β-turn conf.) with the 1597 cm^{-1} band due to the G-band of the carbon layer. Otherwise it is for the C (CVD) + BSA layer, there is an intensive positive band in a different location at (1600, 1552 cm^{-1}) that arises from the carbon G-band and Glu vibrations. The C (CVD) + HSA layer presents a positive asynchronous peak derived of 1635 cm^{-1} (Trp, Arg, His) and 1652 cm^{-1} (α-helix conf.) and 1661 cm^{-1} (β-sheet conf.) of amide I and at ca. 1594 cm^{-1} of the carbon layer G-band.

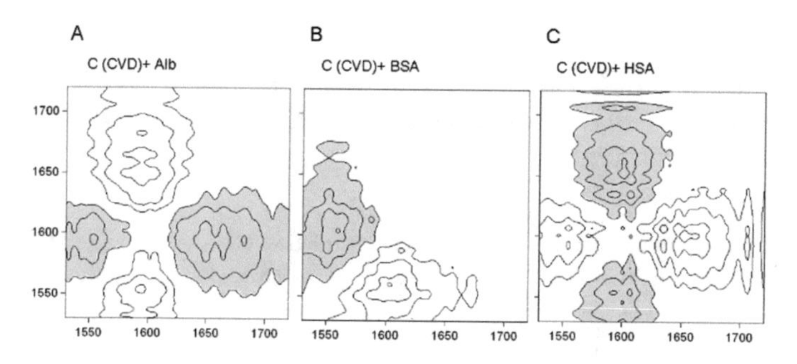

Fig. 5. Asynchronous 2D correlation Raman spectra of the C (CVD) sample incubated with: (A) chicken egg white albumin (Alb); (B) bovine serum albumin (BSA); (C) human serum albumin (HSA); in the wavenumber range of 1530–1720 cm^{-1}; the white and gray color represent positive and negative cross peaks, respectively.

A different picture is observed for the 2D asynchronous maps for the second type carbon nanolayer. The MWCNTs + Alb present a positive asynchronous correlation cross-peak at 1600 cm^{-1} of protein aromatic ring vibrations and 1622 cm^{-1} of Tyr with 1578 cm^{-1} of carbon nanotubes G^{-} vibration. Another tested system MWCNTs + BSA shows intensive negative cross peak −(1630,1602 cm^{-1}) owed to His, Tyr and (G^{+}) carbon band and −(1582, 1603 cm^{-1}) originating from the G^{+}-band of the MWCNTs and the Phe albumin vibrations. The MWCNTs + HSA layer gives a negative asynchronous peak −(1618, 1602 cm^{-1}) due to the Tyr and (G^{+}) carbon band. The other cross-peak originated of the amide I band of 1659 cm^{-1} (β-sheet conf.) and 1687 cm^{-1} (β-turn conf.) with 1601 cm^{-1} (G^{+}) carbon nanotube vibrations.

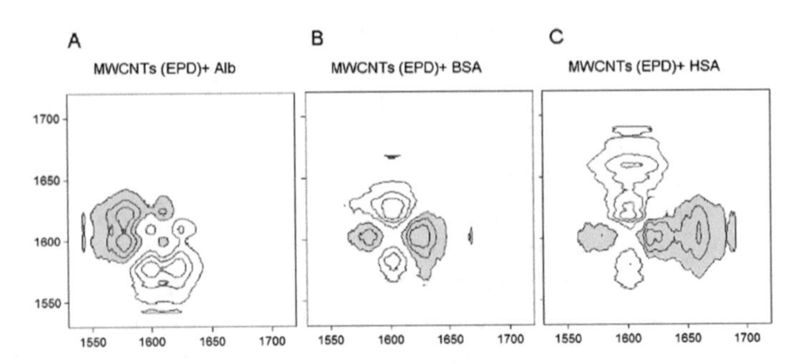

Fig. 6. Asynchronous 2D correlation Raman spectra of the MWCNTs (EPD) sample incubated with: (A) chicken egg white albumin (Alb); (B) bovine serum albumin (BSA); (C) human serum albumin (HSA); in the wavenumber range of 1530–1720 cm^{-1}. The white and gray color represent positive and negative cross peaks, respectively.

The calculated asynchronous cross–peaks are contained in Table 3. The 2D correlation spectroscopy allows to differentiate the adhesion specificity of the selected blood protein to the studied carbon layers.

Table 3. Observed significant asynchronous 2D correlation cross-peaks and their assignments for the C (CVD) and MWCNTs coating incubated in Alb, BSA, HSA in the wavenumber ranges of 1750–1500 cm^{-1} [21–29, 45–48].

2D asynchronous cross-peaks					
C (CVD) + Alb			MWCNTs + Alb		
Assignment	Cross-peaks	Assignment	Assignment	Cross-peaks	Assignment
amide I, α-helix conf.	−(1650,1597)	G-band	Phe, His	+(1600,1578)	G$^-$-band
amide I, β-sheet conf.	−(1665,1597)	G-band	Tyr	+(1622,1578)	G$^-$-band
amide I, β-turn conf.	−(1682,1597)	G-band			
C (CVD) + BSA			MWCNTs + BSA		
G-band	+(1598,1552)	Glu	Phe	+(1603,1583)	G$^+$-band
D′-band	+(1613,1558)	Asp	His	+(1630,1603)	G$^+$-band
C (CVD) + HAS			MWCNTs + HAS		
D′-band	+(1635,1601)	G-band	Tyr	−(1618,1602)	G$^+$-band
D′-band	+(1635,1594)	G-band	amide I, α-helix conf.	−(1660,1602)	G$^+$-band
amide I, α-helix conf.	+(1652,1601)	G-band	amide I, β-turn conf.	−(1687,1601)	G$^+$-band

4 Summary

The results of the performed complex studies of the two types of model carbon coatings exhibit significant dissimilarities regarding their interaction with the chosen blood proteins but also the difference is related to the origin of a protein: whether it is animal or human. Both of the studied carbon layers were incubated with the selected albumins, and the interaction between these two materials is visible by the variation of the contact angle (Fig. 2). They substantially differ in their surface image, the nanotubes form an isotropic fibrous system with a characteristic nanotopography while the C (CVD) pyrolytic surface is smoother (Fig. 2B).

For both types of the studied carbon layers a similar sequence of changes reflecting the occurring interaction with the albumins, which can be estimated by measuring contact angle. For each of the albumin, these differences are clearly marked, and an especially different contact angle is noticed for BSA (Fig. 2 A and B). The C (CVD) layer, which is known to be antithrombogenic, is characterized by high adhesion of protein to the surface while in other cases, the protein layer weakly adheres to the substrate (Table 1).

The results of Raman spectroscopy show that carbon nanolayers interact differently with the selected blood proteins, as indicated by the parameters determined from the Raman spectra, e.g. the position of the characteristic G-bands (Fig. 3). This parameter also allows to reveal the type of interactions and their extent.

The 2D asynchronous maps offer the possibility of determining the type of protein interaction with the surface of the carbon layer. Considering only the most intense cross-peaks you will notice that Alb adopts the structure with a comparable contribution of the α-helix, β-sheet and a sizable portion of β-turn conformation while signals from the individual amino acids Phe, His, Tyr are observed for the MWCNTs layer (Figs. 5A and 6A); [49–51]. For the BSA there is a correlation signal from the individual amino acids: Asp, Glu and Phe, His for C (CVD) and MWCNTs layer, respectively (Figs. 5B and 6B); [48–50]. The 2D correlation spectroscopy does not provide evidence that the secondary structure is mainly α-helix (50–60% in its native state) [50, 52, 53]. HSA acts in an exceptional way with a synthetic nanomaterial, showing equal participation conformation α-helix and also α-helix and β-turn in addition to the Tyr signal for the C (CVD) and MWCNTs layer, respectively [50, 54]. But, the vibrations of the amide I are ahead of the changes in the carbon nanolayer for the C (CVD). The MWCNTs layer shows the opposite order of events (Figs. 5C and 6C). The 2D analysis shows that the amide I bond is modified by the aromatic MWCNTs structure while the interaction on the interphase occurs, what was also noticed for the SWCNTs [55].

5 Conclusions

The albumin conformation is different for the studied surfaces, the amide I band maximum observed for the C (CVD) pyrolytic carbon layer shifts toward the higher vibrations for the MWCNT coating confirming an increase in the β-conformation. The conducted research and spectroscopic characteristics of the studied surfaces nanotopography as a key element in the synthetic surface with blood protein interaction and allow for the explanation of the nature of this process in relation to the type of protein.

The phenomena occurring on the surface of the C pyrolytic carbon with contact with human albumin have a different character than those observed in other cases. It can therefore be assumed that these phenomena, is the nature of conformational changes in HSA and strong adhesion, are responsible for the anti-thrombogenicity of this type of material.

Acknowledgements. This project was financed from the National Science Centre (NCN, Poland) granted on the decision number DEC-2013/09/B/ST8/00146 and UMO-2014/13/B/ST8/01195. AK has been partly supported by the EU Project POWR.03.02.00-00-I004/16.

References

1. Kołodziej, A., Wesełucha-Birczyńska, A., Moskal, P., Stodolak-Zych, E., Dużyja, M., Długoń, E., Sacharz, J., Błażewicz, M.: 2D-Raman correlation spectroscopy recognizes the interaction at the carbon coating and albumin interface. In: Kulczycki, P., Kowalski, P.A., Łukasik, S. (eds.) Contemporary Computational Science, p. 3. AGH-UST Press, Cracow (2018)
2. Cademartiri, L., Ozin, G.A.: Concepts of Nanochemistry. Wiley-VCH, Weinheim (2009)
3. Sahoo, S.K., Parveen, S., Panda, J.J.: The present and future of nanotechnology in human health care. Nanomedicine **3**, 20–31 (2007)
4. Lee, H., Kim, G.: Three-dimensional plotted PCL/β-TCP scaffolds coated with a collagen layer: preparation, physical properties and in vitro evaluation for bone tissue regeneration. J. Mater. Chem. **21**, 6305–6312 (2011)
5. Zhang, L., Webster, T.J.: Nanotechnology and nanomaterials: promises for improved tissue regeneration. Nano Today **4**, 66–80 (2009)
6. Lee, D.-E., Koo, H., Sun, I.-C., Ryu, J.H., Kim, K., Kwon, I.C.: Multifunctional nanoparticles for multimodal imaging and theragnosis. Chem. Soc. Rev. **41**, 2656–2672 (2012)
7. da Rocha, E.L., Porto, L.M., Rambo, C.R.: Nanotechnology meets 3D in vitro models: tissue engineered tumors and cancer therapies. Mater. Sci. Eng., C **34**, 270–279 (2014)
8. Chen, A., Chatterjee, S.: Nanomaterials based electrochemical sensors for biomedical applications. Chem. Soc. Rev. **42**, 5425–5438 (2013)
9. Dash, T.K., Konkimalla, V.B.: Poly-є-caprolactone based formulations for drug delivery and tissue engineering: a review. J. Control. Release **158**, 15–33 (2012)
10. Parikh, R., Dalwadi, S.: Preparation and characterization of controlled release poly-ε-caprolactone microparticles of isoniazid for drug delivery through pulmonary route. Powder Technol. **264**, 158–165 (2014)
11. Shen, Y. (ed.): Functional Polymers for Nanomedicine. RSC Publishing, Cambridge (2013)
12. Chen, L., Han, D., Jiang, L.: On improving blood compatibility: from bioinspired to synthetic design and fabrication of biointerfacial topography at micro/nano scales. Colloids Surf. B **85**, 2–7 (2011)
13. Ritchie, R.O.: Fatigue and fracture of pyrolytic carbon: a damage- tolerant approach to structural integrity and life prediction in "ceramic" heart valve prostheses. J. Heart Valve Dis. **5**(1), 9–31 (1996)

14. Cao, H.: Mechanical performance of pyrolytic carbon in prosthetic heart valve applications. J. Heart Valve Dis. **5**(1), 32–49 (1996)
15. Scholz, M.-S., Blanchfield, J.P., Bloom, L.D., Coburn, B.H., Elkington, M., Fuller, J.D., Gilbert, M.E., Muflahi, S.A., Pernice, M.F., Rae, S.I., Trevarthen, J.A., White, S.C., Weaver, P.M., Bond, I.P.: The use of composite materials in modern orthopaedic medicine and prosthetic devices: a review. Compos. Sci. Technol. **71**, 1791–1803 (2011)
16. Bareket-Keren, L., Hanein, Y.: Carbon nanotube-based multi electrode arrays for neuronal interfacing: progress and prospects. Front. Neural Circuits **6**, 1–16 (2012)
17. Hwang, J.Y., Shin, U.S., Jang, W.C., Hyun, J.K., Wall, I.B., Kim, H.W.: Biofunctionalized carbon nanotubes in neural regeneration: a mini-review. Nanoscale **5**, 487–497 (2013)
18. Silva, G.A.: Neuroscience nanotechnology: progress, opportunities and challenges. Nat. Rev. Neurosci. **7**, 65–74 (2006)
19. Sanchez, V.C., Jachak, A., Hurt, R.H., Kane, A.B.: Biological interactions of graphene-family nanomaterials: an interdisciplinary review. Chem. Res. Toxicol. **25**, 15–34 (2012)
20. Engel, E., Michiardi, A., Navarro, M., Lacroix, D., Planell, J.A.: Nanotechnology in regenerative medicine: the materials side. Trends Biotechnol. **26**, 39–47 (2008)
21. Wesełucha-Birczyńska, A., Frączek-Szczypta, A., Długoń, E., Paciorek, K., Bajowska, A., Kościelna, A., Błażewicz, M.: Application of Raman spectroscopy to study of the polymer foams modified in the volume and on the surface by carbon nanotubes. Vib. Spec. **72**, 50–56 (2014)
22. Wesełucha-Birczyńska, A., Swiętek, M., Sołtysiak, E., Galiński, P., Płachta, Ł., Piekara, K., Błażewicz, M.: Raman spectroscopy and the material study of nanocomposite membranes from poly(ε-caprolactone) with biocompatibility testing in osteoblast-like cells. Analyst **140**, 2311–2320 (2015)
23. Poncin-Epaillard, F., Vrlinic, T., Debarnot, D., Mozetic, M., Coudreuse, A., Legeay, G., El Moualij, B., Zorzi, W.: Surface treatment of polymeric materials controlling the adhesion of biomolecules. J. Funct. Biomater. **3**, 528–543 (2012)
24. Fraczek-Szczypta, A., Długon, E., Wesełucha-Birczyńska, A., Nocuń, M., Błażewicz, M.: Multi walled carbon nanotubes deposited on metal substrate using EPD technique: a spectroscopic study. J. Mol. Struct. **1040**, 238–245 (2013)
25. Benko, A., Przekora, A., Wesełucha-Birczyńska, A., Nocuń, M., Ginalska, G., Błażewicz, M.: Fabrication of multi-walled carbon nanotube layers with selected properties via electrophoretic deposition: physicochemical and biological characterization. Appl. Phys. A **122**, 1–13 (2016)
26. Wesełucha-Birczyńska, A., Stodolak-Zych, E., Turrell, S., Cios, F., Krzuś, M., Długoń, E., Benko, A., Niemiec, W., Błażewicz, M.: Vibrational spectroscopic analysis of ametal/carbon nanotube coating interface and the effect of its interaction with albumin. Vib. Spectrosc. **85**, 185–195 (2016)
27. Wesełucha-Birczyńska, A., Stodolak-Zych, E., Piś, W., Długoń, E., Benko, A., Błażewicz, M.: A model of adsorption of albumin on the implant surface titanium and titanium modified carbon coatings (MWCNT-EPD): 2D correlation analysis. J. Mol. Struct. **1124**, 61–70 (2016)
28. Vajtai, R. (ed.): Springer Handbook of Nanomaterials. Springer, Heidelberg (2013)
29. Ferrari, A.C., Robertson, J.: Interpretation of Raman spectra of disordered and amorphous carbon. Phys. Rev. B **61**, 14095–14107 (2000)
30. Iijima, S.: Helical microtubules of graphitic carbon. Nature **354**, 56–58 (1991)
31. Murr, L.E., Guerrero, P.A.: Carbon nanotubes in wood soot. Atmos. Sci. Lett. **7**, 93–95 (2006)

32. Bang, J.J., Guerrero, P.A., Lopez, D.A., Murr, L.E., Esquivel, E.V.: Carbon nanotubes and other fullerene nanocrystals in domestic propane and natural gas combustion streams. J. Nanosci. Nanotechnol. **4**, 716–718 (2004)
33. Sinha, N., Yeow, J.T.: Carbon nanotubes for biomedical applications. EEE Trans. Nanobiosci. **4**, 180–195 (2005)
34. Zhang, S. (ed.): Biological and Biomedical Coatings Handbook: Applications. CRC Press, Boca Raton (2011)
35. Park, S., Hamad-Schifferli, K.: Nanoscale interfaces to biology. Curr. Opin. Chem. Biol. **14**, 616–622 (2010)
36. Cui, H., Sinko, P.J.: The role of crystallinity on differential attachment/proliferation of osteoblasts and fibroblasts on poly (caprolactone-co-glycolide) polymeric surfaces. Front. Mater. Sci. **6**, 47–59 (2012)
37. Washburn, N.R., Yamada, K.M., Simon Jr., C.G., Kennedy, S.B., Amis, E.J.: High-throughput investigation of osteoblast response to polymer crystallinity: influence of nanometer-scale roughness on proliferation. Biomaterials **25**, 1215–1224 (2004)
38. Anselme, K.: Osteoblast adhesion on biomaterials. Biomaterials **21**, 667–681 (2000)
39. Schaller, J., Gerber, S., Kämfer, U., Lejon, S., Trachsel, C.: Human Blood Plasma Proteins. Wiley, Chichester (2008)
40. Noda, I., Ozaki, Y.: Two-dimensional Correlation Spectroscopy e Applications in Vibrational and Optical Spectroscopy. Wiley, Chichester (2004)
41. Noda, I., Dowrey, A.E., Marcott, C., Story, G.M., Ozaki, Y.: Generalized two-dimensional correlation spectroscopy. Appl. Spectrosc. **54**(7), 236A–248A (2002)
42. Noda, I.: Generalized two-dimensional correlation method applicable to infrared, raman, and other types of spectroscopy. Appl. Spectrosc. **47**, 1329–1336 (1993)
43. Shinzawa, H., Awa, K., Ozaki, Y.: Compression induced morphological and molecular structural changes of cellulose tablets probed with near infrared imaging. J. Near Infrared Spectrosc. **19**, 15–22 (2011)
44. Dshige © Shigeaki Morita, Kwansei-Gakuin University (2004–2005)
45. Ferrari, A.C., Robertson, J.: Raman spectroscopy of amorphous, nanostructured, diamond-like carbon, and nanodiamond. Phil. Trans. R. Soc. Lond. A **362**, 2477–2512 (2004)
46. Dresselhaus, M.S., Dresselhaus, G., Charlier, J.C., Hernández, E.: Electronic, thermal and mechanical properties of carbon nanotubes. Philos. Trans. A Math. Phys. Eng. Sci. **362**, 2065–2098 (2004)
47. Lehman, J.H., Terrones, M., Mansfield, E., Hurst, K.E., Meunier, V.: Evaluating the characteristics of multiwall carbon nanotubes. Carbon **49**, 2581–2602 (2011)
48. Wesełucha-Birczyńska, A., Babeł, K., Jurewicz, K.: Carbonaceous materials for hydrogen storage investigated by 2D Raman correlation spectroscopy. Vib. Spectrosc. **60**, 206–211 (2012)
49. Lewis, J.C., Snell, N.S., Hirschmann, D.J., Fraenkel-Conrat, H.: Amino acid composition of egg proteins. J. Biol. Chem. **186**(1), 23–35 (1950)
50. Tu, A.T.: Raman Spectroscopy in Biology: Principles and Applications. Wiley, New York (1982)
51. Synytsya, A., Judexová, M., Hrubý, T., Tatarkovič, M., Miškovičová, M., Petruželka, L., Setnička, V.: Analysis of human blood plasma and hen egg white by chiroptical spectroscopic methods (ECD, VCD, ROA). Anal. Bioanal. Chem. **405**, 5441–5453 (2013)
52. Anderle, G., Mendelsohn, R.: Thermal denaturation of globular proteins. Fourier transform-infrared studies of the amide III spectral region. Biophys. J. **52**, 69–74 (1987). https://doi.org/10.1016/S0006-3495(87)83189-2
53. Lippert, J.L., Tyminski, D., Desmeules, P.J.: Determination of the secondary structure of proteins by laser Raman spectroscopy. J. Am. Chem. Soc. **98**, 7075–7080 (1976)

54. Meloun, B., Morávek, L., Kostka, V.: Complete amino acid sequence of human serum albumin. FEBS Lett. **58**, 134–137 (1975)
55. Zhong, J., Song, L., Meng, J., Gao, B., Chu, W., Xu, H., Luo, Y., Guo, J., Marcelli, A., Xie, S., Wu, Z.: Bio-nano interaction of proteins adsorbed on single-walled carbon nanotubes. Carbon **47**, 967–973 (2009)

Effect of Elastic and Inelastic Scattering on Electronic Transport in Open Systems

Karol Kulinowski, Maciej Wołoszyn, and Bartłomiej J. Spisak(✉)

Faculty of Physics and Applied Computer Science,
AGH University of Science and Technology,
al. Mickiewicza 30, 30-059 Kraków, Poland
bjs@agh.edu.pl

Abstract. The purpose of this study is to apply the distribution function formalism to the problem of electronic transport in open systems, and numerically solve the kinetic equation with a dissipation term. This term is modeled within the relaxation time approximation, and contains two parts, corresponding to elastic or inelastic processes. The collision operator is approximated as a sum of the semiclassical energy dissipation term, and the momentum relaxation term which randomizes momentum but does not change energy. As a result, the distribution of charge carriers changes due to the dissipation processes, which has a profound impact on the electronic transport through the simulated region discussed in terms of the current–voltage characteristics and their modification caused by the scattering.

Keywords: Kinetic equation · Relaxation time approximation · Scattering processes

1 Introduction

Transport processes exhibit a wide variety of problems which are interesting for fundamental research in different branches of physics and engineering. Especially, a lot of attention is devoted to the electronic transport in solid state systems because they are used as elements of common electronic devices that are subject to progressive miniaturization. Simultaneously, more and more functionality of these devices is expected. These needs stimulate the search for new materials, as well as the development of theoretical and computational techniques or simulations for exploring transport phenomena of particular relevance in these systems.

Full theoretical description of electronic transport in the considered systems should be based on the methods of non-equilibrium statistical mechanics, but such description is an extremely complicated problem [1–4]. Therefore some physical assumptions and approximations are usually applied to simplify this issue [5,6]. One of the most common assumptions is to treat the studied system as an active element of the device connected to large electrodes which play a role

P. Kulczycki et al. (Eds.): ITSRCP 2018, AISC 945, pp. 296–306, 2020.
https://doi.org/10.1007/978-3-030-18058-4_23

of reservoirs of charge and heat. Each of these electrodes is characterized by its own equilibrium distribution function in the Fermi-Dirac form that is specified through the temperature and the chemical potential. For a fixed temperature, the difference between the chemical potentials associated with reservoirs is proportional to the bias voltage which produces the electric field across the studied system and the current of electrons flowing through the system is generated. The real structure of the systems can be defected or it can contain dopants, and moreover a collective thermal vibration of ions generates phonons which interact with the conduction electrons. All these factors have a significant impact on the electronic transport, and a separation of contributions from the different scattering mechanisms states the first step to understand the transport properties of the system.

The aim of this report is to study the influence of the scattering processes of the carriers in a highly defected semiconductor device on its transport characteristics. In such systems, the momentum of the carriers is randomized due to the scattering processes and their energy can be changed as a results of interaction with phonons. For this purpose we solve the kinetic equation with a dissipation term modeled with the relaxation time approximation. Within this approximation we take into account that the dissipation term consists of two parts. One of them describes the momentum relaxation due to the elastic scattering, and the second term describes the momentum and energy relaxation due to inelastic scattering processes.

The paper is organized as follows: in Sect. 2 we present the derivation of the kinetic equation for the distribution function, starting from the modified form of the von Neumann equation for the density operator. Further, we discuss different approximations which are usually made for simplification of the full quantum theory of electronic transport. Sect. 3 contains the discussion of the numerical method that is used to solve the presented problem. In turn, in Sect. 4 we present the results of our calculations and their discussion. This report is concluded in Sect. 5 where we summarize presented results.

2 Theory

It is widely recognized, that the starting point for the quantum theory of electronic transport in solid-state systems is the von Neumann equation, e.g. [7,8],

$$i\hbar \frac{d\hat{\rho}(t)}{dt} = \left[\hat{H}_0, \hat{\rho}(t)\right], \tag{1}$$

where $\hat{H}_0$ is the one-particle Hamiltonian of the unperturbed system in the form $\hat{H}_0 = \hat{p}^2/2m + U(\hat{x})$, here $\hat{p}$ denotes the momentum operator, and $U(\hat{x})$ is the potential energy operator of carriers with the effective mass m. In turn, $\hat{\rho}(t)$ is the one-particle density operator defined by the formula [9],

$$\hat{\rho}(t) = \sum_n p_n \left|\phi_n(t)\right\rangle \left\langle\phi_n(t)\right|, \tag{2}$$

where p_n corresponds the probability of finding the system in a pure state $|\phi_n(t)\rangle$. The density operator can be characterized by its own matrix elements which form the so-called density matrix. For example, the density operator in the position representation is $\rho(\xi,\xi') = \langle\xi|\,\hat{\rho}(t)\,|\xi'\rangle$. The diagonal elements of the density matrix represent the electron's density whereas its off-diagonal elements are responsible for the phase correlations of electrons. Let us note that the application of the von Neumann equation to the description of the conduction electrons dynamics in the considered systems is equivalent to the assumption that the electronic transport is coherent, i.e., quantum interference becomes important ingredient of the dynamical processes description. On the other hand, this coherent dynamics of conduction electrons is often perturbed by different kinds of uncontrollable interactions which destroy the phase coherence of the carriers. One of the examples of this situation in real systems is the interaction of the carriers gas with the phonon gas. In this case, one can observe transfer of energy or momentum between the both subsystems. This transfer is expressed in terms of inelastic or elastic scattering ratio, depending on the responsible physical mechanism. In fact this is a statistical process which introduces some kind of irreversibility to the system, which results in dissipation. These considerations lead to the conclusion that Eq. (1) should contain an additional term which introduces dissipation. On this basis, the von Neumann equation is transformed to the form

$$i\hbar\frac{d\hat{\rho}(t)}{dt} = \left[\hat{H}_0, \hat{\rho}(t)\right] + \hat{D}\left[\hat{\rho}(t)\right], \tag{3}$$

where the operator $\hat{D}\left[\hat{\rho}(t)\right]$ represents the dissipative term. The form of this term is a matter of wide dispute, and several discussions have been reported for this approach, e.g. [10]. Our goal is to write-down Eq. (3) in the phase-space coordinates. For this purpose we transform the density matrix in the position representation $[\xi - \xi']$ into the new position representation $[x - X]$ using the center of mass coordinates: $x = \xi' - \xi$ and $X = (\xi' + \xi)/2$, and then into the mixed representation $[x-p]$ using the Weyl transform [11]. As a result, we obtain the Wigner distribution function (WDF) [12–15],

$$\varrho(x,p,t) = \frac{1}{2\pi\hbar}\int dX \rho\left(x+\frac{X}{2}, x-\frac{X}{2}, t\right)\exp\left[-\frac{ipX}{\hbar}\right], \tag{4}$$

which plays a similar role as the distribution function in the classical statistical mechanics, in the sense that the WDF can be used to calculate expectation value of any dynamical variable as follows

$$\langle A(t)\rangle = \int dxdp\; A_W(p,x)\varrho(x,p;t), \tag{5}$$

where $A_W(p,x)$ is the Weyl symbol of quantum-mechanical operator of a dynamical variable $\hat{A}$ in the position representation,

$$A_W(p,x) = \int dX\,\left\langle x+\frac{X}{2}\right|\hat{A}(\hat{p},\hat{x})\left|x-\frac{X}{2}\right\rangle\exp\left[-\frac{ipX}{\hbar}\right]. \tag{6}$$

The application of the above scheme and the Weyl transform to Eq. (3) leads to the quantum kinetic equation in the Wigner form,

$$\frac{\partial \varrho(x,p,t)}{\partial t} + \frac{p}{m}\frac{\partial \varrho(x,p,t)}{\partial x} = \int dp' \, W(x,p-p')\varrho(x,p',t) + D_W\left[\varrho(x,p,t)\right], \quad (7)$$

where $D_W\left[\varrho(x,p,t)\right]$ is the Weyl symbol of the dissipation term, and the integral kernel $W(x,k)$ is interpreted as the nonlocal potential. Its form is given by the formula

$$W(x,p-p') = \frac{1}{2\pi i\hbar}\int dX \left[U\left(x+\frac{X}{2}\right) - U\left(x-\frac{X}{2}\right)\right] \exp\left[-\frac{i(p-p')X}{\hbar}\right]. \quad (8)$$

We would like to emphasize that the nonlocal potential allows to incorporate the quantum interference effect. In the classical limit the rhs of Eq. (8) can be reduced to the form: $[\partial U(x)/\partial x]\,\partial/\partial p$. In this way Eq. (7) takes the Boltzmann-like form,

$$\frac{\partial \varrho(x,p,t)}{\partial t} + \frac{p}{m}\frac{\partial \varrho(x,p,t)}{\partial x} - \frac{\partial U(x)}{\partial x}\frac{\partial \varrho(x,p,t)}{\partial p} = D_W\left[\varrho(x,p,t)\right]. \quad (9)$$

The Weyl symbol of the dissipation term is still a difficult issue, because its explicit form is unknown, although some results for the Caldeira-Leggett model [16] are known [17]. In this report, we consider the stationary solution of Eq. (9) in the limit of a slowly-varying potential and we apply the simplest model for the dissipation term which is known as the relaxation time approximation [5,6]. In accordance with Ref. [18], we choose this term in the following form,

$$D_W\left[\varrho(x,p)\right] \approx -\gamma_R\left[\varrho(x,p) - \varrho^{eq}(p)\right] - \gamma_M\left[\varrho(x,p) - \varrho(x,-p)\right], \quad (10)$$

where γ_R and γ_M are the strengths of inelastic and elastic scattering, respectively. The first term on rhs of Eq. (10) describes relaxation processes in which momentum as well as energy are changed during the scattering. In turn the second term describes situations in which only the carrier's momentum is changed, whereas energy is conserved. Taking into account all the above assumptions, we can finally write the kinetic equation in the form

$$\frac{p}{m}\frac{\partial \varrho(x,p)}{\partial x} = -\gamma_R\left[\varrho(x,p) - \varrho^{eq}(p)\right] - \gamma_M\left[\varrho(x,p) - \varrho(x,-p)\right]. \quad (11)$$

This equation is numerically solved with the boundary condition which assumes that the states of conduction electrons inflowing to the nanosystem depend on the states of the charge reservoirs as follows [19],

$$\begin{aligned} \varrho(0,p)\Big|_{p>0} &= f^L(E(p)), \\ \varrho(L,p)\Big|_{p<0} &= f^R(E(p)), \end{aligned} \quad (12)$$

where $f^{L(R)}(E(k))$ are the supply functions for the left (L) and right (R) contact in the form

$$f^{L(R)}(E(k)) = \frac{m}{\pi\hbar^2\beta} \ln\left\{\exp\left[-\left(\frac{\hbar^2k^2}{2m} - \mu_F^{L(R)}\right)\beta\right] + 1\right\}, \tag{13}$$

where $p = \hbar k$, $\beta = 1/k_BT$, k_B is the Boltzmann constant, and T is the temperature, while $\mu_F^{L(R)}$ is the Fermi level in the left (right) contact.

A solution of the kinetic equation in the form given by Eq. (11) with the boundary condition in the form (12) allows determining the distribution function for a given value of the applied bias voltage V_B. Then the electronic current, I, as a function of the bias voltage is calculated in accordance to the formula

$$I(V_B) = \frac{e}{2\pi\hbar L}\int dx \int dp \frac{p}{m}\varrho(x,p;V_B). \tag{14}$$

3 Method of calculation

Solution of the kinetic equation (11) may be found numerically using the implicit Euler method, with all the following equations written in the atomic units (a.u.) for which it assumed that $\hbar = e = m_e = 1$, and therefore $p = k$. For this purpose, the phase space is discretized into $N_x \times N_k$ cells, each of them having volume $\Delta_x\Delta_k$, where $\Delta_x = L/N_x$ and $\Delta_k = 2k_{max}/N_k$. In the first step, this procedure requires substituting the derivatives of the distribution function with their three-point discrete approximations resulting from the finite difference method,

$$\frac{d}{dx}\varrho(x,k) \approx \frac{1}{2\Delta_x}(-3\varrho_{i,j} + 4\varrho_{i+1,j} - \varrho_{i+2,j}), \qquad \text{for } k_j > 0, \tag{15}$$

$$\frac{d}{dx}\varrho(x,k) \approx \frac{1}{2\Delta_x}(3\varrho_{i,j} - 4\varrho_{i-1,j} + \varrho_{i-2,j}), \qquad \text{for } k_j < 0, \tag{16}$$

where $\varrho_{i,j} = \varrho(x_i, k_j)$ with $x_i = i\Delta_x$, $i = 0, \ldots, N_x - 1$ and $k_j = [j - (N_k - 1)/2]\Delta_k$, $j = 0, \ldots, N_k - 1$. It should be noted that the above equations take the asymmetric form which is a result of the assumed boundary condition, i.e., the right-hand or the left-hand form is chosen depending on the sign of k since the boundary condition is defined at $x = 0$ for $k > 0$ and at $x = L$ for $k < 0$. Additionally, N_k must be even, so that the mesh points are defined only for positive or negative values of k, and $k = 0$ is omitted as shown in Fig. 1. It also means that if $\varrho_{i,j}$ corresponds to a certain $\varrho(x,k)$, then ϱ_{i,N_k-j-1} corresponds to $\varrho(x,-k)$.

As a result, we obtain the following equations,

$$\varrho_{i,j} = \frac{4\varrho_{i-1,j} - \varrho_{i-2,j} + 2\frac{m\Delta_x}{\hbar k_j}(\gamma_R\varrho_j^{eq} + \gamma_M\varrho_{i,N_k-j-1})}{3 + 2\frac{m\Delta_x}{\hbar k_j}(\gamma_R + \gamma_M)}, \quad \text{for } k_j > 0, \tag{17}$$

$$\varrho_{i,j} = \frac{4\varrho_{i+1,j} - \varrho_{i+2,j} - 2\frac{m\Delta_x}{\hbar k_j}(\gamma_R\varrho_j^{eq} + \gamma_M\varrho_{i,N_k-j-1})}{3 - 2\frac{m\Delta_x}{\hbar k_j}(\gamma_R + \gamma_M)}, \quad \text{for } k_j < 0. \tag{18}$$

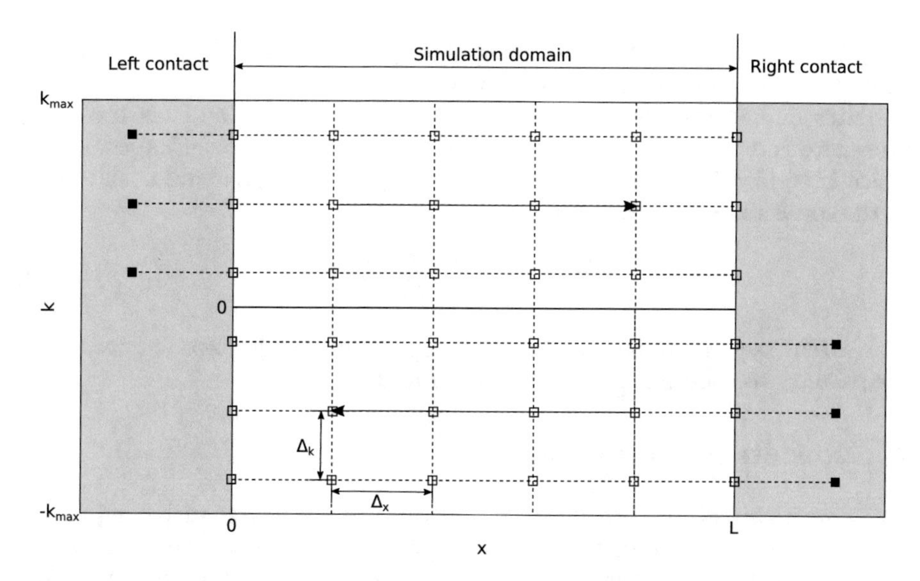

Fig. 1. Discretization scheme for the kinetic equation. Full boxes correspond to the sites at which the fixed boundary condition based on the supply function is applied.

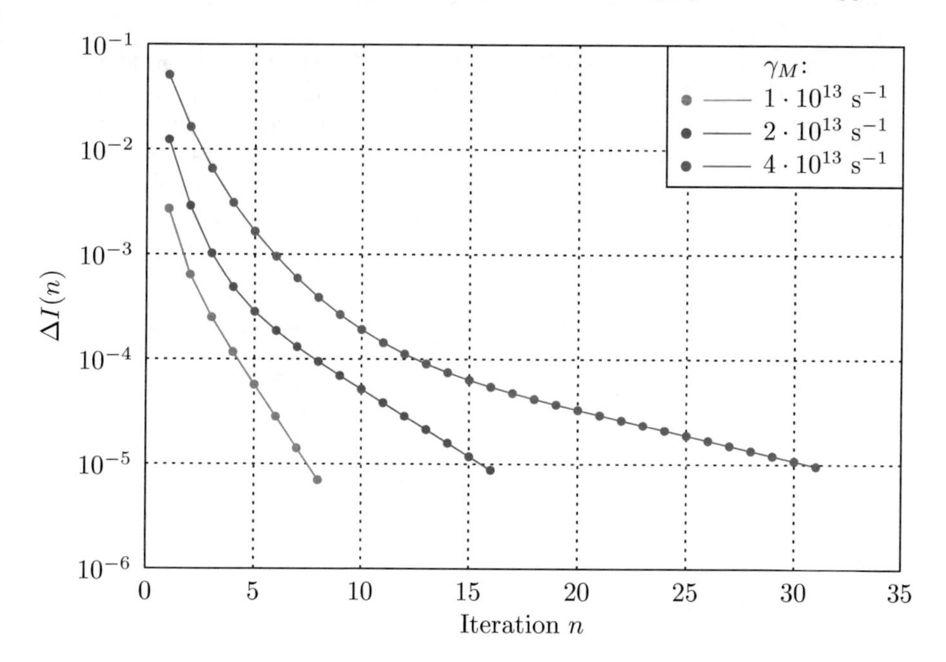

Fig. 2. Illustration of the convergence of the self-consistent procedure in terms of the relative change of the current value between the consecutive iterations. Calculated for $\gamma_M = 1$, 2, and $4 \cdot 10^{13}$ s^{-1} and $\gamma_R = 0$.

They are solved recursively, assuming that ϱ_{i,N_k-j-1} is initially equal to the value of the equilibrium distribution function, $\varrho^{eq}_{N_k-j-1}$.

After finding the values of the distribution function at all sites the whole procedure is repeated in the self-consistent manner until the solution converges, which is verified in terms of the relative change of the current value calculated at the consecutive steps $n-1$ and n,

$$\Delta I(n) = \frac{|I(n) - I(n-1)|}{I(n)}, \tag{19}$$

with $\Delta I(n)$ required to be less than 10^{-5}. Typically $10 \div 40$ steps are needed for the solution to converge, as illustrated in Fig. 2.

4 Results and discussion

The numerical calculations have been performed for a system having length $L = 70$ nm and at temperature $T = 10$ K. The Fermi level at the right contact was assumed to be equal $\mu_F^R = 0.08$ eV, while the Fermi level in the left contact is $\mu_F^L = \mu_F^R + V_B$, where V_B is the applied bias voltage. All calculations were conducted on a computational grid of size $N_x \times N_k$ with $N_x = 40$ and $N_x = 40$, while $\Delta_x = 1.75$ nm and $\Delta_k = 0.02$ nm^{-1}.

Scattering rate gives us information about how frequently electrons participate in the elastic or inelastic scattering processes. The higher scattering rate results therefore in the higher resistance of the system, which reduces the measured electronic current. This dependence, calculated for the considered system, is presented in Fig. 3 which shows how the current–voltage characteristics of the simulated structure reacts to the change of the scattering rates.

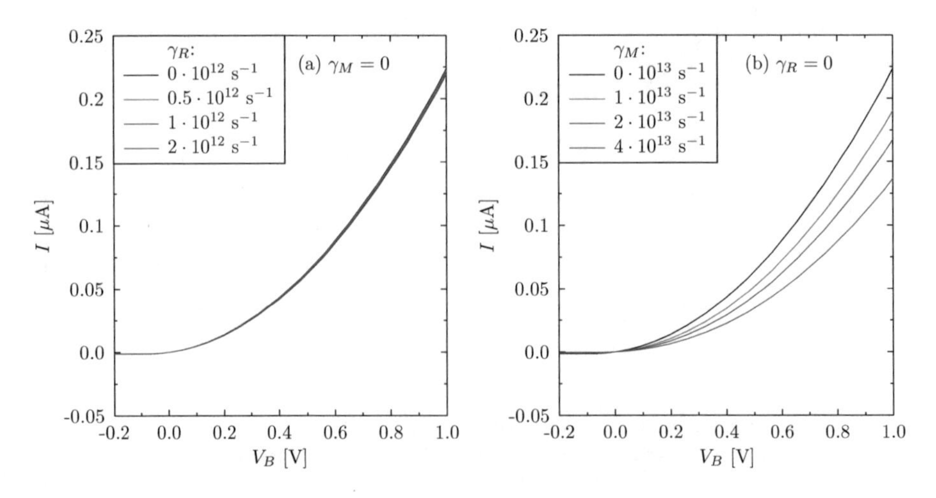

Fig. 3. Current voltage characteristics for (a) different inelastic scattering rates dissipating energy, and (b) various elastic scattering rates responsible for randomization of momentum.

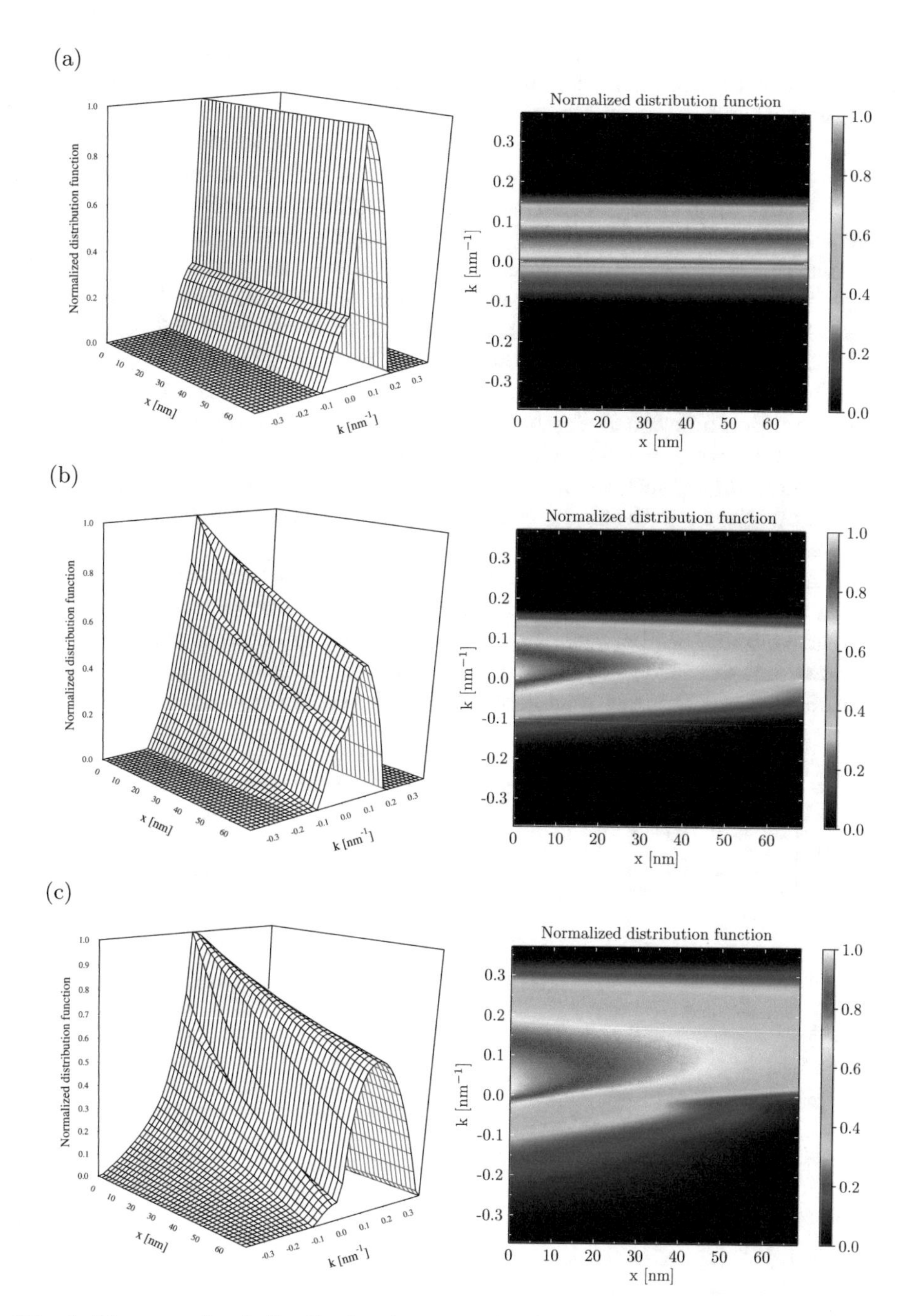

Fig. 4. The normalized distribution function $\varrho(x,k)$ for (a) $\gamma_R = \gamma_M = 0$ at the bias voltage $V_B = 0.2$ V, (b) $\gamma_R = 1 \cdot 10^{12}$ s^{-1} and $\gamma_M = 2 \cdot 10^{13}$ s^{-1} at the bias voltage $V_B = 0.2$ V, and (c) $\gamma_R = 1 \cdot 10^{12}$ s^{-1} and $\gamma_M = 2 \cdot 10^{13}$ s^{-1} at the bias voltage $V_B = 1.0$ V.

The calculated current–voltage characteristics are very similar to the those of the p-n diodes, in particular for the negative bias voltage where the current value is constant and close to zero, independently of the applied bias voltage.

These results also show that the impact of the elastic scattering on the current is very weak, while the influence of the inelastic scattering is much stronger. The reason is that the former does not significantly modify the distribution function, whereas the latter has much stronger influence on the distribution function, and thus also on the current value. Examples of the calculated distribution functions are presented in Fig. 4 which shows the functions $\varrho(x, k)$ that are solutions of the kinetic equation (11). Analysis of those distribution functions clearly shows that larger values of the current are related to bigger carrier density in upper half of phase space, i.e. for $k > 0$. When the non-zero scattering rates are introduced, the distribution function is mostly affected for carriers with positive momentum, while the number of carriers with negative momentum also changes, although not as much. As a result, the introduction of scattering changes the value of the current, and modifies the shape of the distribution function which changes in a much smoother way around $k = 0$.

Influence of the scattering processes on the conductance of the analyzed system can be also illustrated by the relation between the scattering rate and the relative difference between the current flowing in absence of any dissipation processes and the current calculated in presence of elastic or inelastic scattering. Figure 5 confirms that the relative change of the current due to the dissipation

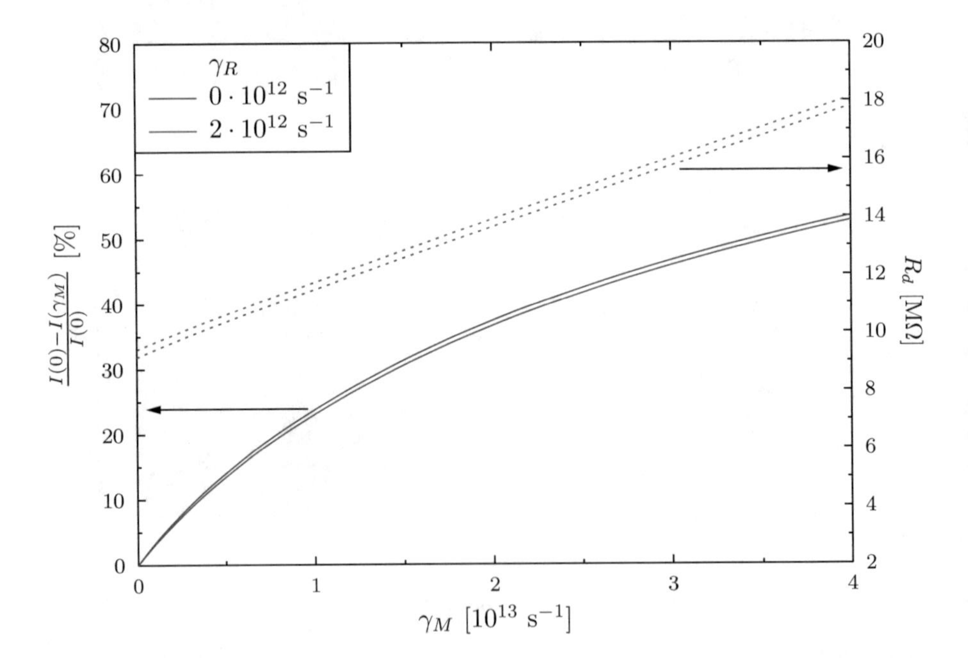

Fig. 5. Relative change of the current (solid lines) and differential resistance (dashed lines) due to influence of the dissipation processes calculated for different scattering rates and the bias voltage $V_B = 0.2$ V.

processes depends on the type of scattering which is taken into account. The relative current change due to the inelastic scattering is a nonlinear, increasing function of γ_M equal up to approximately 50% for $\gamma_M = 4 \cdot 10^{13}$ s^{-1}.

Also the differential resistance, $R_d = dV_B/dI$, is sensitive to the scattering, as shown for $V_B = 0.2$ V in Fig. 5. Its value increases linearly with increasing γ_M, and at all points is slightly larger for greater γ_R.

5 Conclusions

We solved numerically the kinetic equation with the dissipative term which is modeled by the relaxation time approximation, with separate terms describing the relaxation of momentum and relaxation of energy. The numerical solution of the kinetic equation allowed us to find the distribution function. We investigated the modification of the shape of this function due to those two types of scattering. Owing to the knowledge of the distribution functions, also the electronic current was calculated as a function of the bias voltage for different values of the scattering rates corresponding to the term which dissipates energy and to the momentum-relaxation term. Our calculations show that the current–voltage characteristics are considerably modified by the elastic scattering.

Acknowledgement. This work was partially supported by the Faculty of Physics and Applied Computer Science AGH UST statutory tasks within subsidy of Ministry of Science and Higher Education. K.K. has been partly supported by the EU Project POWR.03.02.00-00-I004/16.

The preliminary version of this paper was presented at the 3rd Conference on Information Technology, Systems Research and Computational Physics, 2–5 July 2018, Cracow, Poland [20].

References

1. Fujita, S.: Introduction to Non-Equilibrium Quantum Statistical Mechanics. W. B. Saunders Company, Philadelphia, London (1966)
2. Danielewicz, P.: Ann. Phys. (N.Y.) **152**, 239 (1984). https://doi.org/10.1016/0003-4916(84)90092-7
3. Rammer, J.: Quantum Field Theory of Non-equilibrium States. Cambridge University Press, Cambridge (2007)
4. Schieve, W.C., Horwitz, L.P.: Quantum Statistical Mechanics. Cambridge University Press, Cambridge (2009)
5. Di Ventra, M.: Electrical Transport in Nanoscale Systems. Cambridge University Press, Cambridge (2008)
6. Ferry, D.K., Goodnick, S.M., Bird, J.: Transport in Nanostructures. Cambridge University Press, Cambridge (2009)
7. Kohn, W., Luttinger, J.M.: Phys. Rev. **108**, 590 (1957). https://doi.org/10.1103/PhysRev.108.590
8. Luttinger, J.M., Kohn, W.: Phys. Rev. **109**, 1892 (1958). https://doi.org/10.1103/PhysRev.109.1892

9. Ter Haar, D.: Rep. Prog. Phys. **24**, 304 (1961). https://doi.org/10.1088/0034-4885/24/1/307
10. Chruściński, D., Pascazio, S.: Open Syst. Inf. Dyn. **24**, 1740001 (2017). https://doi.org/10.1142/S1230161217400017
11. Leaf, B.: **9**, 65 (1968). https://doi.org/10.1063/1.1664478
12. Wigner, E.: Phys. Rev. **40**, 749 (1932). https://doi.org/10.1103/PhysRev.40.749
13. Tatarskiĭ, V.I.: Sov. Phys. Usp. **26**, 311 (1983). https://doi.org/10.1070/PU1983v026n04ABEH004345
14. Lee, H.W.: Phys. Rep. **259**, 147 (1995). https://doi.org/10.1016/0370-1573(95)00007-4
15. Schleich, W.P.: Quantum Optics in Phase Space. Wiley, New York (2001)
16. Caldeira, A.O., Leggett, A.J.: Phys. Rev. Lett. **46**, 2114 (1981). https://doi.org/10.1103/PhysRevLett.46.211
17. Zurek, W.H.: Rev. Mod. Phys. **75**, 715 (2003). https://doi.org/10.1103/RevModPhys.75.715
18. Jonasson, O., Knezevic, I.: J. Comput. Electron. **14**, 879 (2015). https://doi.org/10.1007/s10825-015-0734-9
19. Frensley, W.R.: Rev. Mod. Phys. **62**, 745 (1990). https://doi.org/10.1103/RevModPhys.62.745
20. Kulinowski, K., Wołoszyn, M., Spisak, B.J.: In: Kulczycki, P., Kowalski, P.A., Łukasik, S. (eds.) Contemporary Computational Science, p. 4. AGH-UST Press, Cracow (2018)

P$\hbar$ase-Space Approach to Time Evolution of Quantum States in Confined Systems. The Spectral Split-Operator Method

Damian Kołaczek, Bartłomiej J. Spisak(✉), and Maciej Wołoszyn

Faculty of Physics and Applied Computer Science, AGH University of Science and Technology, al. Mickiewicza 30, 30-059 Kraków, Poland
bjs@agh.edu.pl

Abstract. Using the phase space approach, we consider the dynamics of a quantum particle in an isolated confined quantum system with three different potential energy profiles. We solve the Moyal equation of motion for the Wigner function with the highly efficient spectral split-operator method. The main aim of this study is to compare the accuracy of the used algorithm by analysis of the total energy expectation value, in terms of the deviation from its exact value. This comparison is performed for the second and fourth order factorizations of the time evolution operator.

Keywords: Wigner distribution function · Moyal dynamics · Split-operator method

1 Introduction

The theory of quanta revolutionized the way of thinking about physical phenomena in the microscopic world and introduced an abstract mathematical formalism based on some concepts of functional analysis. The structure of the quantum theory is determined by a set of independent axioms which incorporates probabilistic aspect of measurements of observables. Thereby the quantum theory has a statistical character. Usually, we assume that states of a physical system are represented by abstract vectors $|\psi(t)\rangle$ in a Hilbert space. Nevertheless, in some physical situations, description of the system states in terms of the abstract vectors is not sufficient because the states are known only statistically. To meet these challenges, quantum-statistical theory offers a more general approach in which the state of a system is described by the density operator $\hat{\rho}(t)$. Its form, in the spectral representation, is given by a convex combination of the rank-one orthogonal projection operators onto the abstract vectors [1], i.e.,

$$\hat{\rho}(t) = \sum_n p_n \left|\phi_n(t)\right\rangle \left\langle\phi_n(t)\right|, \tag{1}$$

where p_n is the probability of finding the system in a pure state $|\phi_n(t)\rangle$. Time evolution of the density operator is governed by the von Neumann equation

$$i\hbar\frac{d}{dt}\hat{\rho}(t) = \left[\hat{H}(\hat{x},\hat{p}), \hat{\rho}(t)\right], \tag{2}$$

P. Kulczycki et al. (Eds.): ITSRCP 2018, AISC 945, pp. 307–320, 2020.
https://doi.org/10.1007/978-3-030-18058-4_24

where the symbol [,] stands for the commutator of two operators, $\hbar$ is the reduced Planck constant, and i is the imaginary unit. The above equation of motion describes the non-dissipative evolution of the density operator for an isolated (closed) system which is characterized by the one-particle Hamiltonian in the form

$$\hat{H}(\hat{x}, \hat{p}) = \frac{\hat{p}^2}{2m} + U(\hat{x}), \tag{3}$$

where m is the effective mass of a particle, $\hat{x}$ and $\hat{p}$ are the position and the momentum operators, respectively, $[\hat{x}, \hat{p}] = i\hbar\hat{1}$, and $U(\hat{x})$ is the operator of the potential energy. The expectation value of any dynamical observable which is represented by a Hermitian operator $\hat{A}$ is obtained by the formula

$$\langle A(t) \rangle = Tr\left\{\hat{\rho}(t)\hat{A}\right\}, \tag{4}$$

where the symbol $Tr\{\ \}$ indicates the trace of the operator. Hence it can be concluded that the density operator in the quantum-statistical description of the system plays a similar role as the probability distribution function over the phase space in the statistical description of a classical system.

For the first time the relationship between the density operator and function of phase-space variables was established by Wigner [2], who applied the Weyl transform [3,4] to the density operator. As a result, he obtained the following c-number function [5–9]

$$\varrho(x, p, t) = \frac{1}{2\pi\hbar} \int dX \left\langle x + \frac{X}{2} \middle| \hat{\rho}(t) \middle| x - \frac{X}{2} \right\rangle \exp\left[-\frac{ipX}{\hbar}\right], \tag{5}$$

which depends on the position and momentum coordinates, and nowadays is known as the Wigner distribution function (WDF). In fact, this function can take negative values in some regions of the phase space and therefore it is regarded as a quasi-probability density in this space. It is worth to mention that negativity of the WDF exhibits non-classical properties of the state as well as it can be used as an indicator of quantum phenomena in the system [10–14]. It should be also emphasized that the WDF plays a role of a state in the so-called phase-space formulation of quantum mechanics [5,15,16], in which a non-commutative algebra of smooth observables represented by ordinary c-numbers functions is generated by the Groenewold star-product

$$* = \exp\left[i\hbar\left(\overleftarrow{\partial_x}\overrightarrow{\partial_p} - \overleftarrow{\partial_p}\overrightarrow{\partial_x}\right)/2\right], \tag{6}$$

where the arrows indicate in which direction the derivatives act. In practice, this product may be evaluated through translation of the argument of the phase-space function in the following way

$$(f * g)(x, p) = f\left(x + (i\hbar/2)\overrightarrow{\partial_p}, p - (i\hbar/2)\overrightarrow{\partial_x}\right) g(x, p). \tag{7}$$

Hence, one can conclude that the star-product of two such functions is described by a power series in the form

$$(f * g)(x, p) = (fg)(x, p) + \frac{i\hbar}{2}\{f, g\}(x, p) + O(\hbar^2), \tag{8}$$

where the first term in the series is the pointwise product, and the second term is the Poisson bracket. Aforementioned expression suggests that the quantum mechanics in the phase space may be regarded as a deformation theory of classical mechanics [17–19]. Presently, this formulation is exploited in some branches of modern physics and chemistry, i.e., it is applied in quantum-statistical studies of transport processes [20,21], quantum optics [22], or quantum field theory [23] and it is also very inspiring for some branches of mathematics development: deformation of the Lie algebras [24], non-commutative or symplectic geometry [25].

The subject of this report is the exploration of two variants of the spectral split-operator method in application to the phase-space propagation of the quantum state in small confined systems. Originally, the spectral split-operator method was proposed to solve the Schrödinger equation [26], then this method has been adapted to the solution of the Liouville equation [27–29], and it is often used to simulate the dynamics of states in molecular systems. We have performed comparative studies of this numerical method in the second and fourth order. For this purpose we consider the initially localized WDF and its dynamics in the harmonic potential and the class of the power-exponential potentials [30]. This dynamics is generated by the Moyal's equation of motion.

The paper is organized as follows: in Sect. 2 we present theoretical background of quantum mechanics in the phase space, i.e., basic concepts and notations on the Moyal dynamics in the phase space. Then we introduce some elements of the spectral split-operator method of second and fourth order applied to the Moyal equation. Section 3 contains the results of calculations and their discussion. Finally, Sect. 4 contains conclusions and summary of the results.

2 Theory

In the phase-space formulation of quantum mechanics, the considered system is characterized by the Weyl symbol of the Hamiltonian, which can be written in the position representation as follows [5,6],

$$H_W(x,p) = \int dX \left\langle x + \frac{X}{2} \right| \hat{H}(\hat{x},\hat{p}) \left| x - \frac{X}{2} \right\rangle \exp\left[-\frac{ipX}{\hbar}\right], \tag{9}$$

where $\hat{H}(\hat{x},\hat{p})$ corresponds to the hermitian Hamiltonian in the form given by Eq. (3). As it was mentioned at the beginning, the state of the system in the phase space is given by the WDF. In turn, its unitary evolution in time can be described by the equation of motion in the Moyal form [31],

$$\frac{\partial \varrho(x,p;t)}{\partial t} = \hat{\mathcal{L}}_M \varrho(x,p;t), \tag{10}$$

where the operator $\hat{\mathcal{L}}_M$ is given by the Moyal bracket which is defined, for a given Weyl symbol of Hamiltonian, as a skew-symmetric part of the star-product, i.e.,

$$\begin{aligned}\hat{\mathcal{L}}_M \varrho(x,p;t) &= \frac{1}{i\hbar}\left[H_W(x,p) * \varrho(x,p;t) - \varrho(x,p;t) * H_W(x,p)\right] \\ &= \{H_W(x,p), \varrho(x,p;t)\}_M \,.\end{aligned} \tag{11}$$

This means that the Moyal bracket emerges as a generator of quantum dynamics of the state in the phase space. An important property of the Moyal bracket is that it becomes the Poisson bracket in the classical limit ($\hbar \to 0$). It implies that the Moyal equation becomes the Liouville equation for the classical phase-space distribution function.

The Moyal equation can be written in the alternative form as follows [32],

$$i\hbar\partial_t \varrho(x,p,t) = \Big[H_W\left(x + \frac{i\hbar}{2}\overrightarrow{\partial}_p, p - \frac{i\hbar}{2}\overrightarrow{\partial}_x\right) - H_W\left(x - \frac{i\hbar}{2}\overrightarrow{\partial}_p, p + \frac{i\hbar}{2}\overrightarrow{\partial}_x\right)\Big] \varrho(x,p,t), \quad (12)$$

owing to the ansatz given by Eq. (7).

Authors of Ref. [33] defined the Hilbert phase space in which every state of the quantum system, which in general can be mixed, is represented by an abstract vector $|\rho(t)\rangle\rangle$, and its time evolution is given by equation

$$i\hbar\frac{d}{dt}|\rho(t)\rangle\rangle = \left[\frac{\hbar}{m}\hat{p}\hat{\lambda} + U\left(\hat{x} - \frac{\hbar}{2}\hat{\theta}\right) - U\left(\hat{x} + \frac{\hbar}{2}\hat{\theta}\right)\right]|\rho(t)\rangle\rangle, \quad (13)$$

where the operators: $\hat{\lambda}$, $\hat{\theta}$, $\hat{p}$, and $\hat{x}$ belong to a six-operator algebra with respect to the following commutation relations,

$$\left[\hat{x},\hat{\theta}\right] = \hat{0}, \quad [\hat{x},\hat{p}] = \hat{0}, \quad \left[\hat{x},\hat{\lambda}\right] = i\hat{1}, \quad \left[\hat{p},\hat{\theta}\right] = i\hat{1}, \quad \left[\hat{p},\hat{\lambda}\right] = \hat{0}, \quad \left[\hat{\lambda},\hat{\theta}\right] = \hat{0}. \quad (14)$$

Knowing the state at time instant t_0 $|\rho(t_0)\rangle\rangle$, the solution of Eq. (13) for arbitrary $t_0 + \Delta t$ can be written in the form $|\rho(t_0 + \Delta t)\rangle\rangle = \hat{\mathcal{U}}(\Delta t)\,|\rho(t_0)\rangle\rangle$, where

$$\hat{\mathcal{U}}(\Delta t) = \exp\left\{-\frac{i}{\hbar}\left[\frac{\hbar}{m}\hat{p}\hat{\lambda} + U\left(\hat{x} - \frac{\hbar}{2}\hat{\theta}\right) - U\left(\hat{x} + \frac{\hbar}{2}\hat{\theta}\right)\right]\Delta t\right\} \quad (15)$$

is called the time evolution operator. In representation $\langle xp|$ where

$$\hat{x} = x, \quad \hat{p} = p, \quad \hat{\lambda} = -i\partial_x, \quad \hat{\theta} = -i\partial_p, \quad (16)$$

we obtain WDF (5) as it turns out that $\langle xp|\,\rho(t)\rangle\rangle = \varrho(x,p,t)$ and Eq. (13) boils down to Eq. (12).

The time evolution operator (15) can be written in more concise form as follows,

$$\hat{\mathcal{U}}(\Delta t) = \exp\left[-\frac{i}{\hbar}\left(\hat{T} + \hat{U}\right)\Delta t\right]. \quad (17)$$

In this notation, the operator $\hat{T} = (\hbar/m)\hat{p}\hat{\lambda}$ represents the kinetic part, whereas the operator $\hat{U} = U\left(\hat{x} - \hbar\hat{\theta}/2\right) - U\left(\hat{x} + \hbar\hat{\theta}/2\right)$ represents the potential part. Importantly, these two operators do not commute. It should be noticed that in representation $\langle\lambda p|$ where

$$\hat{x} = i\partial_\lambda, \quad \hat{p} = p, \quad \hat{\lambda} = \lambda, \quad \hat{\theta} = -i\partial_p, \quad (18)$$

operator $\hat{T}$ is a multiplication operator. Similarly in representation $\langle x\theta|$ where

$$\hat{x} = x, \quad \hat{p} = i\partial_\theta, \quad \hat{\lambda} = -i\partial_x, \quad \hat{\theta} = \theta, \tag{19}$$

operator $\hat{U}$ is a multiplication operator. Transformations between different representations of the state vector $|\rho(t)\rangle$ are realized by the Fourier transforms, i.e.,

$$\begin{aligned} &\langle x\theta|\, \rho\rangle = \int dp e^{-ip\theta} \langle xp|\, \rho\rangle\,, \quad \langle \lambda p|\, \rho\rangle = \frac{1}{2\pi}\int dx d\theta e^{i(p\theta-\lambda x)} \langle x\theta|\, \rho\rangle\,, \\ &\langle xp|\, \rho\rangle = \frac{1}{2\pi}\int d\theta e^{ix\lambda} \langle \lambda p|\, \rho\rangle\,. \end{aligned} \tag{20}$$

The core of the spectral split operator method is to factorize the time evolution operator (17) as a product of operators dependent only on $\hat{T}$ and only on $\hat{U}$. Each of them can be easily realized numerically as a multiplication in adequate representation and the transformations between various representations can be efficiently realized by fast Fourier transforms. We consider two factorizations of time evolution operator (17), the widely used second order factorization

$$\hat{\mathcal{U}}(\Delta t) = \exp\left[-\frac{i}{2\hbar}\hat{T}\Delta t\right]\exp\left[-\frac{i}{\hbar}\hat{U}\Delta t\right]\exp\left[-\frac{i}{2\hbar}\hat{T}\Delta t\right] + O(\Delta t^3), \tag{21}$$

and the fourth order factorization

$$\begin{aligned} \hat{\mathcal{U}}(\Delta t) = &\exp\left[-\frac{i}{6\hbar}\hat{U}\Delta t\right]\exp\left[-\frac{i}{2\hbar}\hat{T}\Delta t\right]\exp\left[-\frac{2i}{3\hbar}\hat{\tilde{U}}\Delta t\right] \\ &\times \exp\left[-\frac{i}{2\hbar}\hat{T}\Delta t\right]\exp\left[-\frac{i}{6\hbar}\hat{U}\Delta t\right] + O(\Delta t^5), \end{aligned} \tag{22}$$

where $\hat{\tilde{U}} = \hat{U} - \left[\hat{U}, \left[\hat{T}, \hat{U}\right]\right] \Delta t^2/(48\hbar^2)$, which was originally derived in Ref. [34], and then applied to the Liouville equation. Later this factorization has been also used for the Schrödinger equation [35]. It can be clearly seen that the later factorization requires 5/3 times more operations, both multiplications and Fourier transforms. It can be shown that

$$\begin{aligned} \langle x\theta| \left[\hat{U}, \left[\hat{T}, \hat{U}\right]\right] |\rho(t)\rangle = &\frac{\hbar^2}{m}\left[U'\left(x - \frac{\hbar}{2}\theta\right) - U'\left(x + \frac{\hbar}{2}\theta\right)\right] \\ &\times \left[U'\left(x - \frac{\hbar}{2}\theta\right) + U'\left(x + \frac{\hbar}{2}\theta\right)\right] \langle x\theta|\, \rho(t)\rangle\,, \end{aligned} \tag{23}$$

where $U'(x)$ is the first derivative of the potential energy function $U(x)$, so the operator $\left[\hat{U}, \left[\hat{T}, \hat{U}\right]\right]$ is just a multiplication operator in $\langle x\theta|$ representation.

Time shift of WDF from the time instant t_0 to $t_0 + \Delta t$ can be realized in the following steps, accordingly for the second order factorization (21)

$$\begin{aligned} \varrho(x, p, t_0 + \Delta t) \approx &\mathcal{F}^{\lambda\to x} \exp\left[-\frac{i\Delta t}{2m}T(\lambda, p)\right] \mathcal{F}^{\theta\to p}_{x\to\lambda} \exp\left[-\frac{i\Delta t}{\hbar}U(x, \theta)\right] \mathcal{F}^{\lambda\to x}_{p\to\theta} \\ &\times \exp\left[-\frac{i\Delta t}{2m}T(\lambda, p)\right] \mathcal{F}_{x\to\lambda}\varrho(x, p, t_0), \end{aligned} \tag{24}$$

and for the fourth order factorization (22)

$$\begin{aligned}\varrho(x,p,t_0+\Delta t) \approx \mathcal{F}^{\theta\to p} \exp\left[-\frac{i\Delta t}{6\hbar}U(x,\theta)\right] \mathcal{F}_{p\to\theta}^{\lambda\to x} \exp\left[-\frac{i\Delta t}{2m}T(\lambda,p)\right] \mathcal{F}_{x\to\lambda}^{\theta\to p} \\ \times \exp\left[-\frac{2i\Delta t}{3\hbar}\widetilde{U}(x,\theta)\right] \mathcal{F}_{p\to\theta}^{\lambda\to x} \exp\left[-\frac{i\Delta t}{2m}T(\lambda,p)\right] \mathcal{F}_{x\to\lambda}^{\theta\to p} \\ \times \exp\left[-\frac{i\Delta t}{6\hbar}U(x,\theta)\right] \mathcal{F}_{p\to\theta}\varrho(x,p,t_0),\end{aligned} \tag{25}$$

where the ordinary and inverse Fourier transforms are written accordingly in subscript and in superscript.

In numerical calculations we can get the WDF time evolution by successive application of the time evolution operator (15) with given time step Δt. We set the computation box size $[-L_x, L_x] \times [-L_p, L_p]$ and the computation grid size $N_x \times N_p$. We discretize the position and momentum in the following way,

$$x_m = -L_x + m\Delta x, \quad p_n = -L_p + n\Delta p, \tag{26}$$

where $\Delta x = 2L_x/N_x$ and $\Delta p = 2L_p/N_p$. For θ and λ variables we define

$$\Delta\lambda = \frac{\pi}{L_x}, \quad \Delta\theta = \frac{\pi}{L_p}, \quad L_\lambda = \frac{N_x\Delta\lambda}{2}, \quad L_\theta = \frac{N_p\Delta\theta}{2}. \tag{27}$$

Unlike for x and p, the discretizations of variables θ and λ have to be shifted due to the properties of the discrete Fourier transform,

$$\lambda_k = \begin{cases} -L_\lambda + \left(k + \frac{N_x}{2}\right)\Delta\lambda,\ k = 0, 1, ..., \frac{N_x}{2} - 1 \\ -L_\lambda + \left(k - \frac{N_x}{2}\right)\Delta\lambda,\ k = \frac{N_x}{2}, \frac{N_x}{2} + 1, ..., N_x - 1 \end{cases} \tag{28}$$

and

$$\theta_l = \begin{cases} -L_\theta + \left(l + \frac{N_p}{2}\right)\Delta\theta,\ l = 0, 1, ..., \frac{N_p}{2} - 1 \\ -L_\theta + \left(l - \frac{N_p}{2}\right)\Delta\theta,\ l = \frac{N_p}{2}, \frac{N_p}{2} + 1, ..., N_p - 1. \end{cases} \tag{29}$$

For numerical simulation according to Eqs. (24) and (25) one has to discretize functions $T(\lambda,p)$, $U(x,\theta)$ and $\widetilde{U}(x,\theta)$ on appropriate grids and apply the discrete Fourier transforms that are defined as usual, accordingly for ordinary and inverse transform

$$F_k = \sum_{n=0}^{N-1} f_n e^{-i2\pi kn/N}, \quad f_n = \frac{1}{N}\sum_{n=0}^{N-1} F_k e^{i2\pi kn/N}, \tag{30}$$

and for efficient computations should be implemented as fast Fourier transforms.

3 Results and Discussion

3.1 Initial Condition and Considered Potential Energies

The presented algorithm based on the spectral split-operator method for given initial WDF at time instant $t = 0$ allows one to find the time evolution of the

WDF via successive application of time evolution operator. We take the initial condition in the Wigner form of the Gaussian wave packet centered around some point $(0, p_0)$ in the phase space [36],

$$\varrho(x,p;0) = \frac{1}{\pi\hbar} \exp\left\{-\frac{2\delta_x^2(p-p_0)^2}{\hbar^2} - \frac{x^2}{2\delta_x^2}\right\}, \tag{31}$$

where δ_x^2 is the initial variance of the wave packet. Owing to the WDF at successive time instants, we can determine evolution of some dynamical observables that are characteristic for considered systems. In general case, expectation value of any dynamical variable can be calculated in accordance with the formula

$$\langle A(t)\rangle = \int dxdp\, A_W(x,p)\varrho(x,p;t), \tag{32}$$

where $A_W(x,p)$ is the Weyl symbol of quantum-mechanical operator of a dynamical variable $\hat{A}$ in the position representation,

$$A_W(x,p) = \int dX \left\langle x + \frac{X}{2} \middle| \hat{A}(\hat{x},\hat{p}) \middle| x - \frac{X}{2} \right\rangle \exp\left[-\frac{ipX}{\hbar}\right]. \tag{33}$$

From these expressions we can simply derive the formula for temporary changes of the expectation value of the total energy associated with the WDF moving in the considered potential. Inasmuch it can be easily shown that $x_W^l = x^l$ and $p_W^l = p^l$ for arbitrary natural value of l, that formula can be expressed in the following form

$$\langle E(t)\rangle = \int dxdp \left[\frac{p^2}{2m} + U(x)\right] \varrho(x,p;t). \tag{34}$$

Because the analyzed systems are closed, the expectation value of the total energy is constant throughout the motion, and deviations from this value serve us as a measure of algorithm precision. We define the following formulas,

$$\text{error}(E;t,\Delta t) = E^{num}(t,\Delta t) - E^{exact}, \tag{35}$$

$$\text{error}(E;\Delta t) = \max_t |\text{error}(E;t,\Delta t)|, \tag{36}$$

which will be used in further subsections.

Potential energy profiles that are considered are the harmonic oscillator,

$$U_{HO}(x) = \frac{1}{2}m\omega^2 x^2, \tag{37}$$

the Gaussian well,

$$U_G(x) = U_0\left[1 - \exp\left(-\frac{x^2}{2\sigma_x^2}\right)\right] \tag{38}$$

and the power-exponential well,

$$U_{PE}(x) = U_0\left\{1 - \exp\left[-\left(\frac{x}{\sqrt{2}\sigma_x}\right)^{2n}\right]\right\}, \tag{39}$$

which is considered for $n = 2$. It should be noticed that the potential energy $U_G(x)$ is a special case of the power-exponential function (39) for $n = 1$. In Eqs. (37–39) m is the mass of quantum particle, ω is an angular frequency of the harmonic oscillator, U_0 is a depth of the potential well and σ_x is related to the width of the potential well. For the used values of the parameters that will be specified in the following subsections, the potential energy profiles are presented in Fig. 1.

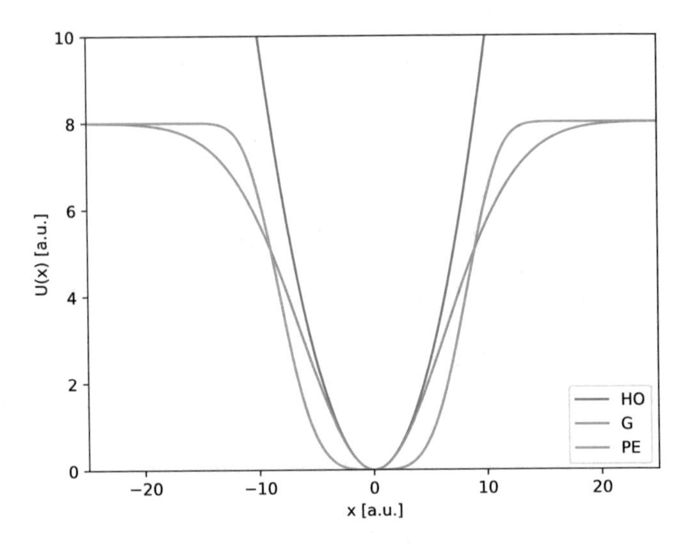

Fig. 1. Considered profiles of the potential energy: harmonic oscillator (HO), Gaussian (G), and power-exponential (PE).

Given the initial WDF (31) and three above forms of the potential energy, one can calculate the analytic forms of the total energy expectation values from Eq. (34),

$$E_{HO}^{exact} = \frac{1}{2m}\left(\frac{\hbar^2}{4\delta_x^2} + p_0^2\right) + \frac{m\omega^2}{2}\delta_x^2, \tag{40}$$

$$E_G^{exact} = \frac{1}{2m}\left(\frac{\hbar^2}{4\delta_x^2} + p_0^2\right) + U_0\left(1 - \sqrt{\frac{\sigma_x^2}{\sigma_x^2 + \delta_x^2}}\right), \tag{41}$$

$$E_{PE}^{exact} = \frac{1}{2m}\left(\frac{\hbar^2}{4\delta_x^2} + p_0^2\right) + U_0\left(1 - \frac{\sigma_x^2}{2\pi\delta_x^2}\exp\left[\frac{\sigma_x^4}{8\delta_x^4}\right]K_{\frac{1}{4}}\left(\frac{\sigma_x^4}{8\delta_x^4}\right)\right), \tag{42}$$

where $K_\nu(x)$ is the modified Bessel function of the second kind.

All calculations were performed in atomic units ($a.u.$), where $\hbar = e = m_e = 1$.

3.2 Harmonic Oscillator Potential

For the potential energy (37) we set $m = 1$ a.u. and $\omega = 1/\sqrt{5}$ a.u. For the initial WDF, the momentum expectation value is $p_0 = 1$ a.u. For such values of parameters the expectation value of the total energy (40) is $E_{HO}^{exact} = 1.025$ a.u. The maxima of the total energy error for both factorizations of the time evolution operator are presented in Fig. 2 as a function of the used time step Δt. It shows that for the shortest time step $\Delta t = 0.05$ a.u. the fourth order method produces error which is five orders of magnitude smaller than for the second order method. Even at the longest tested $\Delta t = 1.6$ a.u. the error of the fourth order method is similar to the error of the second order method with the shortest Δt.

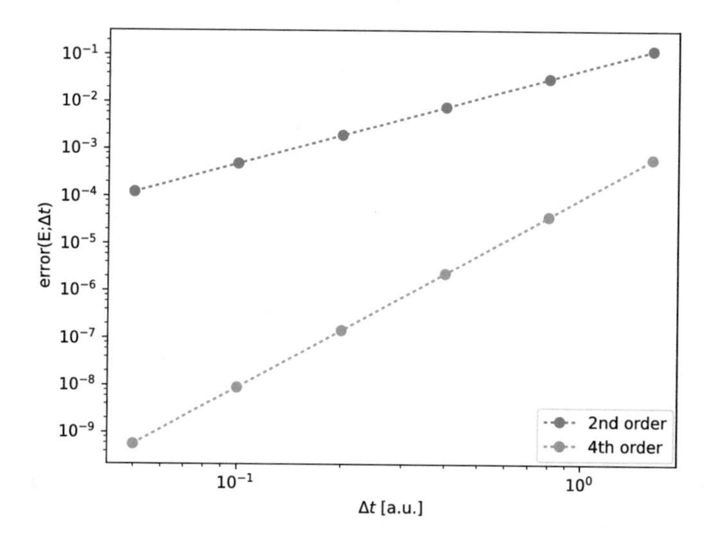

Fig. 2. Maximum value of the error of total energy during simulation for various time steps in the case of the second and the fourth order algorithms applied to potential of harmonic oscillator.

Time dependence of the total energy error for two exemplary time steps for each method is shown in Fig. 3. The only visible difference between plots for both time steps is scaling, but the shape is virtually the same. The same is true for all other considered time steps.

3.3 Gaussian Potential

For the potential energy (38) we set $U_0 = 8$ a.u. and $\sigma_x^2 = 40$ a.u. For the initial WDF, the momentum expectation value is $p_0 = 1$ a.u. For such values of parameters the expectation value of the total energy (41) is $E_g^{exact} = 1.024883$ a.u. The maxima of the total energy error for both factorizations of the time evolution operator presented in Fig. 4 turn out to be pretty similar to the results obtained

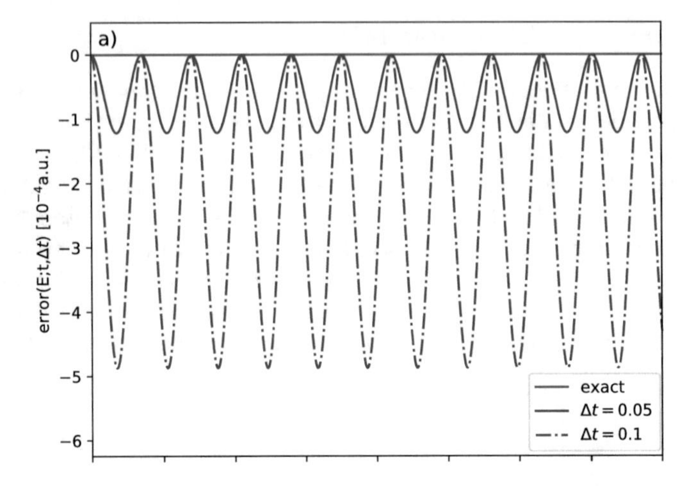

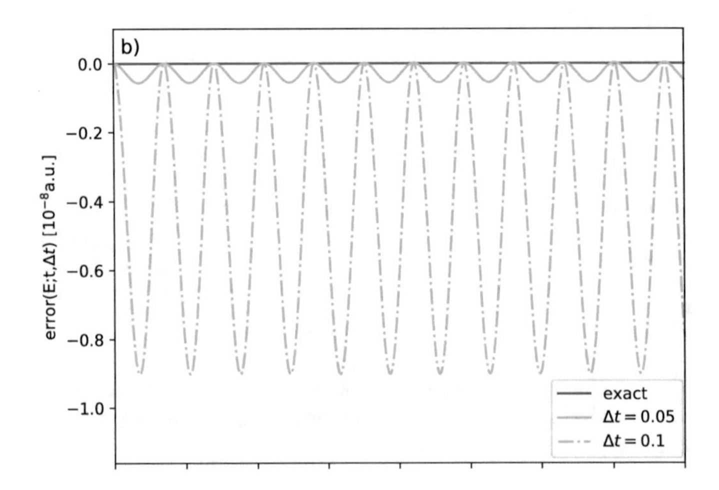

Fig. 3. Total energy error time dependence for time steps: $\Delta t = 0.05$ a.u. (solid line) and $\Delta t = 0.1$ a.u. (dash-dotted line) for (a) the second order algorithm, and (b) the fourth order algorithm. Both for the harmonic oscillator potential.

previously for the harmonic oscillator potential. Analogically, for the time step $\Delta t = 0.05$ a.u. the fourth order method produces error which is five orders of magnitude smaller than the error of the second order method. Also for the longest tested $\Delta t = 1.6$ a.u. the error of the fourth order method is comparable to the error of the second order method for the shortest $\Delta t = 0.5$ a.u.

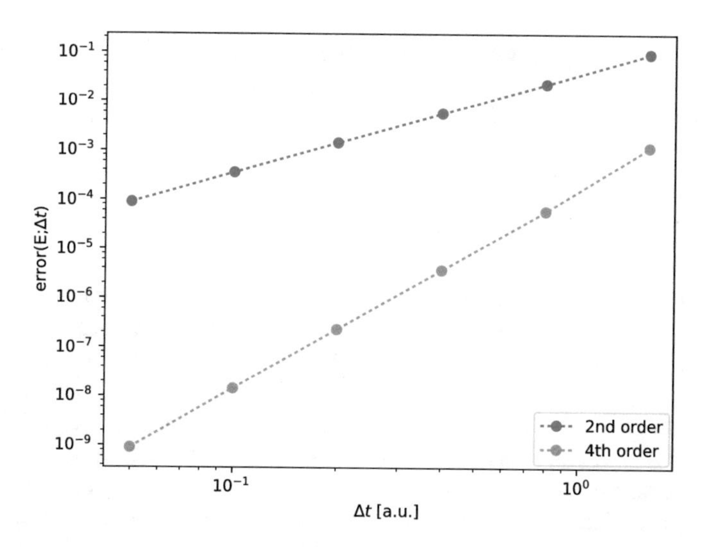

Fig. 4. Maximum value of the error of total energy during simulation for various time steps in the case of the second and the fourth order algorithms applied to Gaussian potential.

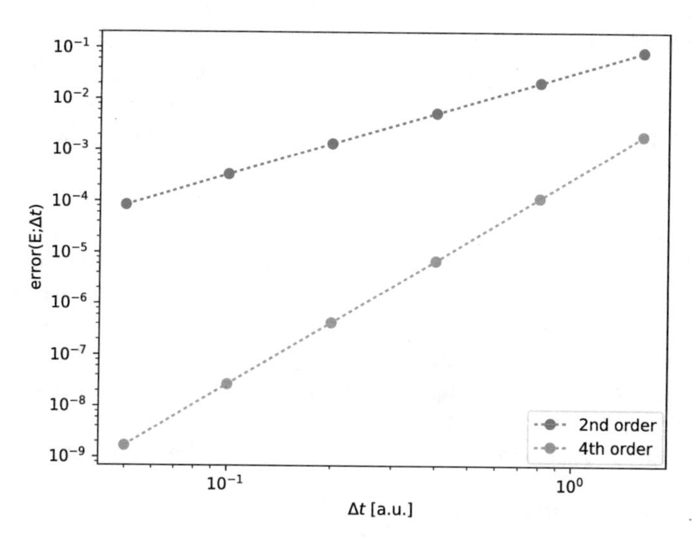

Fig. 5. Maximum value of the error of total energy during simulation for various time steps in the case of the second and the fourth order algorithms applied to power-exponential potential.

3.4 Power-Exponential Potential

For the potential energy (39) we set $U_0 = 8$ a.u. and $\sigma_x^2 = 40$ a.u. For the initial WDF, the momentum expectation value is $p_0 = 1$ a.u. For such values of parameters the expectation value of the total energy (41) is $E_g^{exact} = 1.000234$ a.u. The maxima of the total energy error for both factorizations of the time evolution operator showed in Fig. 5 display very similar behavior to the results presented in Fig. 2 and Fig. 4 for harmonic oscillator potential and Gaussian well potential, respectively.

4 Conclusions

We have studied the phase space dynamics of a quantum particle in three different confined systems according to the Moyal equation determining the time evolution of the Wigner function. We have analyzed precision of the algorithm based on the spectral split-operator method for various time steps and for the second and fourth order factorizations of the time evolution operator. The fourth order factorization has been used for the first time for Wigner function dynamics in the quantum phase space. The influence of the boundaries on the dynamics of the Wigner function has been neglected because the extents of the computational box have been suitably adjusted to avoid this effect. Also the grid size was large enough to keep errors related to the phase space discretization much smaller than errors resulting from the considered lengths of the time step. As a measure of precision, we used the deviation of the total energy from its exact value. For every profile of the potential energy we have obtained similar dependencies of the total energy error as a function of the time step length. These dependencies are very close to linear in log-log scale which means that they are power relations. More precisely, regardless of the potential energy form, the total energy error is approximately proportional to some power of time step which turned out to be equal to the order of used factorization method. For a given time evolution operator factorization and a given potential energy profile, the error is a function displaying very similar shape regardless of the used time step, but it is scaled by approximately the ratio of two considered time steps raised to the power of time evolution operator factorization order. After running all of the calculations we can undoubtedly state that the fourth order factorization scheme of the time evolution operator is far superior to the second order one. It requires only 5/3 times more calculations for single time step but provides several orders of magnitude more precise results for the same time step length.

Acknowledgement. This work was partially supported by the Faculty of Physics and Applied Computer Science AGH UST statutory tasks within subsidy of Ministry of Science and Higher Education. D.K. has been partly supported by the EU Project POWR.03.02.00-00-I004/16.

The preliminary version of this paper was presented at the 3rd Conference on Information Technology, Systems Research and Computational Physics, 2–5 July 2018, Cracow, Poland [37].

References

1. Ter Haar, D.: Rep. Prog. Phys. **24**, 304 (1961). https://doi.org/10.1088/0034-4885/24/1/307
2. Wigner, E.: Phys. Rev. **40**, 749 (1932). https://doi.org/10.1103/PhysRev.40.749
3. Pool, J.C.T.: J. Math. Phys. **7**, 66 (1966). https://doi.org/10.1063/1.1704817
4. Ozorio de Almeida, A.M.: Phys. Rep. **295**, 265 (1998). https://doi.org/10.1016/S0370-1573(97)00070-7
5. Baker, G.A.: Phys. Rev. **109**, 2198 (1958)
6. Tatarskiĭ, V.I.: Sov. Phys. Usp. **26**, 311 (1983). https://doi.org/10.1070/PU1983v026n04ABEH004345
7. Hillery, M., O'Connell, R.F., Scully, M.O., Wigner, E.P.: Phys. Rep. **106**, 121 (1984). https://doi.org/10.1016/0370-1573(84)90160-1
8. Balazs, N.L., Jennings, B.K.: Phys. Rep. **104**, 347 (1984)
9. Lee, H.W.: Phys. Rep. **259**, 147 (1995). https://doi.org/10.1103/PhysRevB.69.073102
10. Benedict, M.G., Czirják, A.: Phys. Rev. A **60**, 4034 (1999). https://doi.org/10.1103/PhysRevA.60.4034
11. Kenfack, A., Życzkowski, K.: J. Opt. B Quantum Semiclass. Opt. **6**, 396 (2004). https://doi.org/10.1088/1464-4266/6/10/003
12. Sadeghi, P., Khademi, S., Nasiri, S.: Phys. Rev. A **82**, 012102 (2010). https://doi.org/10.1103/PhysRevA.82.012102
13. Kenfack, A.: Phys. Rev. A **93**, 036101 (2016). https://doi.org/10.1103/PhysRevA.93.036101
14. Khademi, S., Sadeghi, P., Nasiri, S.: Phys. Rev. A **93**, 036102 (2016). https://doi.org/10.1103/PhysRevA.93.036102
15. Curtright, T.L., Zachos, C.K.: Asia Pac. Phys. Newslett. **01**, 37 (2012). https://doi.org/10.1142/S2251158X12000069
16. Błaszak, M., Domański, Z.: Ann. Phys. **327** (2012). https://doi.org/10.1016/j.aop.2011.09.006
17. Bayen, F., Flato, M., Fronsdal, C., Lichnerowicz, A., Sternheimer, D.: Lett. Math. Phys. **1**, 521 (1977). https://doi.org/10.1007/BF00399745
18. Bayen, F., Flato, M., Fronsdal, C., Lichnerowicz, A., Sternheimer, D.: Ann. Phys. **111**, 61 (1978). https://doi.org/10.1016/0003-4916(78)90224-5
19. Bayen, F., Flato, M., Fronsdal, C., Lichnerowicz, A., Sternheimer, D.: Ann. Phys. **111**, 111 (1978). https://doi.org/10.1016/0003-4916(78)90225-7
20. Xue, Y., Prodan, E.: Phys. Rev. B **86**, 155445 (2012). https://doi.org/10.1103/PhysRevB.86.155445
21. Leung, B., Prodan, E.: J. Phys. A **46**, 085205 (2013). https://doi.org/10.1088/1751-8113/46/8/085205
22. Isar, A., Scheid, W.: Phys. A **335**, 79 (2004). https://doi.org/10.1016/j.physa.2003.12.017
23. Lechner, G.: Commun. Math. Phys. **312**, 265 (2012). https://doi.org/10.1007/s00220-011-1390-y
24. Delius, G.W., Hüffmann, A.: J. Phys. A **29**, 1703 (1996). https://doi.org/10.1088/0305-4470/29/8/018
25. Castellani, L.: Class. Quantum Grav. **17**, 3377 (2000). https://doi.org/10.1088/0264-9381/17/17/301
26. Feit, M.D., Fleck, J.A., Steiger, A.: J. Comput. Phys. **47**, 412 (1982). https://doi.org/10.1016/0021-9991(82)90091-2

27. Torres-Vega, G., Frederick, J.H.: Phys. Rev. Lett. **67**, 2601 (1991). https://doi.org/10.1103/PhysRevLett.67.2601
28. Dattoli, G., Giannessi, L., Ottaviani, P.L., Torre, A.: Phys. Rev. E **51**, 821 (1995)
29. Gómez, E.A., Thirumuruganandham, S.P., Santana, A.: Comput. Phys. Commun. **185**, 136 (2014)
30. Ciurla, M., Adamowski, J., Szafran, B., Bednarek, S.: Phys. E **15**, 261 (2002). https://doi.org/10.1016/S1386-9477(02)00572-6
31. Hiley, B.J.: J. Comput. Electron. **14**, 869 (2015). https://doi.org/10.1007/s10825-015-0728-7
32. Kubo, R.: J. Phys. Soc. Japan **19**, 2127 (1964). https://doi.org/10.1143/JPSJ.19.2127
33. Bondar, D.I., Cabrera, R., Zhdanov, D.V., Rabitz, H.A.: Phys. Rev. A **88**, 052108 (2013). https://doi.org/10.1103/PhysRevA.88.052108
34. Chin, S.A.: Phys. Lett. A **226**, 344 (1997). https://doi.org/10.1016/S0375-9601(97)00003-0
35. Chin, S.A., Chen, C.R.: J. Chem. Phys. **117**, 1409 (2002). https://doi.org/10.1063/1.1485725
36. Kaczor, U., Klimas, B., Szydłowski, D., Wołoszyn, M., Spisak, B.: Open Phys. **14**, 354 (2016). https://doi.org/10.1515/phys-2016-0036
37. Kołaczek, D., Spisak, B.J., Wołoszyn, M.: In: Kulczycki, P., Kowalski, P.A., Łukasik, S. (eds.) Contemporary Computational Science, p. 5. AGH-UST Press, Cracow (2018)

Probability Measures and Projections on Quantum Logics

Oľga Nánásiová[1(✉)], Viera Čerňanová[2], and Ľubica Valášková[3]

[1] Institute of Computer Science and Mathematics, Slovak University of Technology, Ilkovičova 3, 812 19 Bratislava, Slovakia
olga.nanasiova@stuba.sk

[2] Department of Mathematics and Computer Science, Faculty of Education, Trnava University, Priemyselná 4, 918 43 Trnava, Slovakia
vieracernanova@hotmail.com

[3] Department of Mathematics and Descriptive Geometry, Slovak University of Technology, Radlinského 11, 810 05 Bratislava, Slovakia
lubica.valaskova@stuba.sk

Abstract. The present paper is devoted to modelling of a probability measure of logical connectives on a quantum logic (QL), via a G-map, which is a special map on it. We follow the work in which the probability of logical conjunction, disjunction and symmetric difference and their negations for non-compatible propositions are studied.

We study such a G-map on quantum logics, which is a probability measure of a projection and show, that unlike classical (Boolean) logic, probability measure of projections on a quantum logic are not necessarilly pure projections.

We compare properties of a G-map on QLs with properties of a probability measure related to logical connectives on a Boolean algebra.

Keywords: Logical connectives · Orthomodular lattice · Quantum logic · Probability measure · State

1 Introduction

The primary version of this paper was presented at the 3rd Conference on Information Technology, Systems Research and Computational Physics, 2–5 July 2018, Cracow, Poland [1].

The problem of modelling of probability measures for logical connectives of non-compatible propositions started by publishing the paper Birkhoff, von Neumann [2]. Quantum logic allows to model situations with non-compatible events (events that are not simultaneously measurable). Methods of quantum logic appear in data processing, economic models, and in other domains of application e.g. [2,6,10,26].

Calculus for non-compatible observables has been described in [9], while modelling of logical connectives in terms of their algebraic properties and algebraic structures can be found in [5,7,20].

P. Kulczycki et al. (Eds.): ITSRCP 2018, AISC 945, pp. 321–330, 2020.
https://doi.org/10.1007/978-3-030-18058-4_25

The present paper follows up the work [17], where the authors studied logical connectives: conjuction, disjunction, and symmetric difference together with their negations, from the perspective of a probability measure. An overview of various insights into this issue is provided in [24].

The paper is organized as follows. Section 2 reminds some basic notions and their properties. A special function that associates a probability measure to some logical connectives on a quantum logic is defined and studied in Sects. 3 and 4. In the last Sect. 5 properties of a G-map are compared with properties of a probability measure related to logical connectives on a Boolean algebra.

2 Basic Definitions and Properties

In the first part of this section, we recall fundamental notions: orthomodular lattice, compatibility, orthogonality, state, and their basic properties. For more details, see [4,22]. In the second subsection, we recall some situations with two-dimensional states allowing to model a probability measure of logical connectives in the case of non-compatible events [9,12–18,25].

2.1 Quantum Logic

Definition 1. *An orthomodular lattice (OML) is a lattice L with 0_L and 1_L as the smallest and the greatest element, respectively, endowed with a unary operation $a \mapsto a'$ that satisfies: (i) $a'' := (a')' = a$; (ii) $a \leq b$ implies $b' \leq a'$; (iii) $a \vee a' = 1_L$; (iv) $a \leq b$ implies $b = a \vee (a' \wedge b)$ (the orthomodular law).*

Definition 2. *Elements a, b of an orthomodular lattice L are called*

- *orthogonal if $a \leq b'$; (notation $a \perp b$);*
- *compatible if $a = (a \wedge b) \vee (a \wedge b')$; (notation $a \leftrightarrow b$).*

Definition 3. *A state on an OML L is a function $m : L \to [0, 1]$ such that*

(i) $m(1_L) = 1$;

(ii) $a \perp b$ implies $m(a \vee b) = m(a) + m(b)$.

Note that the notions *state* and *probability measure* are closely tied, and it is clear that $m(0_L) = 0$.

There exist three kinds of OMLs: without any state, with exactly one state and with infinite number of states (see e.g. [21]). The first and the second type of OLMs as a basic structure are not suitable to build a generalized probability theory. The last type of OMLs, which has infinite number of states is considered in the present paper.

Definition 4. *An OML L with infinite number of states is called a quantum logic (QL).*

When studying states on a quantum logic, one can meet some problems, that do not exist on a Boolean algebra. It means, that some of basic properties of probability measures are not necessarilly satisfied for non-compatible random events. Here are some of them: Bell-type inequalities (e.g. [10,11,23,25]), Jauch-Piron state, (e.g. [3,8]), problems of pseudometric (see [17]).

2.2 Probability Measures of Logical Connectives on QLs

In the paper [13], the notion of *a map for simultaneous measurements (an s-map)* on a QL has been introduced. This function is a measure of conjunction even for non-compatible propositions, see e.g. [24].

A map $p : L \times L \to [0,1]$ is called a *map for simultaneous measurements* (abbr. *s-map*) if the following conditions hold:

(s1) $p(1_L, 1_L) = 1$;
(s2) if $a \perp b$ then $p(a,b) = 0$;
(s3) if $a \perp b$ then for any $c \in L$:
$p(a \vee b, c) = p(a,c) + p(b,c)$ and $p(c, a \vee b) = p(c,a) + p(c,b)$.

The following properties of s-map have been proved:
Let $p : L \times L \to [0,1]$ be an s-map and $a, b, c \in L$. Then

1. if $a \leftrightarrow b$ then $p(a,b) = p(a \wedge b, a \wedge b) = p(b,a)$;
2. if $a \leq b$ then $p(a,b) = p(a,a)$;
3. if $a \leq b$ then $p(a,c) \leq p(b,c)$ and $p(c,a) \leq p(c,b)$ for any $c \in L$;
4. $p(a,b) \leq \min(p(a,a), p(b,b))$;
5. the map $m_p : L \to [0,1]$ defined as $m_p(a) = p(a,a)$ is a state on L, induced by p.

The property 1. shows that s-maps can be seen as providing probabilities of 'virtual' conjunctions of propositions, even non-compatible ones, for in the case of compatible propositions the value $p(a,b)$ coincides with the value that a state m_p generated by p takes on the meet $a \wedge b$, which in this case really represents conjunction of a and b [24].

On the other hand, the identity $p(a,b) = p(b,a)$ may not be true in general. So an s-map can be used for describing of stochastic causality [9,15]. Moreover, for any $a \in L$: $m_p(a) = p(a,a) = p(1_L, a) = p(a, 1_L)$.

Measures of logical connectives disjunction (*j-map*) and symmetric difference (*d-map*) are studied on a QL e.g. [14,17].

Let L be a QL. A map $q : L \times L \to [0,1]$ is called a *join map* (abbr. *j-map*) if the following conditions hold:

(j1) $q(0_L, 0_L) = 0, \quad q(1_L, 1_L) = 1$;
(j2) if $a \perp b$ then $q(a,b) = q(a,a) + q(b,b)$;
(j3) if $a \perp b$ then for any $c \in L$:
$q(a \vee b, c) = q(a,c) + q(b,c) - q(c,c)$ and $q(c, a \vee b) = q(c,a) + q(c,b) - q(c,c)$.

If p is an s-map on a QL, m_p is a state induced by p and $q_p : L \times L \to [0,1]$ such that for any $a, b \in L$ $q_p(a,b) = m_p(a) + m_p(b) - p(a,b)$, then q_p is a j-map[1].

Let L be a QL. A map $d : L \times L \to [0,1]$ is called a *difference map* or simply *d-map*[2] if the following conditions hold:

(d1) $d(a,a) = 0$ for any $a \in L$, and $d(1_L, 0_L) = d(0_L, 1_L) = 1$;
(d2) if $a \perp b$ then $d(a,b) = d(a, 0_L) + d(0_L, b)$;
(d3) if $a \perp b$ then for any $c \in L$:
$d(a \vee b, c) = d(a,c) + d(b,c) - d(0_L, c)$ and $d(c, a \vee b) = d(c,a) + d(c,b) - d(c, 0_L)$.

3 Special Bivariables Maps on QLs

In [17], special bivariables maps G satisfying $G(0_L, 1_L) = G(1_L, 0_L)$ have been introduced. The following definition brings an extended version of G-map.

Definition 5. *Let L be a QL. A map $G : L \times L \to [0,1]$ is called a* G-map *if the following holds:*

(G1) if $a, b \in \{0_L, 1_L\}$ then $G(a,b) \in \{0,1\}$;
(G2) if $a \perp b$ then $G(a,b) = G(a, 0_L) + G(0_L, b) - G(0_L, 0_L)$;
(G3) if $a \perp b$ then for any $c \in L$:

$$G(a \vee b, c) = G(a,c) + G(b,c) - G(0_L, c)$$
$$G(c, a \vee b) = G(c,a) + G(c,a) - G(c, 0_L).$$

A G-map enables modelling probability of logical connectives even for non-compatible propositions.

Lemma 1. *Let $G : L \times L \to [0,1]$ be a G-map, where L is a QL. Then for $a \leftrightarrow b$ it holds*

$$G(a,b) = G(a \wedge b, a \wedge b) + G(a \wedge b', 0_L) + G(0_L, a' \wedge b) - 2G(0_L, 0_L).$$

Proof. Consider a, b compatible. Then $a = (a \wedge b) \vee (a \wedge b')$ and $b = (a \wedge b) \vee (a' \wedge b)$. Since $a \wedge b$, $a \wedge b'$ are orthogonal, it follows immediately from Definition 5:

$$\begin{aligned} G(a,b) &= G\left((a \wedge b) \vee (a \wedge b'),\, b\right) = G\left(a \wedge b,\, b\right) + G\left(a \wedge b',\, b\right) - G(0_L,\, b) \\ &= G\left(a \wedge b,\, a \wedge b\right) + G\left(a \wedge b,\, a' \wedge b\right) - G(a \wedge b,\, 0_L) \\ &\quad + G\left(a \wedge b',\, 0_L\right) + G\left(0_L,\, b\right) - G(0_L,\, 0_L) - G\left(0_L,\, b\right) \\ &= G\left(a \wedge b,\, a \wedge b\right) + G\left(a \wedge b,\, 0_L\right) + G\left(0_L,\, a' \wedge b\right) - G\left(0_L,\, 0_L\right) \\ &\quad - G(a \wedge b,\, 0_L) + G\left(a \wedge b',\, 0_L\right) - G\left(0_L,\, 0_L\right) \\ &= G(a \wedge b, a \wedge b) + G(a \wedge b', 0_L) + G(0_L, a' \wedge b) - 2G(0_L, 0_L). \end{aligned}$$ □

[1] It is easy to see that if $a \leftrightarrow b$, then $q_p(a,b) = m_p(a) + m_p(b) - m_p(a \wedge b) = m_p(a \vee b)$ which explains its name.

[2] If $a \leftrightarrow b$, then $d(a,b) = m_d(a \triangle b) = m_d(a \wedge b') + m_d(a' \wedge b)$, where m_d is a state induced by d.

Proposition 1. *Let $G : L \times L \to [0,1]$ be a G-map, where L is a QL. Then the map $G' = 1 - G$ is a G-map.*

Proof. It suffices to verify the rules (G2) and (G3) from Definition 5, because (G1) is obvious.
(G2): Consider $a, b \in L$ such that $a \perp b$. Then

$$\begin{aligned} G'(a,b) &= 1 - G(a,b) = 1 - (G(a,0_L) + G(0_L,b) - G(0_L,0_L)) \\ &= 1 - G(a,0_L) + 1 - G(0_L,b) - 1 + G(0_L,0_L) \\ &= G'(a,0_L) + G'(0_L,b) - G'(0_L,0_L). \end{aligned}$$

(G3): Consider $a, b, c \in L$, where $a \perp b$. Then

$$\begin{aligned} G'(a \vee b, c) &= 1 - G(a \vee b, c) = 1 - (G(a,c) + G(b,c) - G(0_L,c)) \\ &= 1 - G(a,c) + 1 - G(b,c) - 1 + G(0_L,c) \\ &= G'(a,c) + G'(b,c) - G'(0_L,c). \end{aligned}$$

The proof of the second identity is similar. □

There are sixteen families Γ_i, $(i = 1, ..., 16)$ of maps G according to values in vertices $(1_L, 1_L)$, $(1_L, 0_L)$, $(0_L, 1_L)$, $(0_L, 0_L)$. Eight of them with $G(1_L, 0_L) = G(0_L, 1_L)$ are studied in [17]. See Table 1.

Table 1. Results from the paper [17]

	Γ_1	Γ_2	Γ_3	Γ_4	Γ_5	Γ_6	Γ_7	Γ_8
$G(0_L, 0_L)$	0	0	0	0	1	1	1	1
$G(1_L, 0_L)$	0	0	1	1	1	0	0	1
$G(0_L, 1_L)$	0	0	1	1	1	0	0	1
$G(1_L, 1_L)$	0	1	1	0	0	0	1	1
Probability of	0_L	$a \wedge b$	$a \vee b$	$(a \Leftrightarrow b)'$	$a' \vee b'$	$a' \wedge b'$	$a \Leftrightarrow b$	1_L
		$a \leftrightarrow b$	$a \leftrightarrow b$	$a \leftrightarrow b$	$a \leftrightarrow b$	$a \leftrightarrow b$	$a \leftrightarrow b$	

Family Γ_2 is the set of all s-maps (measures of conjuntion), Γ_3 the set of all j-maps (measures of disjunction), and Γ_4 is that of all d-maps (measures of symmetric difference) on a QL (see [17] for more details). In the present paper a map G generating a measure of projection on a QL is studied.

4 Probability Measures of Projections on QLs

This part is devoted to $\Gamma_9 - \Gamma_{12}$ with values in the vertices shown in the Table 2. As Γ_9 and Γ_{10} are analogical, only Γ_9 is studied in detail. Moreover: $G \in \Gamma_{11}$ iff $1 - G \in \Gamma_9$, and $G \in \Gamma_{12}$ iff $1 - G \in \Gamma_{10}$ (Proposition 1 and Table 2).

Table 2. $\Gamma_9 - \Gamma_{12}$ values in vertices

	Γ_9	Γ_{10}	Γ_{11}	Γ_{12}
$G(0_L, 0_L)$	0	0	1	1
$G(0_L, 1_L)$	0	1	1	0
$G(1_L, 0_L)$	1	0	0	1
$G(1_L, 1_L)$	1	1	0	0

Lemma 2. *Let L be a QL and $G \in \Gamma_9$. Then for any $a, b \in L$ it holds*

1. $G(1_L, a) = 1$, $G(0_L, a) = 0$;
2. $G(a, 0_L) = G(a, a) = G(a, 1_L)$;
3. $G(a, 0_L) = \frac{1}{2}(G(a, b) + G(a, b'))$;
4. $G(a, 0_L) = \frac{1}{n}\sum_{i=1}^{n} G(a, b_i)$, *where $b_1, \cdots, b_n$ is an orthogonal partition of unity 1_L.*

Proof. 1. Let $G \in \Gamma_9$ and $a \in L$. Then from

$$1 = G(1_L, 1_L) = G(1_L, a) + G(1_L, a') - G(1_L, 0_L) = G(1_L, a) + G(1_L, a') - 1$$
$$2 = G(1_L, a) + G(1_L, a')$$
$$0 = G(0_L, 1_L) = G(0_L, a) + G(0_L, a') - G(0_L, 0_L) = G(0_L, a) + G(0_L, a'),$$

and taking into account that $G : L^2 \to [0, 1]$, it follows that $G(1_L, a) = 1$ and $G(0_L, a) = 0$.

2. Let $G \in \Gamma_9$ and $a \in L$. Then

$$G(1_L, 0_L) = G(a, 0_L) + G(a', 0_L) - G(0_L, 0_L) = G(a, 0_L) + G(a', 0_L)$$
$$G(1_L, a) = G(a, a) + G(a', a) - G(0_L, a)$$
$$= G(a, a) + G(a', 0_L) + G(0_L, a) - G(0_L, 0_L)$$

As $G(1_L, 0_L) = G(1_L, a)$ and $G(0_L, a) = G(0_L, 0_L) = 0$, one obtains $G(a, 0_L) = G(a, a)$. Moreover,

$$G(a, 1_L) = G(a, a) + G(a, a') - G(a, 0_L)$$
$$= G(a, a) + G(a, 0_L) + G(0_L, a') - G(0_L, 0_L) - G(a, 0_L)$$
$$= G(a, a).$$

3. Let $G \in \Gamma_9$ and $a, b \in L$. Then

$$G(a, 1_L) = G(a, b) + G(a, b') - G(a, 0_L)$$
$$G(a, 1_L) + G(a, 0_L) = G(a, b) + G(a, b')$$
$$2G(a, 0_L) = G(a, b) + G(a, b').$$

4. To prove this identity, it suffices to set $1_L = \bigvee_{i=1}^{n} b_i$ and follow the method used in the previous item. □

Proposition 2. *Let L be a QL, and $G \in \Gamma_9$. Then for any $a, b \in L$ it holds*

1. *If $a \leftrightarrow b$ then $G(a,b) = G(a,0_L)$.*
2. *For any choice of b, the map $m_b : L \to [0,1]$: $m_b(a) = G(a,b)$ is a state on L.*

Proof. Let $G \in \Gamma_9$.

1. Consider $a, b \in L, a \leftrightarrow b$. Then according to Lemmas 1 and 2

$$G(a,b) = G(a \wedge b, a \wedge b) + G(a \wedge b', 0_L) + G(0_L, a' \wedge b) - 2G(0_L, 0_L)$$
$$G(a,b) = G(a \wedge b, 0_L) + G(a \wedge b', 0_L) - G(0_L, 0_L)$$
$$G(a,b) = G(a, 0_L).$$

2. Consider $b \in L$, and define $m_b : L \to [0,1]$: $m_b(a) = G(a,b)$. Then

$$m_b(1_L) = G(1_L, b) = 1,$$

and for arbitrary $x, y \in L$, $x \perp y$, it holds

$$m_b(x \vee y) = G(x \vee y, b) = G(x,b) + G(y,b) - G(0_L, b) = m_b(x) + m_b(y).$$

Therefore m_b is a state on L. □

From Proposition 2 it follows that any $G \in \Gamma_9$ is a probability measure of the projection onto the first coordinate. Analogical properties are fullfiled for any $G \in \Gamma_{10}$, which is a probability measure of the projection onto the second coordinate.

If L is a Boolean algebra, then for any $G \in \Gamma_9$ ($G \in \Gamma_{10}$) it holds $G(a,b) = G(a,0_L)$ ($G(a,b) = G(0_L,b)$) for all $a, b \in L$. If L is a QL but not a Boolean algebra, then the equality does not hold in general, as illustrates the following example.

Table 3. G-maps from Γ_9 on a horizontal sum of Boolean algebras

	a	a'	b	b'	0_L	1_L
a	α	α	r_1	r_2	α	α
a'	$1-\alpha$	$1-\alpha$	$1-r_1$	$1-r_2$	$1-\alpha$	$1-\alpha$
b	u_1	u_2	β	β	β	β
b'	$1-u_1$	$1-u_2$	$1-\beta$	$1-\beta$	$1-\beta$	$1-\beta$
0_L	0	0	0	0	0	0
1_L	1	1	1	1	1	1

Table 4. States on L

	a	a'	b	b'	0_L	L
m_b	r_1	$1-r_1$	β	$1-\beta$	0	1
m_0	α	$1-\alpha$	β	$1-\beta$	0	1

Example 1. Consider $L = \{0, 1, a, a', b, b'\}$, a horizontal sum of Boolean algebras $\mathcal{B}_a = \{0, 1, a, a'\}$, $\mathcal{B}_b = \{0, 1, b, b'\}$. Consider $r_1, r_2, u_1, u_2 \in [0, 1]$. Every $G \in \Gamma_9$ can be fully defined by Table 3, where $\alpha = \frac{1}{2}(r_1 + r_2)$, $\beta = \frac{1}{2}(u_1 + u_2)$ according to Lemma 2. If $r_1 \neq r_2$ then $G(a, b) \neq G(a, 0_L)$.

From Table 3, one can extract all states on L, related to the choice of r_1, r_2, u_1, u_2. Each column in the Table 3 represents a state on L. As example, m_b and m_0 are in Table 4.

Definition 6. *Let $G \in \Gamma_9$. The map G is called a pure projection if $G(a, b) = G(a, 0_L)$ for any $a, b \in L$.*

Proposition 3. *For every s-map p there exists a G–map $G_p \in \Gamma_9$ such that $G_p(a, b) = G_p(a, 0_L)$.*

Proof. Set $G_p(a, b) = p(a, b) + p(a, b') = p(a, a)$, where p is an arbitrary s-map. Then $G_p \in \Gamma_9$ and $G_p(a, b) = G_p(a, 0_L)$ for any $b \in L$. □

The results for $\Gamma_9 - \Gamma_{12}$ are summarized in Table 5.

Table 5. Results for $\Gamma_9 - \Gamma_{12}$

	Γ_9	Γ_{10}	Γ_{11}	Γ_{12}
probability of	a	b	a'	b'

5 Summary

Two issues related to the G-map on a quantum logic arised: existence of this function and its properties on a QL. The existence of s-map has been solved in [12,13].

Some features of G-map are summarized in the following:

- The classes Γ_2, Γ_3, Γ_5, Γ_6 are mutually isomorphic.
- Each probability measure on $\mathcal{B}$ induces a pseudometric. It means, that for any probability measure m, the map d_m: $d_m(a, b) = m(a \wedge b') + m(a' \wedge b)$ is a pseudometric on $\mathcal{B}$ induced by m.
  If $p \in \Gamma_2$ and $d_p(a, b) = p(a, b') + p(a', b)$, then $d_p \in \Gamma_4$ and it does not have to be a pseudometric on the quantum logic.

- Let L be a QL, m be a state and p be an s-map. Then the first Bell-type inequality (1) is not necessarily fulfilled and the version (2) is always fulfilled.

$$m(a) + m(b) - m(a \wedge b) \leq 1 \tag{1}$$
$$p(a,a) + p(b,b) - p(a,b) \leq 1 \tag{2}$$

The second Bell-type innequality (3) is not necessarily fulfilled and the version (4) is fulfilled for every s-map, which induces a pseudometric on L [25].

$$m(a) + m(b) + m(c) - m(a \wedge b) - m(a \wedge c) - m(c \wedge b) \leq 1 \tag{3}$$
$$p(a,a) + p(b,b) + p(c,c) - p(a,b) - p(a,c) - p(c,b) \leq 1 \tag{4}$$

- Analogically implication (5) (Jauch-Piron state, see e.g. [3,8]) can be violated on a QL L but implication (6) is always valid

$$m(a) = m(b) = 1 \quad \Rightarrow \quad m(a \wedge b) = 1 \tag{5}$$
$$p(a,a) = p(b,b) = 1 \quad \Rightarrow \quad p(a,b) = 1, \tag{6}$$

and moreover for any $c \in L$ $p(a,c) = p(c,a) = p(c,c)$.
- The classes Γ_9 – Γ_{12}: On a Boolean algebra, every projection is a pure projection. On a quantum logic, a G-map is not necessarilly a pure projection, see Example 1.

The modeling of random events using of quantum logics allows to test, among other things, stochastic causality.

Acknowledgement. The author (O. Nánásiová) would like to thank for the support of the VEGA grant agency by means of grant VEGA 1/0710/15 and VEGA 1/0159/17 and the author (L. Valášková) would like to thank for the support of VEGA 1/0420/15.

References

1. Nánásiová, O., Čerňanová, V., Valášková, Ľ.: Probability measures and projections on quantum logics. In: Kulczycki, P., Kowalski, P.A., Łukasik, S. (eds.) Contemporary Computational Science, p. 78. AGH-UST Press, Cracow (2018)
2. Birkhoff, G., von Neumann, J.: The logic of quantum mechanics. Ann. Math. **37**(4), 823–843 (1936). second series
3. Bunce, L.J., Navara, M., Pták, P., Maitland Wright, D.: Quantum logics with Jauch-Piron states. Q. J. Math. **36**(3), 261–271 (1985)
4. Dvurečenskij, A., Pulmannová, S.: New Trends in Quantum Structures. Springer, Dordrecht (2000). ISBN 978-94-017-2422-7
5. Dvurečenskij, A., Pulmannová, S.: Connection between joint distribution and compatibility. Rep. Math. Phys. **19**(3), 349–359 (1984)
6. Sozzo, S.: Conjunction and negation of natural concepts: a quantum-theoretic modeling S Sozzo. J. Math. Psychol. **66**, 83–102 (2015)
7. Herman, L., Marsden, L., Piziak, R.: Implication connectives in orthomodula lattices. Notre Dame J. Formal Logic **XVI**(3), 305–326 (1975)

8. Jauch, J.M., Piron, C.: On the structure of quantal proposition systems. Helv. Phys. Acta **42**, 842–848 (1969)
9. Kalina, M., Nánásiová, O.: Calculus for non-compatible observables, construction through conditional states. Int. J. Theor. Phys. **54**(2), 506–518 (2014)
10. Khrennikov, A.Y.: EPR-Bohm experiment and Bell's inequality: quantum physics meets probability theory. TMF **157**(1), 99–115 (2008). (Mi tmf6266)
11. Khrennikov, A.: Violation of Bell's inequality and non-Kolmogorovness. In: Accardi, L., et al. (eds.) Foundations of Probability and Physics-5. American Institute of Physics, Mellville (2009)
12. Nánásiová, O.: Principle conditionig. Int. J. Theor. Phys. **43**(7–8), 1757–1768 (2004)
13. Nánásiová, O.: Map for simultaneous measurements for a quantum logic. Int. J. Theor. Phys. **42**(9), 1889–1903 (2003)
14. Nánásiová, O., Drobná, E., Valášková, Ľ.: Quantum logics and bivariable functions. Kybernetika **46**(6), 982–995 (2010)
15. Nánásiová, O., Khrennikov, A.: Representation theorem of observables on a quantum system. Int. J. Theor. Phys. **45**(3), 469–482 (2006)
16. Nánásiová, O., Pykacz, J.: Modelling of uncertainty and bi-variable maps. J. Electr. Eng. **67**(3), 169–176 (2016)
17. Nánásiová, O., Valášková, Ľ.: Maps on a quantum logic. Soft Comput. **14**(10), 1047–1052 (2010)
18. Nánásiová, O., Valášková, Ľ.: Marginality and triangle inequality. Int. J. Theor. Phys. **49**(12), 3199–3208 (2010)
19. Pavičić, M., Megill, N.D.: Is quantum logic a logic? In: Engesser, K., Gabbay, D., Lehmann, D. (eds.) Handbook of Quantum Logic and Quantum Structures, pp. 23–47. Elsevier, Amsterdam (2009)
20. Pavičić, M.: Classical logic and quantum logic with multiple and common lattice models. Hindawi Publishing Corporation Advances in Mathematical Physics volume 2016, Article ID 6830685, 12 pages (2016)
21. Pavičić, M.: Exhaustive generation of orthomodular lattices with exactly one nonquantum state. Rep. Math. Phys. **64**, 417–428 (2009)
22. Pták, P., Pulmannová, S.: Orthomodular Structures as Quantum Logics. Springer, Netherlands (1991)
23. Pitovsky, I.: Quantum Probability-Quantum Logic. Springer, Berlin (1989)
24. Pykacz, J., Frackiewicz, P.: The problem of conjunction and disjunction in quantum logics. Int. J. Theor. Phys. **56**(12), 3963–3970 (2017)
25. Pykacz, J., Valášková, L., Nánásiová, O.: Bell-type inequalities for bivariate maps on orthomodular lattices. Found. Phys. **45**(8), 900–913 (2015)
26. Sergioli, G., Bosyk, G.M., Santucci, E., Giuntini, R.: A quantum-inspired version of the classification problem. Int. J. Theor. Phys. **56**, 3880–3888 (2017). https://doi.org/10.1007/s10773-017-3371-1

Statistical Analysis of Models' Reliability for Punching Resistance Assessment

Jana Kalická, Mária Minárová(✉), Jaroslav Halvoník, and Lucia Majtánová

Slovak University of Technology, Bratislava, Slovakia
{jana.kalicka,maria.minarova,jaroslav.halvonik,lucia.majtanova}@stuba.sk

Abstract. The paper deals with statistical analysis of engineering data set. The purpose of analysis is to stipulate suitability of formulas that compete for being involved in prepared EuroCode that will be valid from 2020. Authors dispose with a sufficient numbers of lab tests. Having input geometrical and physical parameters of each experiment at hand, the corresponding theoretical value is computed by using three formulas provided by three models. Case by case, the ratio between measured and theoretical value reveal the safety immediately: greater then one means safety, less then one means failure. This ratio stands as the one parametric dimensionless statistical variable which is analysed afterwards.

The preliminary version of this paper was presented at the 3rd Conference on Information Technology, Systems Research and Computational Physics, 2–5 July 2018, Cracow, Poland [3].

Keywords: Punching reliability · Kolmogorov - Smirnov test · Tuckey's fence · Quartiles · Coefficient of variance · Data mining

1 Flat Slabs Punching Resistance

Punching is the most often failure of flat slabs. It is very dangerous due to its brittle character. Brittle character means that the failure spreads very quickly towards all directions causing a progressive collapse. Indeed, the slabs have to be build up with respect to prior investigation based significantly on experimental results. The investigation aiming to the constructions failure avoiding together with economical aspects concerned resulted in normative prescriptions ensuring the their further safety. The adequate models built up and implemented, finally resulted in normative formulas that should ensure the safety of the designed constructions.

1.1 Three Normative Forms for Reliability Computation

The first mode, fully empirical, was introduced in Model Code in 1990, [8]. It is based solely on the statistical analysis. It stipulates the shear stress dependence on the reinforcement ratio ρ and on the concrete compressive stress f_{ck}, formula (1). Later the nonlinear elasticity theory was employed in the investigation and

P. Kulczycki et al. (Eds.): ITSRCP 2018, AISC 945, pp. 331–338, 2020.
https://doi.org/10.1007/978-3-030-18058-4_26

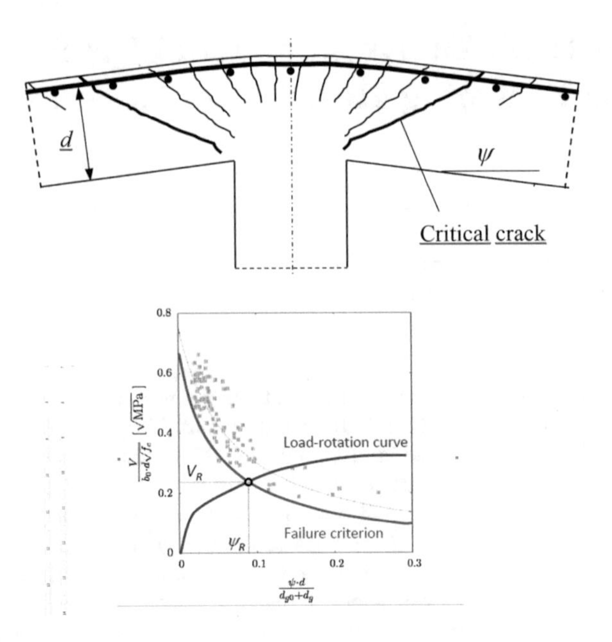

Fig. 1. Theory of critical shear crack, [1]

the Critical shear theory was developed. The theory was introduced by Muttoni and Schwartz in [4] and upgraded by Mutttoni in [5], resulting in Model Code 2010 normative form (2), [6]. The model is more complex as it inter alia includes greater number of geometrical and physical parameters, e.g. longitudinal shear reinforcement, and magnitude of the aggregate. Lately, due to some simplification effort, the third model EC (2017) was developed, [7]. Afterwards it was included to the second generation of Euro Code of the second generation, EC2 (Fig. 1).

Remark. Each model consists in one principal formula quantifying the shear stress resistance $V_{Rd,c}$; for better clarity sake, some auxiliary forms are provided.

Model EC2 (2004)

$$V_{Rd,c} = \frac{C_{Rk,c}}{\gamma_c} k(100\rho f_{ck})^{\frac{1}{3}} u_1 d \tag{1}$$

with $C_{Rk,c}[MPa]$ being an empirical factor, $\gamma_c[-]$ partial safety factor, $k[-]$ size effect factor, $k = 1+(200[mm]/d[mm])^{\frac{1}{2}}$, $d[m]$ effective depth of the slab, i.e. the vertical distance from the bottom of the slab up to the reinforcement placement, $u_1[m]$ basic control perimeter at the distance $2d$ from the axis of the column,

$$\rho = (\rho_x \rho_y)^{\frac{1}{2}} \tag{2}$$

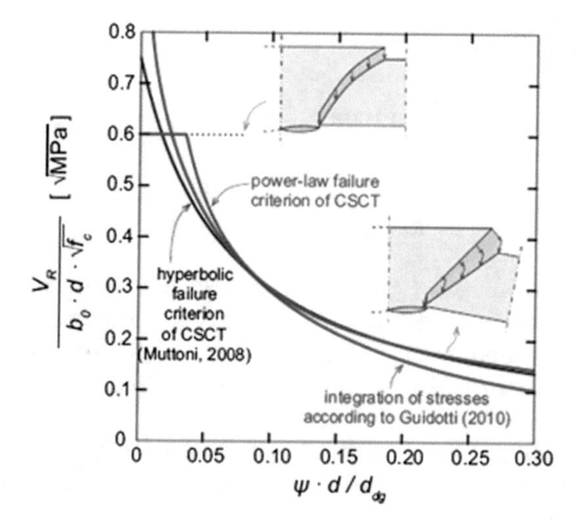

Fig. 2. Graphical performance of failure criterion as stipulated by different models

Herein, ρ_x and $\rho_y[-]$ are the reinforcement ratios in x and y direction respectively:

$$\begin{aligned} \rho_x &= \tfrac{A_{sx}}{d_x b} \\ \rho_y &= \tfrac{A_{sy}}{d_y b} \end{aligned} \tag{3}$$

with A_{sx} and A_{sy} being the areas of reinforcement in x and y direction respectively, b longitude of specimen, set to 1 (Fig. 2).

Model MC (2010)

$$V_{Rd,c} = k_\psi \frac{\sqrt{f_{ck}}}{\gamma_c} b_0 d_v \tag{4}$$

$$\begin{aligned} k_\psi &= \tfrac{1}{1.5+0.9k_{dg}\psi d} \\ \psi &= \tfrac{r_s}{d}\tfrac{f_y d}{E_s}\left(\tfrac{m_{Sd}}{m_{Rd}}\right)^{\frac{3}{2}} \end{aligned} \tag{5}$$

with $d_v[m]$ being the effective depth of the slab, usually $d_v = d$, $b_0[m]$ the length of control perimeter at the distance $d_v/2$, $k_{dg}[m]$ factor involving maximal aggregate magnitude $d_g[mm]$: $k_{dg} = \frac{32}{16+d_g}$, $r_s[m]$ distance from axis of the column to the line of contraflexure of radial bending momentums, $f_{yd}[MPa]$ yielded strength of principal reinforcement, $m_{Sd}[m^{-1}]$ average design bending capacity per unit length, $m_{Rd}[m^{-1}]$ average design bending capacity per unit length.

Model EC2 (2017)

$$V_{Rd,c} = \min\left\{\frac{k_b}{\gamma_c}\left(100\rho f_{ck}\frac{d_g}{a_v}\right)^{\frac{1}{3}} b_0 d_v, \frac{0.6}{\gamma_c}\sqrt{f_{ck}} b_0 d_v\right\} \tag{6}$$

$$k_b = \max\left\{1, \sqrt{8\mu\frac{d}{b_0}}\right\} \tag{7}$$

with $a_v[m]$ being shear span ($\geq 2.5d$), geometric average of shear spans in both orthogonal directions, μ parameter accounting shear force and bending momentums in the shear region, in case of indoor column without unbalanced momentum it is set to 8. Accordingly, for normal weight concrete it is taken $d_g = 32\,\text{mm}$.

Remark. These three models performed by formulas for $V_{Rd,c}$ computation, compete for becoming normative in the on-coming unified European norm that is intended to be valid from 2020.

2 Data Description

The laboratory punching resistance tests releasing the data of shear resistance in a lot of variations were executed through the decades. The data, i.e. results of tests together with input geometrical and physical parameters, were collected in databases. Nowadays there exist one large database where great number of input and output experimental data are recorded. From its inception, with months and years lapsing, the observations were recorded in more and more details. As the oldest experiments included are from 1938, the amount of checked input parameters has been enhanced as well as the outputs recorded. That is why, due to lack of some parameters (prevalently in older experiments), not each lab test can be involved in data analysis carried out. The database still enhances by new experiment items. Some of them are executed at our university. With aim of reliable comparison among three mentioned models, the statistical analysis is carried out that validates the match between each theoretical and experimental values. Due to the reasons mentioned above, in our analysis 404 items of EC2004, 385 items of Model Code 2010 and 385 items of Euro Code 2017 are taken into account.

There are a lot of characteristics within the data to be analysed. Three of them, the most influencing, are withdrawn as being of our further interest:

- effective depth $d[m]$, whereby $d \in [50, 660.9]$
- reinforcement ratio $\rho[-]$, whereby $\rho \in [0.0025, 0.0702]$
- cylindrical stiffness of concrete f_{ck}, whereby $f_{ck} \in [9.282, 119]$.

All values (3 theoretical and 1 experimental) are included in the database. In accordance with civil engineering practice, not differences, but the ratios between experimental value V_{test} and corresponding theoretical value $V_{Rd,c}$ (V_{model}) will be treated, namely in three cases of $V_{Rd,c}$: that yielded form (1), from (4) or from (7), respectively. Thus the ratio $\frac{V_{test}}{V_{Rd,c}}$ stands as the statistical variable to be handled. The item $\frac{V_{test}}{V_{Rd,c}}$ enhances the five-dimensional vector $(d, \rho, f_{ck}, V_{Rd,c}, V_{test})$ belonging to each model, by one. Moreover, it is worth to note that the ratio above 1 means safety, under 1 means failure. The statistical investigation and data mining is exerted.

3 Graphical Analysis

By using graphical outputs of statistical analysis result, a synoptic comparison of the three models can be performed. Graphical analysis involves histograms depicting the probability distribution of ratio values, box-and-whiskers diagrams demonstrating the overall distribution of data set, quartile distribution as well and detecting the outstanding data. All of three data sets involve outstanding values. By using Tuckey's fence (8) we detect and omit the outstanding data (outliers) from the next analysis.

$$[Q_1 - k(Q_3 - Q_1, Q_3 + k(Q_3 - Q_1))] \tag{8}$$

with Q_1 being the first (lower) quartile, Q_3 the third (upper) quartile, k coefficient of outlying, usually $k = 1.5$ for outliers, $k = 3$ for far out values. From Figs. 3, 4 and 5 it is evident that model EC2 (2017) is not sufficiently safe. The Table 1 affirms that even the mean value is below one. Histograms refer to the normality of data distribution in all three cases both before and after excluding the outstanding data. From the point of view of civil engineering practice it is interesting to trace the outstanding data with regard to the particular input parameters. In Fig. 5 we provide an example of such an approach. It is apparent that the most outstanding data are situated in the interval of the most used values of f_{ck}, i.e. $[20, 40]$. The normality is affirmed by Kolmogorov-Smirnov test, see Sect. 4.

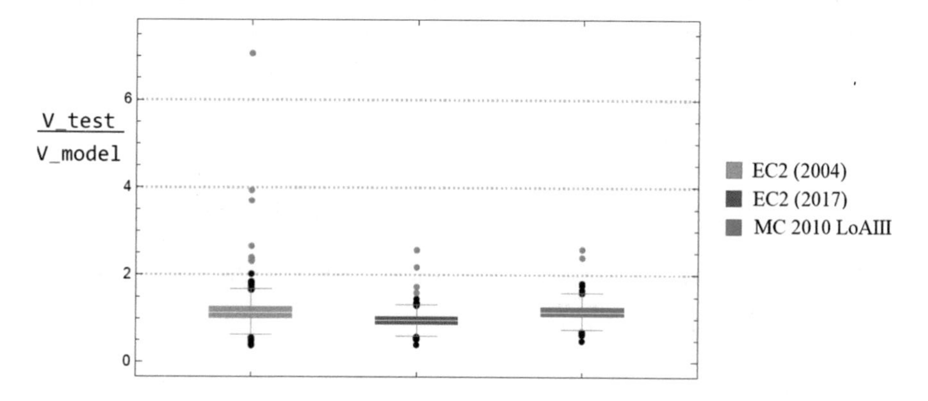

Fig. 3. Box plot of ratio experimental/theoretical value of punching resistance EC2 (2004), EC2(2017), Model Code (2010), original data

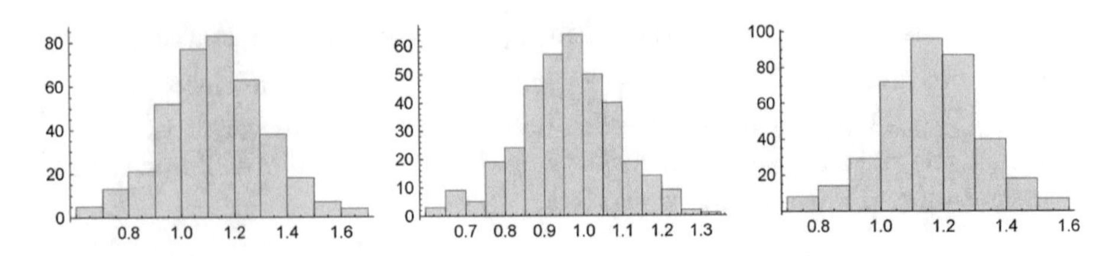

Fig. 4. Histograms of ratio experimental/theoretical value of punching resistance EC2 (2004), EC2 (2017), Model Code (2010), outstanding data excluded

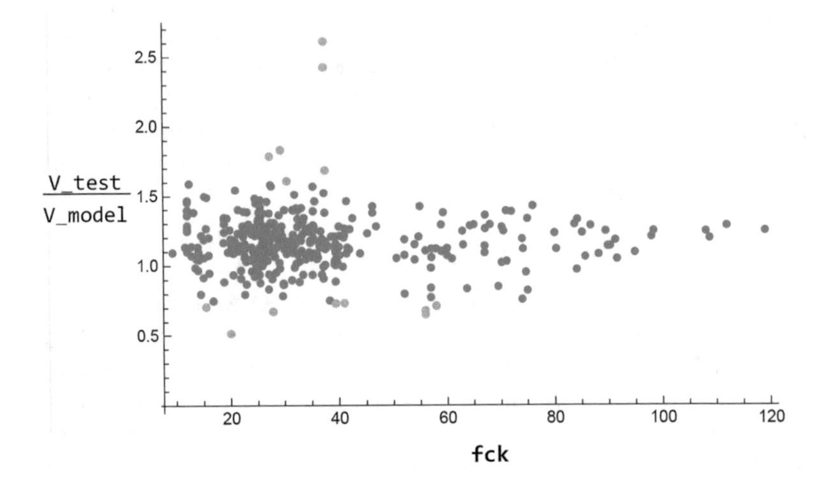

Fig. 5. Model Code (2010), outstanding data (in yellow color) with regard to f_{ck}

4 Computational Data Analysis

Computational analysis begins with stipulation of the basic statistical characteristics computation, e.g. estimate of the mean and standard deviation.

$$\hat{\mu} = \bar{x} = \frac{1}{n}\sum_{i=1}^{n} x_i, \quad \hat{\sigma} = s_x = \sqrt{\frac{1}{n-1}\sum_{i=1}^{n}(x_i - \bar{x})^2}, \quad V_x = \frac{\bar{x}}{s_x} \tag{9}$$

with $x_i = \frac{V_{test}}{V_{model}}$, n number of measurements. For the civil engineers, also 0.05 quantile and safe and unsafe zone splitting is interesting.

Normality of data distribution of all three models is affirmed by the Kolmogorov - Smirnov test (significance level $\alpha = 0.05$).

Table 1. Statistical parameters of particular models

Model	Median	Average	Standard deviation	Coefficient of variation	0.05 quantile	$P(\frac{V\_test}{V\_model} \leq 1)$
EC2 (2004)	1.12054	1.12842	0.186157	0.164971	0.809558	0.238845
EC2 (2017)	0.958613	0.96237	0.124694	0.12957	0.760439	0.627072
MC2010	1.15752	1.16523	0.160081	0.137381	0.884772	0.137466

5 Results Interpretation

The statistical analysis demonstrates that best match between theoretical and experimental value of reliability within the required input data range is achieved by using Model Code (2010). It is sufficiently safe and economically optimal at the same time (Fig. 6).

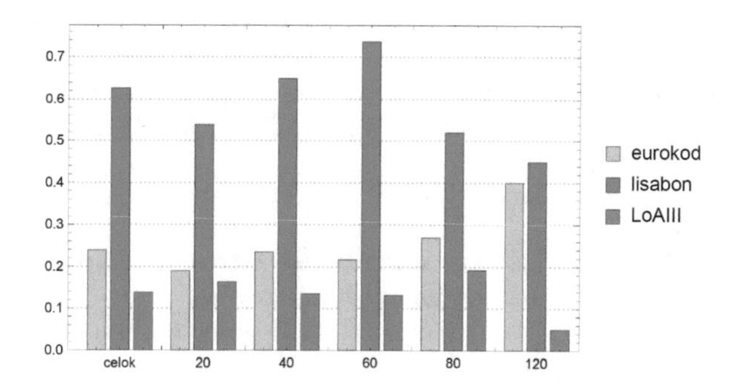

Fig. 6. Amount of data below 1, i.e. on the side of unsafety, in particular three cases

6 Conclusion

The importance of the statistical analysis of data and data mining in engineering practice is indisputable. Validation of three models is influencing factor of future normative formulas utilization in civil building practice. It contributes to the higher safety of construction having economical optimality in mind.

Acknowledgement. This work was supported by grants APVV-14-0013, VEGA 1/0810/16 and VEGA-1/0420/15.

References

1. EN1992-1-1 Design of Concrete Structures, Part 1-1 General Rules and Rules for Buildings, May 2004
2. Fédération Internationale du Béton (fib), Model Code 2010 - Final draft, vol. 1, Fédération Internationale du Béton, Bulletin 65, Lausanne, Switzerland, vol. 2 (2012)
3. Kalická, J., Minárová, M., Halvoník, J., Majtánová, L.: Statistical analysis of models' reliability for punching resistance assessment. In: Kulczycki, P., Kowalski, P.A., Łukasik, S. (eds.) Contemporary Computational Science, p. 79. AGH-UST Press, Cracow (2018)
4. Muttoni, A., Schwartz, J.: Behaviour of beams and punching in slabs without shear reinforcement. In: IABSE Colloquium, Switzerland, Zurich, vol. 62, pp. 703–708 (1991)
5. Muttoni, A., Fernández, Ruiz M.: Shear strength of members without transverse reinforcement as function of critical shear crack width. ACI Struct. J. **105**(2), 163–172 (2008)
6. Muttoni, A., Fernández Ruiz, M.: The levels-of-approximation approach in MC 2010: applications to punching shear provisions. Struct. Concr. **13**(1), 32–41 (2012)
7. Muttoni, A., Fernández Ruiz, M.: The critical shear crack theory for punching design: From mechanical model to closed-form design expressions. Punching shear of structural concrete slabs, Honoring Neil M. Hawkins ACI-fib International Symposium, pp. 237-252, April 2017
8. Zsutty, T.: Beam shear strength prediction by analysis of existing data. ACI J. **65**(11), 943–951 (1968)
9. The Wolfram Language: Fast Introduction for Programmers. http://www.wolfram.com/language/fast-introduction-for-programmers/en/
10. Mendelhall, W., Sincich, T.: Statistics for the Engineering and Computer Sciences. Dellen Publishing Company, San Francisco (1988)

On Persistence of Convergence of Kernel Density Estimates in Particle Filtering

David Coufal[1,2](✉)

[1] Czech Academy of Sciences, Institute of Computer Science, Pod Vodárenskou věží 2, 182 07 Praha 8, Czech Republic
david.coufal@cs.cas.cz

[2] Faculty of Mathematics and Physics, Department of Probability and Mathematical Statistics, Charles University, Sokolovská 83, 186 75 Praha 8, Czech Republic

Abstract. A sufficient condition is provided for keeping the character of the filtering density in the filtering task. This is done for the Sobolev class of filtering densities. As a consequence, estimating the filtering density in particle filtering persists its convergence at any time of filtering. Specifying the condition complements previous results on using the kernel density estimates in particle filtering.

Keywords: Particle filtering · Kernel density estimates · Convergence

1 Introduction

The task of filtering is the task of finding the optimal estimate of the signal's state, which is not directly observable, using available indirect observations [9, 10]. Particle filtering is an effective tool for solving the filtering task in non-linear and/or non-Gaussian settings when a closed-form solution is not available [7,8].

Formulating the filtering task in the context of probability theory leads to determining the conditional distribution of the signal's state conditioned on observations. This conditional distribution is called the *filtering distribution*. When the filtering distribution cannot be specified exactly, we are interested in its approximation. This may be accomplished by using particle filtering.

During its operation, a particle filter [8,10] generates a set of points, called *particles*, from the signal's state space. The set of particles then constitutes an empirical measure that approximates the filtering distribution. This measure has the form of a weighted sum of Dirac measures located at the individual particles. Estimating the filtering distribution's generalized moments boils down to weighted summation of moments' defining functions over the individual particles.

Provided the number of generated particles goes to infinity, the convergence results for particle filters [4,5,7] ensure that the estimates converge to the corresponding theoretical values.

Although the particle filters allow for approximation of the generalized moments, the question of approximating the filtering distribution's density - the *filtering density* is the subject of both theoretical and practical interest [2,5].

P. Kulczycki et al. (Eds.): ITSRCP 2018, AISC 945, pp. 339–346, 2020.
https://doi.org/10.1007/978-3-030-18058-4_27

It is natural to use samples from particle filtering to approximate the filtering density. The problem here, however, is that the samples do not represent an i.i.d. samples from the filtering distribution. There is a weak dependence among particles due to the resampling step in the particle filtering algorithm [7]. Hence it might be disputable to use the standard methodologies for density approximation in particle filtering as they draw on the i.i.d. character of the sample [11–13]. In spite of this obstacle, however, it can be shown that the standard kernel density estimation methodology is still applicable here [2,5].

In [2], the convergence of the kernel density estimates to the filtering density has been proved using techniques of Fourier analysis [12]. The result also provides the upper bounds on the error of estimation. This has been further extended for partial derivatives of the filtering density along with providing the corresponding lower bounds in [1].

In the kernel estimates convergence theorem, there is assumed that the filtering density has certain character, namely the Sobolev one [2]. Whilst this assumption can be checked directly for the initial density of the signal process, this is not the case for other operation times as we do not have at our disposal an explicit representation of the filtering density. The purpose of this paper is to provide a handy tool for checking persistence of the Sobolev character of the filtering density over time, and, as the consequence, to ensure that the kernel estimates converge through whole operation time of the filter.

The rest of the paper is organized as follows. The next section describes in more details the particle filtering algorithm and introduces the notions important for presenting the main result. The third section delivers the announced sufficient condition on persistence of the Sobolev character and the fourth section concludes the paper.

The preliminary version of this paper was presented at the 3rd Conference on Information Technology, Systems Research and Computational Physics, 2–5 July 2018, Cracow, Poland [3].

2 Preliminaries

Here we provide a very basic overview of particle filtering and kernel density estimation along with relevant notions so that the main result could be presented in the proper context.

2.1 Filtering Task

The probabilistic formulation of the filtering task reads as follows. Let $\{\boldsymbol{X}\}_{t=0}^{\infty}$, $\{\boldsymbol{Y}\}_{t=1}^{\infty}$ be two stochastic processes with continuous state space, defined on a common probability space $(\Omega, \mathcal{A}, P)$. The first process $\{\boldsymbol{X}\}_{t=0}^{\infty}$, $\boldsymbol{X}_t : (\Omega, \mathcal{A}) \to (\mathbb{R}^d, \mathcal{B}(\mathbb{R}^d))$, $t \in \mathbb{N}_0$, $d \in \mathbb{N}$ is called the *signal process* and represents a generally inhomogeneous Markov chain. Its behaviour is determined by the initial distribution of $\boldsymbol{X}_0$ and the set of transition kernels $K_{t-1} : \mathcal{B}(\mathbb{R}^d) \times \mathbb{R}^d \to [0,1]$, $t \in \mathbb{N}$ where $\mathcal{B}(\mathbb{R}^d)$ denotes the standard Borel σ-algebra on $\mathbb{R}^d$. The second process,

$\boldsymbol{Y}_t : (\mathbb{R}, \mathcal{B}(\mathbb{R}^d)) \rightarrow (\mathbb{R}^{d_y}, \mathcal{B}(\mathbb{R}^{d_y}))$, $t \in \mathbb{N}$, $d_y \in \mathbb{N}$ is the *observation process*. Its values are given by transforming the signal states and consequent corruption with an independent noise: $\boldsymbol{Y}_t = h_t(\boldsymbol{X}_t) + \boldsymbol{V}_t$, $t \in \mathbb{N}$. The transformation functions $h_t : \mathbb{R}^d \rightarrow \mathbb{R}^{d_y}$, $t \in \mathbb{N}$ as well as the distributions of the noise random variables $\boldsymbol{V}_t : (\Omega, \mathcal{A}) \rightarrow (\mathbb{R}^{d_y}, \mathcal{B}(\mathbb{R}^{d_y}))$, $t \in \mathbb{N}$ are assumed to be known.

The general solution to the filtering task is the specification of the *filtering distribution* $\pi_t(d\boldsymbol{x}_t) = P(\boldsymbol{X}_t \in d\boldsymbol{x}_t | \boldsymbol{Y}_1, \ldots, \boldsymbol{Y}_t) = P(\boldsymbol{X}_t \in d\boldsymbol{x}_t | \boldsymbol{Y}_{1:t})$. More specifically, assuming that all the involved distributions have densities with respect to the corresponding (in the sense of the dimension) Lebesgue measures, we are interested in the specification of the density of π_t. Naturally, this density is called the *filtering density* and denoted $p(\boldsymbol{x}_t|\boldsymbol{y}_{1:t})$, i.e., $\pi_t(d\boldsymbol{x}_t) = p(\boldsymbol{x}_t|\boldsymbol{y}_{1:t})\, d\boldsymbol{x}_t$. For other distributions we assume that $\pi_0(d\boldsymbol{x}_0) = p_0(\boldsymbol{x}_0)\, d\boldsymbol{x}_0$, $K_{t-1}(d\boldsymbol{x}_t|\boldsymbol{x}_{t-1}) = K_{t-1}(\boldsymbol{x}_t|\boldsymbol{x}_{t-1})\, d\boldsymbol{x}_t$ and $P(\boldsymbol{V}_t \in d\boldsymbol{v}_t) = g_t^v(\boldsymbol{v}_t)\, d\boldsymbol{v}_t$ with g_t^v being bounded and strictly positive, i.e., $||g_t^v||_\infty < \infty$ and $g_t^v(\boldsymbol{v}_t) > 0$. In the filtering equations below, we also consider the conditional density of $P(\boldsymbol{Y}_t \in d\boldsymbol{y}_t | \boldsymbol{X}_t = \boldsymbol{x}_t)$ which writes $g_t(\boldsymbol{y}_t|\boldsymbol{x}_t) = g_t^v(\boldsymbol{y}_t - h_t(\boldsymbol{x}_t))$. See [2], for more details.

Though the joint density of states and observations $p(\boldsymbol{x}_{0:t}, \boldsymbol{y}_{1:t})$ can be specified in the closed form [2,7,10], this is not the case for the filtering density due to coping with a generally intractable integration term when going from the joint $p(\boldsymbol{x}_{0:t}, \boldsymbol{y}_{1:t})$ to the marginal $p(\boldsymbol{x}_t|\boldsymbol{y}_{1:t})$. A certain progress, however, can be made towards this direction by writing down the following recurrence equations, called the *filtering equations* [2,10]:

$$p(\boldsymbol{x}_t|\boldsymbol{y}_{1:t-1}) = \int K_{t-1}(\boldsymbol{x}_t|\boldsymbol{x}_{t-1}) p(\boldsymbol{x}_{t-1}|\boldsymbol{y}_{1:t-1})\, d\boldsymbol{x}_{t-1}, \tag{1}$$

$$p(\boldsymbol{x}_t|\boldsymbol{y}_{1:t}) = \frac{g_t(\boldsymbol{y}_t|\boldsymbol{x}_t) p(\boldsymbol{x}_t|\boldsymbol{y}_{1:t-1})}{\int g_t(\boldsymbol{y}_t|\boldsymbol{x}_t) p(\boldsymbol{x}_t|\boldsymbol{y}_{1:t-1})\, d\boldsymbol{x}_t}, \quad t \in \mathbb{N}. \tag{2}$$

The filtering equations provide an insight into the filtering density's development over time. The development is split into two steps. The first step is called the prediction step and the second the update step. Yet, despite this breakdown the formulas do not generally leads to a closed-form specification of the filtering density. Lacking a convenient representation for the filtering distribution, particle filtering is used to get its approximation in practice [2,5,7].

2.2 Particle Filtering

Particle filtering employs the Markovian structure of the filtering task and generates the set of particles $\{\boldsymbol{x}_t^i \in \mathbb{R}^d\}_{i=1}^n$ at every step of its operation $t = 1, \ldots, T$. The number of generated particles $n \in \mathbb{N}$ is a design parameter in the particle filtering algorithm and $T \in \mathbb{N}$ is a finite computation horizon.

The particle filtering algorithm is internally split into several steps that partially follows from the specification of the filtering equations [7,8]. The filtering distribution is primarily approximated by a non-uniformly weighted empirical measure. To prevent the algorithm from degenerating, the resampling step is introduced. Several resampling schemes has been proposed [6]. Whilst resampling improves stability of the algorithm, it also brings a weak dependence among generated particles. In effect, the empirical measure cannot be considered as an i.i.d. sample from the filtering distribution.

The set of resampled particles at the output of the algorithm constitutes the uniformly weighted empirical measure $\pi_t^n(d\boldsymbol{x}_t) = \frac{1}{n}\sum_{i=1}^n \delta_{\boldsymbol{x}_t^i}(d\boldsymbol{x}_t)$ where $\delta_{\boldsymbol{x}_t^i}(d\boldsymbol{x}_t)$ is the Dirac measure located at the particle $\boldsymbol{x}_t^i$. This empirical measure approximates the filtering distribution. Indeed, several convergence results has been proved [4,7], which ensures the reasonability of the particle filtering algorithm from the theoretical point of view.

Here we only mention the result stating the L_2 convergence [4]. It reads as that for any bounded, generally complex-valued function $f : \mathbb{R}^d \to \mathbb{C}$, it holds that $\lim_{n\to\infty} \mathbb{E}[|\pi_t^n f - \pi_t f|^2] = 0$ where $\pi_t^n f = \int f \, d\pi_t^n = \frac{1}{n}\sum_i f(\boldsymbol{x}_t^i)$ and $\pi_t f = \int f \, d\pi_t$.

For detailed reviews of the particle filtering algorithm see, for example, [2,5,7,8].

2.3 Kernel Density Estimates in Particle Filtering

The purpose of introducing kernel estimates to particle filtering is to reconstruct the filtering density using the standard methodology of kernel density estimation [11–13]. A straightforward application of this methodology leads to the estimates

$$\hat{p}_t^n(\boldsymbol{x}) = \frac{1}{nh^d}\sum_{i=1}^n K\left(\frac{\boldsymbol{x} - \boldsymbol{x}_t^i}{h}\right), \quad t = 1, \ldots, T \tag{3}$$

where $K : \mathbb{R}^d \to \mathbb{R}$ is a suitable function called the *kernel* and $h > 0$ is the *bandwidth.*

Nevertheless, the convergence results on the kernel density estimates cannot be directly transferred to particle filtering as the set of generated particles $\{\boldsymbol{x}_t^i\}_{t=1}^n$ does not constitute a random sample from the filtering distribution. As explained, the i.i.d. character is violated due to the resampling step in the particle filtering algorithm. Yet, restricting on the certain class of densities and kernels it can be shown that the mean integrated squared error (MISE) of the kernel estimates (3) is upper bounded in dependence on the number of generated particles, which eventually leads to the convergence result as desired.

In order to present the corresponding theorem we introduce the following notions of the *Sobolev class* of functions and the *order of the kernel.*

Definition 1. *Let $\beta \geq 1$ be an integer and $L > 0$ a real. The Sobolev class of functions $\mathcal{P}_{S(\beta,L)}^{\propto}$ consists of $L_1(\mathbb{R}^d)$ integrable functions $f : \mathbb{R}^d \to \mathbb{R}$ satisfying*

$$\int ||\boldsymbol{\omega}||^{2\beta} |\mathcal{F}[f](\boldsymbol{\omega})|^2 \, d\boldsymbol{\omega} \leq (2\pi)^d L^2 \tag{4}$$

where $||\cdot||$ is the Euclidean norm and $\mathcal{F}[f] : \mathbb{R}^d \to \mathbb{C}$ is the Fourier transform of f. $f \in L_1(\mathbb{R}^d)$ is called β-Sobolev if $f \in \mathcal{P}_{S(\beta,L)}^{\propto}$. If $f \in L_1(\mathbb{R}^d)$ is known to be a density we use the notation $f \in \mathcal{P}_{S(\beta,L)}$.

To introduce the order of the kernel, we use differential operator D^α with multi-index $\alpha = (\alpha_1, \alpha_2, \ldots, \alpha_d)$, $\alpha_i \in \mathbb{N}_0$. For a suitably differentiable function $f : \mathbb{R}^d \to \mathbb{R}$, one has $D^\alpha f = \partial^{|\alpha|} f / \partial^{\alpha_1} x_1 \cdots \partial^{\alpha_d} x_d$ with $|\alpha| = \sum_{i=1}^d \alpha_i$ being the order of the derivative.

Definition 2. *Let $\ell \geq 1$ be an integer. We say that the kernel $K \colon \mathbb{R}^d \to \mathbb{R}$ is of order ℓ, if K is $L_1(\mathbb{R}^d) \cap L_2(\mathbb{R}^d)$ integrable, its Fourier transform $K_{\mathcal{F}}(\boldsymbol{\omega}) = \mathcal{F}[K](\boldsymbol{\omega})$ is real, satisfies $K_{\mathcal{F}}(\mathbf{0}) = 1$ and has all partial derivatives $D^\alpha K_{\mathcal{F}}$ up to the ℓ-th order such that $D^\alpha K_{\mathcal{F}}(\mathbf{0}) = 0$ for all $|\alpha| = 1, \ldots, \ell$.*

Under the assumption that the order of the kernel fits the Sobolev character of the filtering density, i.e., if $\ell = \beta$, we have the following theorem proved in its general version for derivatives in [1]; and in [2] when no differentiation is considered, i.e., for $|\alpha| = 0$.

Theorem 1. *In the filtering problem, let $\{\pi_t\}_{t=0}^T$, $\{D^\alpha p_t\}_{t=0}^T$, $T \in \mathbb{N}$ be the sequences of filtering distributions and partial derivatives of the corresponding filtering densities for some multi-index $\alpha = (\alpha_1, \ldots, \alpha_d)$, $|\alpha| \in \mathbb{N}_0$, i.e., $p_t(\boldsymbol{x}_t) = p(\boldsymbol{x}_t | \boldsymbol{y}_{1:t})$ and $\pi_t(d\boldsymbol{x}_t) = p_t(\boldsymbol{x}_t)\, d\boldsymbol{x}_t$. Let $D^\alpha p_t$, $t \in \{0, \ldots, T\}$ be β-Sobolev for some $\beta \in \mathbb{N}$ and $L_{t,\alpha} > 0$, i.e., $D^\alpha p_t \in \mathcal{P}^\propto_{S(\beta, L_{t,\alpha})}$. Let $\{\pi_t^n\}_{t=1}^T$, $\{D^\alpha \hat{p}_t^n\}_{t=1}^T$, $n \in \mathbb{N}$ be the sequences of empirical measures from particle filtering and partial derivatives of related kernel density estimates (3) with the bandwidth varying as $h(n) = a n^{-\frac{1}{2\beta + d + 2|\alpha|}}$ for some $a > 0$. Let the kernel K employed in the estimates be of order β and $||D^\alpha K||_2^2 < \infty$. Then we have the following upper bounds on the MISE of $D^\alpha \hat{p}_t^n$ for $t \in \{1, \ldots, T\}$:*

$$\mathbb{E}\left[\int (D^\alpha \hat{p}_t^n(\boldsymbol{x}_t) - D^\alpha p_t(\boldsymbol{x}_t))^2 \, d\boldsymbol{x}_t\right] \leq C_{t,\alpha}^2 \cdot n^{-\frac{2\beta}{2\beta + d + 2|\alpha|}} \tag{5}$$

where $C_{t,\alpha}$ is constant w.r.t. the number of particles n.

Under the respective assumptions, the theorem says that for each $t = 1, \ldots, T$ the derivatives of the kernel density estimates (3) converge to the derivatives of the filtering densities in the MISE as n goes to infinity. The rate of convergence is $O\left(n^{-\frac{2\beta}{2\beta + d + 2|\alpha|}}\right)$. The dimension d and the order of derivative $|\alpha|$ increases the upper bound, whilst the Sobolev character β makes it reduced. Now we discuss the persistence of convergence.

3 Sobolev Character of Derivatives of Filtering Densities

This section deal with the substantial assumption in Theorem 1 that $D^\alpha p_t \in \mathcal{P}^\propto_{S(\beta, L_{t,\alpha})}$, i.e., that derivatives of the fitering density satisfy the inequality (4). This might be directly verified for p_0, but a straightforward verification for higher time instants $t > 1$ is inconvenient. Here we present a tool for making this more comfortable. It corresponds to a condition on the Fourier transform of the transition kernels in the signal process. This condition then ensures the Sobolev character of the derivatives as required. The theorem bellow is an extension of Theorem 5.2 in [2].

To present the theorem let us recall the prediction and update formulas (1) and (2), respectively, describing the evolution of the filtering density over time. The equations can be written in a more concise form as

$$\overline{p}_t(\boldsymbol{x}_t) = \int K_{t-1}(\boldsymbol{x}_t|\boldsymbol{x}_{t-1})p_{t-1}(\boldsymbol{x}_{t-1})\, d\boldsymbol{x}_{t-1}, \quad p_t(\boldsymbol{x}_t) = \frac{g_t(\boldsymbol{x}_t)\overline{p}_t(\boldsymbol{x}_t)}{\overline{\pi}_t g_t}.$$

In the formulas, $\overline{p}_t(\boldsymbol{x}_t)$ is the abbreviation for the density of the prediction distribution, i.e., $\overline{p}_t(\boldsymbol{x}_t) = p(\boldsymbol{x}_t|\boldsymbol{y}_{1:t-1})$, $g_t(\boldsymbol{x}_t)$ is the shortcut for the conditional density $g_t(\boldsymbol{y}_t|\boldsymbol{x}_t)$ and $\overline{\pi}_t g_t = \int g_t(\boldsymbol{x}_t)\overline{p}_t(\boldsymbol{x}_t)\, d\boldsymbol{x}_t$.

Theorem 2. *In the filtering problem, let* $D^\alpha p_0 \in \mathcal{P}^{\propto}_{S(\beta,L_{0,\alpha})}$. *Let* $\{K_{t-1}, t \in \mathbb{N}\}$ *be the set of the transition kernels, and* $\{D^\alpha K_{t-1}, t \in \mathbb{N}\}$ *be the set of its partial derivatives. Let* $\{\mathcal{F}[D^\alpha K_{t-1}](\boldsymbol{\omega}|\boldsymbol{x}_{t-1}), t \in \mathbb{N}\}$ *be the set of the corresponding conditional Fourier transforms, i.e.,*

$$\mathcal{F}[D^\alpha K_{t-1}](\boldsymbol{\omega}|\boldsymbol{x}_{t-1}) = \int e^{\mathrm{i}\langle \boldsymbol{\omega}, \boldsymbol{x}_t\rangle} D^\alpha K_{t-1}(\boldsymbol{x}_t|\boldsymbol{x}_{t-1})\, d\boldsymbol{x}_t.$$

For all $t \in \mathbb{N}$, *let* $\mathcal{F}[D^\alpha K_{t-1}]$ *be bounded by some function* $K_b^\alpha : \mathbb{R}^d \to \mathbb{C}$ *in such a way that for any* $\boldsymbol{x}_{t-1} \in \mathbb{R}^d$ *and* $\boldsymbol{\omega} \in \mathbb{R}^d$,

$$|\mathcal{F}[D^\alpha K_{t-1}](\boldsymbol{\omega}|\boldsymbol{x}_{t-1})| \leq |K_b^\alpha(\boldsymbol{\omega})|.$$

Let K_b^α *be* β*-Sobolev for some* $L_{K_b^\alpha} > 0$, *i.e.,* $K_b^\alpha \in \mathcal{P}^{\propto}_{S(\beta,L_{K_b^\alpha})}$. *Then* $D^\alpha p_t \in \mathcal{P}^{\propto}_{S(\beta,L_{t,\alpha})}$, $t \in \mathbb{N}$ *with the recurrence for* $L_{t,\alpha}$ *written as* $L_{t,\alpha} = ||g_t||_\infty L_{K_b^\alpha}/\overline{\pi}_t g_t$.

Proof. The theorem holds for $D^\alpha p_0$ by the assumption. From the prediction formula, multiplying both sides of the prediction formula by the complex exponential, we get

$$e^{\mathrm{i}\langle \boldsymbol{\omega}, \boldsymbol{x}_t\rangle}\, \overline{p}_t(\boldsymbol{x}_t) = e^{\mathrm{i}\langle \boldsymbol{\omega}, \boldsymbol{x}_t\rangle} \int K_{t-1}(\boldsymbol{x}_t|\boldsymbol{x}_{t-1})p_{t-1}(\boldsymbol{x}_{t-1})\, d\boldsymbol{x}_{t-1}.$$

By integration, the left-hand side just gives the characteristic function $\overline{\psi}_t(\boldsymbol{\omega})$ of $\overline{p}_t(\boldsymbol{x}_t)$, i.e.,

$$\overline{\psi}_t(\boldsymbol{\omega}) = \int e^{\mathrm{i}\langle \boldsymbol{\omega}, \boldsymbol{x}_t\rangle} \overline{p}_t(\boldsymbol{x}_t)\, d\boldsymbol{x}_t.$$

The right-hand side has then form

$$\begin{aligned}
&\int\int e^{\mathrm{i}\langle \boldsymbol{\omega}, \boldsymbol{x}_t\rangle} K_{t-1}(\boldsymbol{x}_t|\boldsymbol{x}_{t-1})p_{t-1}(\boldsymbol{x}_{t-1})\, d\boldsymbol{x}_{t-1}\, d\boldsymbol{x}_t \\
&= \int p_{t-1}(\boldsymbol{x}_{t-1}) \left(\int e^{\mathrm{i}\langle \boldsymbol{\omega}, \boldsymbol{x}_t\rangle} K_{t-1}(\boldsymbol{x}_t|\boldsymbol{x}_{t-1})\, d\boldsymbol{x}_t \right) d\boldsymbol{x}_{t-1} \\
&= \int p_{t-1}(\boldsymbol{x}_{t-1})\mathcal{F}[K_{t-1}](\boldsymbol{\omega}|\boldsymbol{x}_{t-1})\, d\boldsymbol{x}_{t-1}.
\end{aligned}$$

Multiplying by $(-\mathrm{i})^{|\alpha|}(\omega_1^{\alpha_1}\cdot\dots\cdot\omega_d^{\alpha_d})$ we move both sides to the Fourier transforms of the corresponding partial derivatives. That is,

$$(-\mathrm{i})^{|\alpha|}(\omega_1^{\alpha_1}\cdot\dots\cdot\omega_d^{\alpha_d})\overline{\psi}_t(\boldsymbol{\omega}) = (-\mathrm{i})^{|\alpha|}(\omega_1^{\alpha_1}\cdot\dots\cdot\omega_d^{\alpha_d})\mathcal{F}[\overline{p}_t] = \mathcal{F}[D^\alpha\overline{p}_t]$$

and

$$\int p_{t-1}(\boldsymbol{x}_{t-1})(-\mathrm{i})^{|\alpha|}(\omega_1^{\alpha_1}\cdot\dots\cdot\omega_d^{\alpha_d})\mathcal{F}[K_{t-1}](\boldsymbol{\omega}|\boldsymbol{x}_{t-1})\,d\boldsymbol{x}_{t-1} = \\ = \int p_{t-1}(\boldsymbol{x}_{t-1})\mathcal{F}[D^\alpha K_{t-1}](\boldsymbol{\omega}|\boldsymbol{x}_{t-1})\,d\boldsymbol{x}_{t-1}.$$

Multiplication by the corresponding complex conjugates gives the expression

$$|\mathcal{F}[D^\alpha\overline{p}_t]|^2 = \left|\int p_{t-1}(\boldsymbol{x}_{t-1})\mathcal{F}[D^\alpha K_{t-1}](\boldsymbol{\omega}|\boldsymbol{x}_{t-1})\,d\boldsymbol{x}_{t-1}\right|^2.$$

The Jensen's inequality and the boundedness of $\mathcal{F}[D^\alpha K_{t-1}]$ further gives

$$|\mathcal{F}[D^\alpha\overline{p}_t]|^2 \le \left(\int |D^\alpha\mathcal{F}[K_{t-1}](\boldsymbol{\omega}|\boldsymbol{x}_{t-1})|\,p_{t-1}(\boldsymbol{x}_{t-1})\,d\boldsymbol{x}_{t-1}\right)^2 \\ \le \left(|K_b^\alpha(\boldsymbol{\omega})|\int p_{t-1}(\boldsymbol{x}_{t-1})\,d\boldsymbol{x}_{t-1}\right)^2 = |K_b^\alpha(\boldsymbol{\omega})|^2.$$

Thus,

$$\int ||\boldsymbol{\omega}||^{2\beta}|D^\alpha\mathcal{F}[\overline{p}_t]|^2\,d\boldsymbol{\omega} \le \int ||\boldsymbol{\omega}||^{2\beta}|K_b^\alpha(\boldsymbol{\omega})|^2 \le (2\pi)^d L^2_{K_b^\alpha}.$$

The above formula shows that $D^\alpha\overline{p}_t \in \mathcal{P}^\propto_{S(\beta,L_{K_b^\alpha})}$ for any $t\in\mathbb{N}$. We proceed with specifying the Sobolev constant $L_{t,\alpha}$ for the partial derivative $D^\alpha p_t$.

The g_t function reads as $g_t(\boldsymbol{x}_t) = g_t(\boldsymbol{y}_t|\boldsymbol{x}_t) = g_t^v(\boldsymbol{y}_t - h(\boldsymbol{x}_t))$. As the densities of the noise terms g_t^v are considered bounded and strictly positive in Sect. 2.1, we have $\sup_{\boldsymbol{x}_t,\boldsymbol{y}_t}\{g_t^v(\boldsymbol{y}_t - h(\boldsymbol{x}_t))\} = ||g_t||_\infty < \infty$ and $0 < \overline{\pi}_t g_t < \infty$.

Again, multiplying the update formula by the complex exponential, integrating, multiplying by $(-\mathrm{i})^{|\alpha|}(\omega_{i_1}^{\alpha_1}\cdot\dots\cdot\omega_{i_d}^{\alpha_d})$ and the respective conjugates we move to the Fourier transforms of the partial derivatives and get

$$\begin{aligned}(\overline{\pi}_t g_t)\,p_t(\boldsymbol{x}_t) &= g_t(\boldsymbol{x}_t)\,\overline{p}_t(\boldsymbol{x}_t),\\ (\overline{\pi}_t g_t)\int e^{\mathrm{i}\langle\boldsymbol{\omega},\boldsymbol{x}_t\rangle}p_t(\boldsymbol{x}_t)\,d\boldsymbol{x}_t &= \int e^{\mathrm{i}\langle\boldsymbol{\omega},\boldsymbol{x}_t\rangle}g_t(\boldsymbol{x}_t)\,\overline{p}_t(\boldsymbol{x}_t)\,d\boldsymbol{x}_t,\\ (\overline{\pi}_t g_t)\,(-\mathrm{i})^{|\alpha|}(\omega_1^{\alpha_1}\cdot\dots\cdot\omega_d^{\alpha_d})\,\psi_t(\boldsymbol{\omega}) &\le ||g_t||_\infty\,(-\mathrm{i})^{|\alpha|}(\omega_1^{\alpha_1}\cdot\dots\cdot\omega_d^{\alpha_d})\,\overline{\psi}_t(\boldsymbol{\omega}),\\ ||\boldsymbol{\omega}||^{2\beta}\,(\overline{\pi}_t g_t)^2\,|\mathcal{F}[D^\alpha p_t]|^2 &\le ||\boldsymbol{\omega}||^{2\beta}\,||g_t||^2_\infty\,|\mathcal{F}[D^\alpha\overline{p}_t]|^2,\\ (2\pi)^{-d}\int ||\boldsymbol{\omega}||^{2\beta}|\mathcal{F}[D^\alpha p_t]|^2\,d\boldsymbol{\omega} &\le \frac{||g_t||^2_\infty L^2_{K_{b,\alpha}}}{(\overline{\pi}_t g_t)^2} = L^2_{t,\alpha},\end{aligned}$$

which concludes the proof. □

4 Conclusions

In the paper, we investigated the persistence of convergence of kernel density estimates in particle filtering. We provided the sufficient condition on the Fourier transform of transition kernels in the signal process guaranteeing that the Sobolev character of the filtering density does not change over time. As the consequence, the convergence of the kernel density estimates is maintained over the whole operation time of the filter.

Acknowledgements. This work was supported by programme CZ.02.1.01/0.0/0.0/16_013/0001787 (OP VVV) of the Ministry of Education, Youth and Sport of the Czech Republic, institutional support RVO:67985807 and grant SVV No. 260454 of Charles University.

References

1. Coufal, D.: Convergence rates of kernel density estimates in particle filtering. Stat. Probab. Lett. (submitted)
2. Coufal, D.: On convergence of kernel density estimates in particle filtering. Kybernetika **52**(5), 735–756 (2016)
3. Coufal, D.: On persistence of convergence of kernel density estimates in particle filtering. In: Łukasik, S., Kulczycki, P., Kowalski, P.A. (eds.) Contemporary Computational Science, p. 91. AGH-UST Press, Cracow (2018)
4. Crisan, D., Doucet, A.: A survey of convergence results on particle filtering methods for practitioners. IEEE Trans. Signal Process. **50**(3), 736–746 (2002)
5. Crisan, D., Míguez, J.: Particle-kernel estimation of the filter density in state-space models. Bernoulli **20**(4), 1879–1929 (2014)
6. Douc, R., Cappé, O., Moulines, E.: Comparison of resampling schemes for particle filtering. In: 4th International Symposium on Image and Signal Processing and Analysis (ISPA), pp. 64–69 (2005)
7. Doucet, A., de Freitas, N., Gordon, N. (eds.): Sequential Monte Carlo Methods in Practice. Springer, New York (2001)
8. Doucet, A., Johansen, A.M.: A tutorial on particle filtering and smoothing: fifteen years later. In: Crisan, D., Rozovskii, B. (eds.) The Oxford Handbook of Nonlinear Filtering. Oxford University Press, Oxford (2011)
9. Fristedt, B., Jain, N., Krylov, N.: Filtering and Prediction: A Primer. American Mathematical Society, Providence (2007)
10. Särkkä, S.: Bayesian Filtering and Smoothing. Cambridge University Press, Cambridge (2013)
11. Silverman, B.W.: Density Estimation for Statistics and Data Analysis. Chapman and Hall/CRC, London (1986)
12. Tsybakov, A.B.: Introduction to Nonparametric Estimation. Springer, New York (2009)
13. Wand, M.P., Jones, M.C.: Kernel Smoothing. Chapman and Hall/CRC, London (1995)

Multidimensional Copula Models of Dependencies Between Selected International Financial Market Indexes

Tomáš Bacigál[1(✉)], Magdaléna Komorníková[1], and Jozef Komorník[2]

[1] Slovak University of Technology, 810 05 Bratislava, Slovakia
{Tomas.Bacigal,Magdalena.Komornikova}@stuba.sk
[2] Comenius University, 820 05 Bratislava, Slovakia
Jozef.Komornik@fm.uniba.sk

Abstract. In this paper we focus our attention on multi-dimensional copula models of returns of the indexes of selected prominent international financial markets. Our modeling results, based on elliptic copulas, 7-dimensional vine copulas and hierarchical Archimedean copulas demonstrate a dominant role of the SPX index among the considered major stock indexes (mainly at the first tree of the optimal vine copulas). Some interesting weaker conditional dependencies can be detected at it's highest trees. Interestingly, while global optimal model (for the whole period of 277 months) belong to the Student class, the optimal local models can be found (with very minor differences in the values of GoF test statistic) in the classes of vine and hierarchical Archimedean copulas. The dominance of these models is most striking over the interval of the financial market crisis, where the quality of the best Student class model was providing a substantially poorer fit.

Keywords: International financial market indexes · Elliptic copulas · Vine copulas · Hierarchical Archimedean copulas

1 Introduction

In this paper we apply multi-dimensional copula to model dependence among returns of selected prominent indexes of international financial markets. The following indexes were considered (with months' values from the time interval 31.1.1995 – 31.1.2018): SPX (Standard and Poor's Index is designed to measure performance of the broad US economy through changes in the aggregate market value of 500 stocks representing all major industries), DAX (The German Stock Index is a total return index of 30 selected German blue chip stocks traded on the Frankfurt Stock Exchange), UKX (The FTSE 100 Index is a capitalization-weighted index of the 100 most highly capitalized companies traded on the London Stock Exchange), NKY (The Nikkei-225 Stock Average is a price-weighted average of 225 top-rated Japanese companies listed in the First Section of the

P. Kulczycki et al. (Eds.): ITSRCP 2018, AISC 945, pp. 347–360, 2020.
https://doi.org/10.1007/978-3-030-18058-4_28

Tokyo Stock Exchange), HSI (The Hang Seng Index is a free-float capitalization-weighted index of a selection of companies from the Stock Exchange of Hong Kong), LEGATRUU (The Bloomberg Barclays Global Aggregate Bond Index is a flagship measure of global investment grade debt from twenty-four local currency markets), SPGSCITR (The S&P GSCI Total Return Index in USD is widely recognized as the leading measure of general commodity price movements and inflation in the world economy).

The paper is organized as follows. The second section is devoted to a brief overview of the theory of vine and hierarchical Archimedean copulas and methodology of copula fitting to multi-dimensional time series. The third section contains application to real data modeling. Finally we discuss results and conclude.

The preliminary version of this paper was presented at the 3rd Conference on Information Technology, Systems Research and Computational Physics, 2–5 July 2018, Cracow, Poland [2].

2 Theory

Copula represents a multivariate distribution that captures the dependence structure between/among random variables leaving alone their marginal distributions. Due to Sklar [23]

$$F(x_1, ..., x_n) = C\left[F_1(x_1), ..., F_n(x_n)\right],$$

where F is joint cumulative distribution function of random vector $(X_1, ..., X_n)$, F_i is marginal cumulative distribution function of X_i, and $C : [0,1]^n \rightarrow [0,1]$ is a copula which is a n-increasing function with 1 as neutral element and 0 as annihilator, see e.g. monograph Nelsen (2006) [17]. Besides three fundamental copulas

$$M(x_1, ..., x_n) = \min(x_1, ..., x_n), \quad W(x_1, x_2) = \max(x_1 + x_2 - 1, 0),$$

$$\Pi(x_1, ..., x_n) = \prod_{i=1}^{n} x_i,$$

which model perfect positive dependence, perfect negative dependence (not applicable for $n > 2$) and independence, respectively, there exist numerous parametric classes, such as Archimedean, Extreme-Value and elliptical copulas. Within the last one there belong such important parametric families as *Gaussian* copulas

$$C_G(x_1, ..., x_n) = \Phi\left[\Phi_1^{-1}(x_1), ..., \Phi_n^{-1}(x_n)\right]$$

and *Student* t-copulas

$$C_t(x_1, ..., x_n) = t\left[t_1^{-1}(x_1), ..., t_n^{-1}(x_n)\right],$$

(where Φ and t are joint distribution functions of multivariate normal and Student t distributions, similarly Φ_i^{-1} and t_i^{-1}, $i = 1, ..., n$ are univariate quantile functions related to X_i), able to flexibly describe dependence in multidimensional random vector.

2.1 Vine Copulas

An n-dimensional regular vine tree structure $\mathcal{S} = \{T_1, ..., T_n\}$ is a sequence of $n-1$ linked trees with properties (see [3,4]):

- Tree T_1 is a tree on nodes 1 to n;
- Tree T_j has $n+1-j$ nodes and $d-j$ edges;
- Edges in tree T_j become nodes in tree T_{j+1};
- Two nodes in tree T_{j+1} can be joined by an edge only if the corresponding edges in tree T_j share a node.

Vine copulas [1,4,22] are constructed as lego using bivariate copulas as the construction blocks in a vine tree structure, which can be estimated (by default following the correlation strength ordering), visualized and interpreted. We outline the construction of three-dimensional probability density function f

$$
\begin{aligned}
f(x_1, x_2, x_3) &= f_1(x_1) \cdot f_{2|1}(x_1, x_2) \cdot f_{3|12}(x_1, x_2, x_3) \\
&= f_1(x_1) \cdot c_{12}\left[F_1(x_1), F_2(x_2)\right] \cdot f_2(x_2) \qquad (1) \\
&\quad \cdot c_{31|2}\left[F_{x_3|x_2}(x_2, x_3), F_{x_1|x_2}(x_1, x_2)\right] \cdot c_{23}\left[F_2(x_2), F_3(x_3)\right] \cdot f_3(x_3)
\end{aligned}
$$

where f_i is a (marginal) probability density function of $X_i,\ i = 1, 2, 3,$

$$
f_{i|j}(x_i, x_j) = \frac{f(x_i, x_j)}{f_j(x_j)}
$$

is conditional density function of X_i given X_j. A copula density c_{ij} couples X_i and X_j while $c_{ij|k}$ couples bivariate conditional distributions of $X_i|X_k$ and $X_j|X_k, i, j, k \in \{1, 2, 3\}, i \neq j \neq k \neq i$. Finally,

$$
F_{x_i|x_j} = \frac{\partial C_{ij}\left[F_i(x_i), F_j(x_j)\right]}{\partial F_j(x_j)}
$$

is a conditional cumulative distribution function of X_i given X_j.

The construction (1) is one of the three possible pair-copula decompositions, which, graphically, are both [5] (see Fig. 1)

- canonical (C-) vine trees: each tree has a unique node connected to $d-j$ edges (use only star like tree - useful for ordering by importance);
- drawable (D-) vine trees: no node is connected to more than 2 edges (use only path like trees - useful for temporal ordering of variables);

and see, e.g., [4] for a natural generalization in higher dimensions, in which C-vines and D-vines are just small subsets of a more general class - regular vines (see [3]).

Besides the well-known 2-dimensional product copula and elliptical copulas (Gaussian and Student), as a building block of vine copulas we utilized also numerous 2-dimensional families of Archimedean and Extreme-value copulas, as well as their rotations, described below in the section Methods.

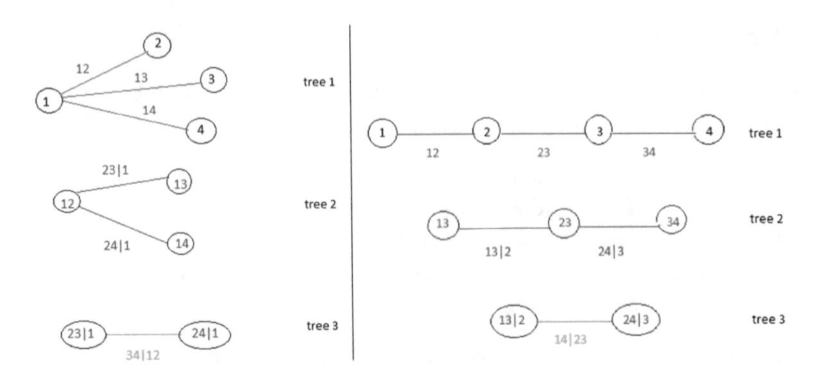

Fig. 1. C-Vine tree (left) and D-Vine tree (right) [5]

From the graph theory perspective, vine trees can be seen as repeatedly applied line graphs (starting with T_1), for instance [16] use concept of line graphs in his algorithm to generate regular vines, see also references therein.[1]

2.2 Hierarchical Copulas

Another class suitable for modeling multidimensional stochastic dependence are hierarchical Archimedean copulas. Their basic building blocks – Archimedean copulas – are defined (for any dimension n) by formula

$$C(x_1, \ldots, x_n) = \phi(\phi^{-1}(x_1) + \ldots + \phi^{-1}(x_n))$$

where the so-called generator $\phi\colon [0,\infty) \searrow [0,1]$ satisfies boundary conditions $\phi(0) = 1$, $\phi(\infty) = 0$ (strict Archimedean copulas) and absolute monotonicity (for further details see [14]). Such a construction is analytically convenient and very flexible in bivariate setting, however it is too restrictive in higher dimensions since the whole dependence structure is rendered by a single univariate function, and – moreover – it is exchangeable.

Hierarchical Archimedean copulas (HAC) overcome this problem by nesting simple Archimedean copulas. Since the general multivariate structure is notationally too complex, we illustrate the principle in four dimensions. For example, fully nested HAC (Fig. 2, left) can be given by

$$\begin{aligned} C_s(x_1, \ldots, x_4) &= C_3\Big(C_2\big(C_1(x_1,x_2),x_3\big),x_4\Big) \\ &= \phi_3\left(\phi_3^{-1} \circ \phi_2\Big(\phi_2^{-1} \circ \phi_1(\phi_1^{-1}(x_1) + \phi_1^{-1}(x_2)) + \phi_2^{-1}(u_3)\Big) + \phi_1^{-1}(u_4)\right), \quad (2) \end{aligned}$$

[1] Another, a rather exotic result is given by [25] who showed conditions under which the so-called Kautz graph (such a line graph of a complete digraph that has diameter 2) is a Cayley graph (an important class because of symmetry and easiness of their generation).

where $C_i, j = 1, \ldots, n-1$ are Archimedean copulas with their corresponding generators ϕ_j and $s = (((1,2),3),4)$ the nesting structure. An example of partially nested Archimedean copula (Fig. 2, right) is given by

$$C_s(x_1, \ldots, x_4) = C_3\Big(C_1(x_1, x_2), C_2(x_3, x_4)\Big), \quad \text{where } s = ((1,2),(3,4)). \tag{3}$$

Fully and partially nested Archimedean copulas form a class of hierarchical Archimedean copulas which can adopt arbitrarily complex structure s, generally $s = (\ldots, (i_a, i_b), i_c, \ldots)$, where $i. \in \{1, \ldots, n\}$ is reordering of the indices of variables with $a, b, c \in \{1, \ldots, n \mid a \neq b \neq c\}$, see, e.g., [9,12,19]. This makes it a very flexible yet parsimonious distribution model. The generators within a single HAC can come either from a single generator family or from different families. In the first case there is required complete monotonicity of composition $\phi_i^{-1} \circ \phi_j$, $(i \neq j)$, which imposes some constraints on their parameters, see sufficient conditions given by [13]. For majority of generators HAC requires decreasing parameters from top to bottom in its hierarchy. In the case of different generator families, the condition of complete monotonicity is not always fulfilled. The software implementation in R, the *HAC* package [19] which we use in our study, considers only single-parameter generators from the same family. Then the whole distribution is specified with at most $n-1$ parameters which can be seen as an alternative to covariance driven models, as remarked in [19], nevertheless, besides the parameters also structure s needs to be estimated. As there are already $n!/k!$ possibilities of combining n variables to fully nested HAC with k-dimensional AC on its lowest level, the greedy approach to structure estimation would be unreasonable even in moderate dimensions, therefore HAC package offers computationally efficient recursive procedure suggested by [18].

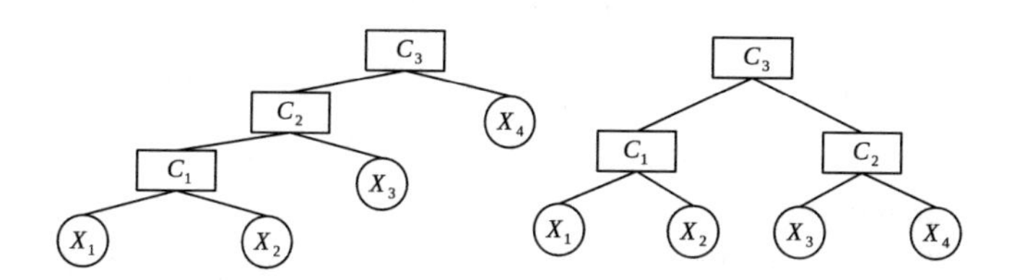

Fig. 2. Fully nested and partially nested Archimedean copulas structure.

2.3 Methods

Within the considered classes of 2-dimensional copulas as well as n-dimensional elliptical copulas, the optimal models were selected using the Maximum likelihood estimation (MLE) method. Recall that for given m observations $\{X_{j,i}\}_{i=1,\ldots,m}$ of j-th random variable X_j, $j = 1, ..., n$, the parameters θ of all copulas under consideration were estimated by maximizing the likelihood function

$$L(\theta) = \sum_{i=1}^{m} \log\left[c_\theta(U_{1,i}, \ldots, U_{n,i})\right], \tag{4}$$

where c_θ denotes density of a parametric copula family C_θ, and

$$U_{j,i} = \frac{1}{m+1} \sum_{k=1}^{m} \mathbf{1}(X_{j,k} \leq X_{j,i}), \quad i = 1, ..., m,$$

are so-called pseudo-observations. The higher dimensional structures of vine and HAC copulas were estimated as described in [6] and [19], respectively.

Goodness-of-fit was performed by a test proposed by Genest et al. [8] and based on empirical copula process using Cramer-von Misses test statistic

$$S_{CM} = \sum_{i=1}^{m} \left[C_\theta(U_{1,i}, \ldots, U_{n,i}) - C_m(U_{1,i}, \ldots, U_{n,i})\right]^2 \tag{5}$$

with empirical copula $C_m(\mathbf{x}) = \frac{1}{m} \sum_{i=1}^{m} \prod_{j=1}^{n} \mathbf{1}(X_{j,i} \leq x_j)$ and indicator function $\mathbf{1}(A) = 1$ whenever A is true, otherwise $\mathbf{1}(A) = 0$.

All calculations were done in R [20] with the specific help of packages *copula* [10], *VineCopula* [21], and *HAC* [19]. Because their goodness-of-fit methods, including the Cramer-von Misses metric, are not directly comparable and computing the values of vine copula cumulative distribution function involves integration over 7-dimensional space, we rather approximated C_θ from (5) by empirical copula of the random samples generated from the corresponding copulas each counting 100 000 realizations.

Besides the usual parametric families of Archimedean class such as Gumbel, Clayton, Frank, Joe, copulas BB1, BB6, BB7, BB8 and Tawn copulas (see e.g. [11,15,17,24]) in bivariate case we used also their rotations C_α by angle α defined

$$C_{90}(x_1, x_2) = x_2 - C(1 - x_1, x_2),$$

$$C_{180}(x_1, x_2) = x_1 + x_2 - 1 + C(1 - x_1, 1 - x_2) \quad \text{survival copula},$$

$$C_{270}(x_1, x_2) = x_1 - C(x_1, 1 - x_2),$$

that are implemented in the package *VineCopula.*

As a preliminary analysis of dependence between random variables, we employ classical measures of dependence such as Pearson's and Kendall's correlation coefficients, moreover to inspect the conditional (in)dependence (which is further investigated parametrically with vines) the partial correlation matrix comes handy. Given a Pearson's correlation matrix Σ, the partial correlation between variables X_i, X_j conditional on all the other pairs in vector $(X_1, \ldots, X_n)$ can be computed

$$\rho_{ij|-ij} = \frac{-p_{ij}}{\sqrt{p_{ii} p_{jj}}}$$

where p_{ij} $(i, j = 1, \ldots, n)$ are elements of the matrix $P = \Sigma^{-1}$. Recall that partial correlation is a measure of the strength and direction of a linear relationship between two continuous random variables that takes into account (removes) the influence of some other continuous random variables.

Conditional correlations are important, e.g., (a) when building (gaussian) graphical models, where insignificant connections are removed to obtain more parsimonious model, as well as (b) to better understand the structure of estimated vine copula.

3 Results

All indexes are computed in terms of returns

$$return_i = \frac{index_i - index_{i-1}}{index_{i-1}}, \quad i = 2, 3, ..., n.$$

Before further analysis, we filtered all considered time series of returns by ARIMA-GARCH filters [7]. For all investigated series of returns, the best filters were identified (by the system Mathematica, Version 11) in the class GARCH(1, 1).

The obtained residuals have pairwise Kendall correlation coefficients τ in the interval (−0.025, 0.648), maximal value was achieved for the couple SPX–UKX, see Table 1.

Table 1. Kendall's correlation coefficients for the residuals

	SPX	DAX	UKX	NKY	HSI	LEG	SPG
SPX	1.	0.577	0.648	0.520	0.298	0.050	0.359
DAX	0.577	1	0.603	0.232	0.196	−0.025	0.027
UKX	0.648	0.603	1.	0.361	0.328	0.017	0.141
NKY	0.520	0.232	0.361	1.	0.247	0.162	0.420
HSI	0.298	0.196	0.328	0.247	1.	0.047	0.212
LEG	0.050	−0.025	0.017	0.162	0.047	1.	0.111
SPG	0.359	0.027	0.141	0.420	0.212	0.111	1.

Figure 3 reveals conditional correlations, showing that the relations of (filtered) returns of SPX–UKX, SPX–DAX, SPX–HSI and DAX–UKX attain the largest values, which is in accordance with corresponding strongest dependencies between couples in the first tree of the optimal global vine copula in Table 2.

Subsequently, to those residuals we applied linear transformations in order to map them into unit interval. Results served as inputs to calculations of 7-dimensional copula models.

We extended our analysis by examining development of the Kendall's correlations. We have chosen frequency of calculations of Kendall's correlation coefficients over the intervals of 72 months overlapping by 36 months with the neighboring intervals, the last time period spans only 60 months. For each of the

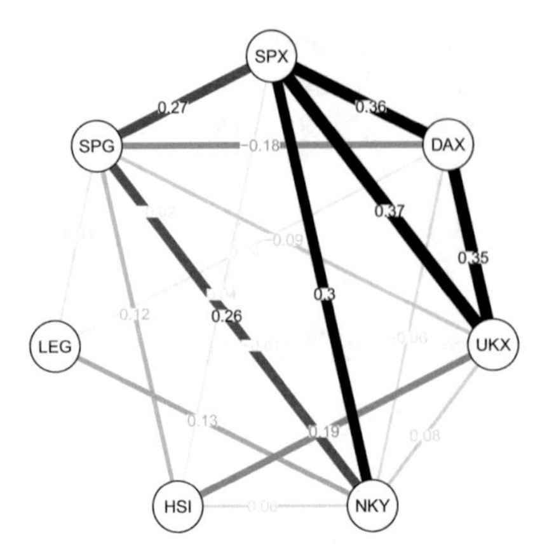

Fig. 3. Partial correlation coefficients for the residuals

couples of considered indexes, we calculated a sequence of 7 local Kendall correlation coefficients on individual local time intervals. We can see (in Fig. 4) that 8 of considered 21 couples of indexes have most values of local Kendall correlation coefficients significantly positive.

3.1 Global Models

The best 7-dimensional vine copula (based on forward selection of trees and AIC criterion for pair-copulas, see [21]) is summarized in Table 2. We observe that at the lowest tree there are modeled stronger links between SPX with UKX, DAX and HSI. It illustrates a very strong international position of the US economy. (It is also interesting to realize that there exist historically strong ties of HSI to the increasingly influential Chinese economy.) At the second tree, we can clearly observe a modest dependence between UKX and DAX, conditioned on SPX. All other elements of the second tree are clearly weaker. Interestingly, at the very last tree, a slight dependence (negative) between LEGATRUU with SPGSCITR, conditioned on all considered stock indexes can be observed.

The best HAC is shown in Fig. 5 (left).

According to the GoF test, the best models for the investigated data are in the class of Student t-copula followed by HAC (see Table 3).

3.2 Modeling Development Dependencies by Local Models

We continued by searching models for the 7 time intervals described above (for which sequence of Kendalls correlation coefficient was calculated). A best vine copula was identified (tree structure) for each interval but estimated (in the same structure) also for all the other intervals. This way we got the selection of 5 different, locally best fitting vine copula structures and their corresponding sequences of estimated vine copulas. Through almost whole considered time

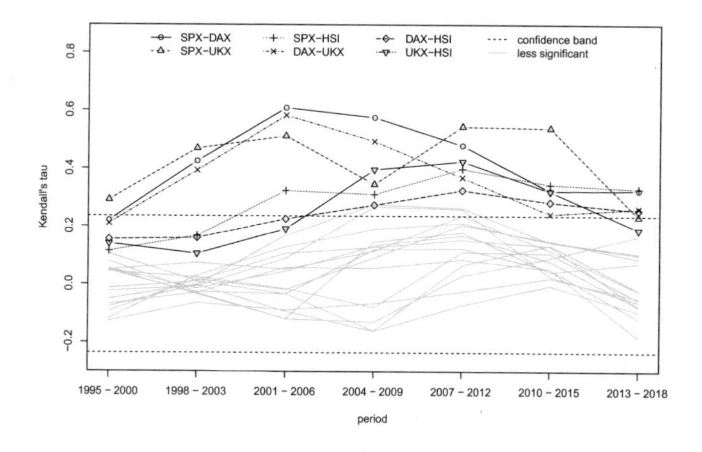

Fig. 4. Evolution of Kendall's τ for all couples of the (filtered) returns

Table 2. The summary of the best 7-dimensional vine copula (I1 = SPX, I2 = DAX, I3 = UKX, I4 = NKY, I5 = HSI, I6 = LEGATRUU, I7 = SPGSCITR)

Tree	Edge	Family	par_1	par_2	τ
1	I4 - I7	t	−0.13	6.04	−0.08
	I2 - I4	SC	0.32	0.00	0.14
	I1 - I2	t	0.61	2.80	0.42
	I1 - I3	t	0.62	2.75	0.43
	I5 - I1	t	0.43	3.19	0.28
	I6 - I5	G	1.09	0.00	0.08
2	I2 - I7; I4	t	−0.01	5.83	−0.01
	I1 - I4; I2	J	1.11	-	0.06
	I3 - I2; I1	t	0.26	3.87	0.17
	I5 - I3; I1	t	0.12	3.76	0.08
	I6 - I1; I5	Tawn	17.67	0.01	0.01
3	I1 - I7; I2 - I4	SJ	1.08	0.00	0.05
	I3 - I4; I1 - I2	t	0.07	7.19	0.05
	I5 - I2; I3 - I1	t	0.09	5.18	0.05
	I6 - I3; I5 - I1	I	-	-	0.00
4	I3 - I7; I1 - I2 - I4	I	-	-	0.00
	I5 - I4; I3 - I1 - I2	Tawn90	−9.33	0.00	0.00
	I6 - I2; I5 - I3 - I1	t	−0.10	6.14	−0.06
5	I5 - I7; I3 - I1 - I2 - I4	I	-	-	0.00
	I6 - I4; I5 - I3 - I1 - I2	G	1.06	0.00	0.06
6	I6 - I7; I5 - I3 - I1 - I2 - I4	F	−0.81	0.00	−0.09

type: R-vine logLik: 257.52 AIC: −455.03 BIC: −346.42

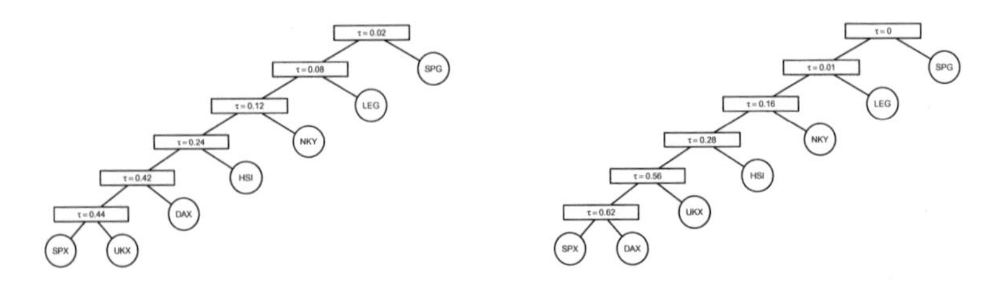

Fig. 5. The best global Gumbel HAC copula (left) and the best Gumbel HAC copula H3 for development (right)

Table 3. GoF test statistics for all four global multi-dimensional models

	Normal copula	Student t-copula	HAC	vine Copula
GoF test statistics	0.0220	**0.0149**	**0.0185**	0.0277

period, the best vine copula was V5 (see Figs. 7 and 8). Similarly we estimated a sequence of 7 elliptic and 7 HAC. Among the elliptic copulas, t-copula was mostly better (except for the last 2 subintervals). We have selected HAC from the classes Gumbel, Clayton, Frank, Joe and Ali-Mikhail-Haq. Throughout all considered time intervals, the best model among them was Gumbel HAC H3 closely followed by Joe HAC H4. The corresponding GoF test statistic (for the best copulas in each class) is displayed in Fig. 6 and it shows slightly superior performance of hierarchical Archimedean copula H3 (Gumbel) and vine copula V5 over elliptical copulas throughout the whole analyzed period. We see that (in most individual time intervals) the difference between the best vine class copulas and the best optimal HAC class copulas are almost negligible.

The graphical visualization of the optimal vine copula in terms of trees is presented on Fig. 7 while the densities of the copula building blocks are depicted on Fig. 8 as contour plot.

The graphical visualization of the optimal Gumbel HAC H3 is presented on Fig. 5 (right).

Here come two interesting observations. First, the breath-taking performance of HAC considering its parsimony: for illustration take now only bivariate copulas used in vines and HAC copulas, then number of parameters needed for construction of n-dimensional normal copula are $n(n-1)/2$, vine copulas $n+(n-1)+\ldots+1$ but HAC copulas only $n-1$! It is true that when (conditional) independence takes place in the random vector, vine copula gets significantly reduced, however in our particular case as for global copula 20 parameters are involved comparing to 6 of HAC, and as for the optimal evolving copula the vine structure contains 8 parameters.

Second, unlike elliptical copulas, the best vine and HAC copulas reveals some asymmetry with respect to tail behavior, and while vines are better for directly displaying conditional relationships, hierarchical Archimedean copulas shows clusters of random variables in somewhat clearer way.

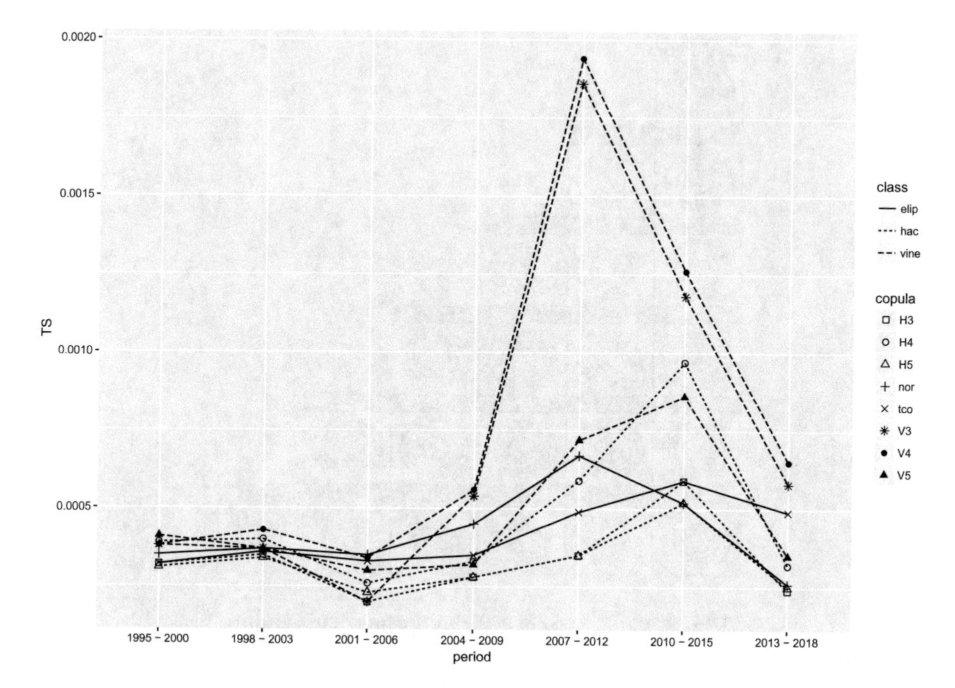

Fig. 6. Evolution of GoF test statistic for the best copulas in each class

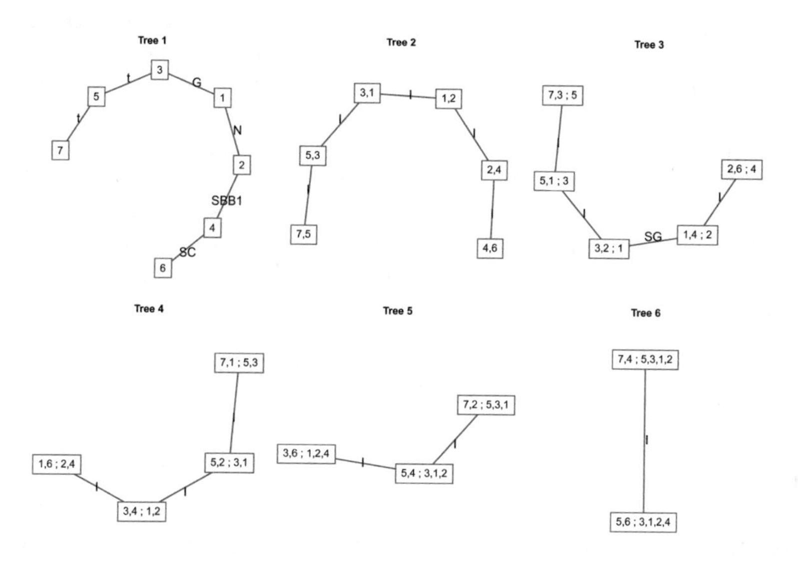

Fig. 7. The tree structure of the optimal vine copula with pair copula family indicated on edges.

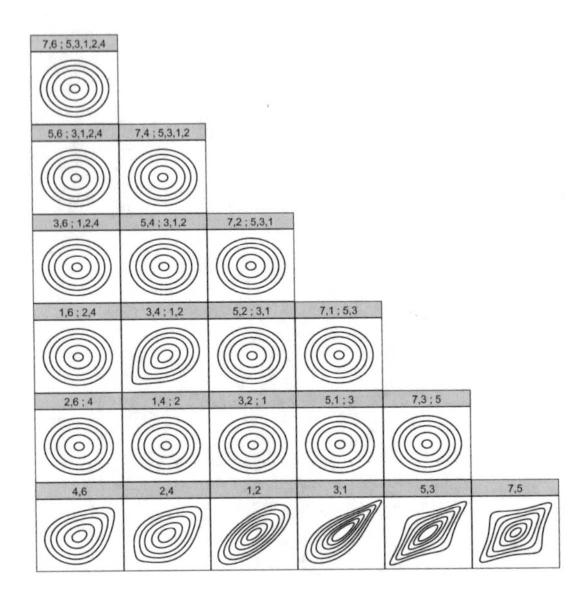

Fig. 8. Contour plots of the optimal vine copula

4 Conclusion and Future Work

Modeling dependencies between international financial market indexes is interesting and important for investors, risk managers and policy makers. Application of more dimensional copulas is bringing a new insight and experience for modeling activities.

Interestingly, while global optimal model (for the whole period of 277 months) belong to the Student class, the optimal local models can be found (with very minor differences in the values of GoF test statistic) in the classes of vine copulas and HAC. The dominance of these models is most striking over the interval of the financial market crisis, where the quality of the best Student class model was providing a substantially poorer fit.

Acknowledgement. The support of the grants APVV-14-0013 and VEGA 1/0420/15 is kindly announced.

References

1. Aas, K., Czado, C., Frigessi, A., Bakken, H.: Pair-copula constructions of multiple dependence. Technical report AMBA/24/06. Norwegian Computing Center, Oslo, Norway (2006)
2. Bacigál, T., Komorníková, M., Komorník, J.: Multidimensional copula models of dependencies between selected international financial market indexes. In: Kulczycki, P., Kowalski, P.A., Łukasik, S. (eds.) Contemporary Computational Science, p. 92. AGH-UST Press, Cracow (2018)

3. Bedford, T., Cooke, R.M.: Vines: a new graphical model for dependent random variables. Ann. Stat. **30**(4), 1031–1068 (2002)
4. Czado, C.: Pair-copula constructions of multivariate copulas. In: Jaworski, P., Durante, F., Härdle, W., Rychlik, T. (eds.) Copula Theory and Its Applications, pp. 93–109. Springer, Heidelberg (2010)
5. Czado, C.: Vine copulas and their applications to financial data. AFMathConf 2013, Brussels, Technische Universität München (2013)
6. Dissmann, J.F., Brechmann, E.C., Czado, C., Kurowicka, D.: Selecting and estimating regular vine copulae and application to financial returns. Comput. Stat. Data Anal. **59**(1), 52–69 (2013)
7. Franses, P.H., Dijk, D.: Non-linear Time Series Models in Empirical Finance. Cambridge University Press, Cambridge (2000)
8. Genest, C., Rémillard, B., Beaudoin, D.: Goodness-of-fit tests for copulas: a review and a power study. Insur. Math. Econ. **44**(2), 199–213 (2009)
9. Hofert, M.: Construction and sampling of nested Archimedean copulas. In: Jaworski, P., Durante, F., Härdle, W., Rychlik, T. (eds.) Copula Theory and Its Applications, pp. 147–160. Springer, Heidelberg (2010)
10. Hofert, M., Kojadinovic, I., Maechler, M., Yan, Y.: copula: Multivariate Dependence with Copulas. R package version 0.999-18 (2017). https://CRAN.R-project.org/package=copula
11. Joe, H.: Families of m-variate distributions with given margins and $m(m-1)/2$ bivariate dependence parameters. In: Rűschendorf, L., Schweitzer, B., Taylor, M.D. (eds.) Distributions with Fixed Marginals and Related Topics, pp. 120–141. Institute of Mathematical Statistics, Beachwood (1996)
12. Joe, H.: Dependence modeling with copulas. No. 134 in monographs on statistics and applied probability (2015)
13. McNeil, A.J.: Sampling nested Archimedean copulas. J. Stat. Comput. Simul. **78**(6), 567–581 (2008)
14. McNeil, A.J., Nešlehová, J.: Multivariate Archimedean copulas, d-monotone functions and ℓ_1 norm symmetric distributions. Ann. Stat. **37**(5b), 3059–3097 (2009)
15. Mendes, B., de Melo, E., Nelsen, R.: Commun. Stat. Simul. Comput. **36**, 997–1017 (2007)
16. Morales-Nápoles, O.: Counting vines. In: Dependence Modeling: Vine Copula Handbook, pp. 189–218 (2010)
17. Nelsen, R.B.: An Introduction to Copulas. Springer Series in Statistics, 2nd edn. Springer, New York (2006)
18. Okhrin, O., Okhrin, Y., Schmid, W.: On the structure and estimation of hierarchical Archimedean copulas. J. Econom. **173**(2), 189–204 (2013)
19. Okhrin, O., Ristig, A.: Hierarchical Archimedean copulae: the HAC package. J. Stat. Softw. **58**(4), 1–20 (2014). http://www.jstatsoft.org/v58/i04/
20. R Core Team: R: a language and environment for statistical computing. R Foundation for Statistical Computing, Vienna, Austria (2015). https://www.R-project.org/
21. Schepsmeier, U., Stoeber, J., Brechmann, E.C., Graeler, B., Nagler, T., Erhardt, T.: VineCopula: statistical inference of vine copulas. R package version 1.6–1 (2015). http://CRAN.R-project.org/package=VineCopula
22. Schirmacher, D., Schirmacher, E.: Multivariate dependence modeling using pair-copulas. Technical report, 14–16 (2008)
23. Sklar, A.: Fonctions de répartition a n dimensions et leurs marges. Publ. Inst. Statist. Univ. Paris **8**, 229–231 (1959)

24. Tawn, J.A.: Bivariate extreme value theory: models and estimation. Biometrika **75**, 397–415 (1988)
25. Ždímalová, M., Staneková, L.: Which Faber-Moore-Chen digraphs are Cayley digraphs? Discrete Math. **310**(17–18), 2238–2240 (2010)

New Types of Decomposition Integrals and Computational Algorithms

Adam Šeliga(✉)

Faculty of Civil Engineering, Slovak University of Technology,
Radlinského 11, 810 05 Bratislava, Slovakia
adam.seliga@stuba.sk

Abstract. In this paper we define two new types of decomposition integrals, namely the chain and the min-max integral and prove some of their properties. Their superdecomposition duals are also mentioned. Based on the wide applicability of decomposition integrals, some computational algorithms and their complexity are discussed.

Keywords: Decomposition integrals · Non-linear integrals · Computational complexity

1 Introduction

Theory of integration plays an important role in almost every subject of mathematics. First ideas behind this theory come back to Eudoxus circa 370 BC. Later, in the 17th century, Newton and Leibniz developed framework called infinitesimal calculus which was formalized using limits by Riemann in the 19th century. Riemann integral is a linear operator on the space of continuous functions. Later, new integrals that were non-linear appeared, e.g. the Sugeno integral [9] and the Choquet integral [1].

Recently, decomposition integrals found wide applicability in many fields of mathematics, to note few, in economics, game theory, or multicriteria optimization. This is due the fact that decomposition integrals are built upon capacities, which are set functions similar to measures, but they allow to include subjective reasoning, such as cooperative and inhibitory effects, to considered processes.

The aim of this contribution is to define new types of decomposition integrals and discuss their properties. Dual superdecomposition integrals are also presented. Furthermore, the computational complexity of some algorithms is discussed too.

The preliminary version of this paper was presented at the 3rd Conference on Information Technology, Systems Research and Computational Physics, 2–5 July 2018, Cracow, Poland [7].

P. Kulczycki et al. (Eds.): ITSRCP 2018, AISC 945, pp. 361–371, 2020.
https://doi.org/10.1007/978-3-030-18058-4_29

2 Decomposition Integrals

In this paper, without loss of generality, we will consider a fixed finite space $X = \{1, 2, \ldots, n\}$ with $n \in \mathbb{N}$. A capacity is any set function $\mu\colon \mathrm{Pow}(X) \to [0, \infty[$ that is monotone, i.e. $A \subseteq B \subseteq X$ implies $\mu(A) \leq \mu(B)$, and grounded, i.e. $\mu(\emptyset) = 0$. A chain on X is any sequence $\{A_i\}_{i=1}^{k}$ such that $\emptyset \neq A_1 \subsetneq \cdots \subsetneq A_k \subseteq X$. The class of all capacities is denoted by $\mathcal{M}$, and the class of all functions with domain X and co-domain $[0, \infty[$ by $\mathcal{F}$. A decomposition system $\mathcal{H}$ is any subset of $\mathrm{Pow}(\mathrm{Pow}(X) \setminus \{\emptyset\})$ that is non-empty, i.e. $\mathcal{H} \neq \emptyset$.

Definition 1. *A decomposition integral [2, 5] with respect to a decomposition system $\mathcal{H}$ is mapping $I_{\mathcal{H}}\colon \mathcal{F} \times \mathcal{M} \to [0, \infty[$ such that*

$$I_{\mathcal{H}}(f, \mu) = \bigvee \left\{ \sum_{i \in J} a_i \mu(A_i)\colon \{A_i\}_{i \in J} \in \mathcal{H}, a_i \geq 0 \ \forall i \in J, \sum_{i \in J} a_i \mathbf{1}_{A_i} \leq f \right\}. \quad (1)$$

Based on the choice of $\mathcal{H}$ in (1) we get different decomposition integrals.

Example 1. If $\mathcal{H}_1$ is the set of all singleton set systems, we speak about the Shilkret integral [8], i.e.

$$\mathrm{Sh}(f, \mu) = \bigvee \{\mu(A) \min f(A)\colon A \in \mathrm{Pow}(X) \setminus \{\emptyset\}\}.$$

Note that we use the abbreviate notation $\min f(A) = \min\{f(x)\colon x \in A\}$. If $\mathcal{H}_2$ consists only of partitions of X then we speak about the Pan integral [10], i.e.

$$\mathrm{Pan}(f, \mu) = \bigvee \left\{ \sum_{i \in J} \mu(A_i) \min f(A_i)\colon \{A_i\}_{i \in J} \text{ is partition of } X \right\}.$$

In case that $\mathcal{H}_3$ is class of all chains on X, then the corresponding integral is the Choquet integral [1], i.e.

$$\mathrm{Ch}(f, \mu) = \int_0^\infty \mu(f \geq t) \, \mathrm{d}t,$$

Conceding that $\mathcal{H}_4 = \mathrm{Pow}(\mathrm{Pow}(X) \setminus \{\emptyset\})$, we get the concave integral $\mathrm{con}(f, \mu)$ introduced by Lehrer [3]. Note that the choice $\mathcal{H} = \{\mathrm{Pow}(X) \setminus \{\emptyset\}\}$ also yields to the concave integral.

Definition 2. *A superdecomposition integral [4] with respect to a decomposition system $\mathcal{H}$ is mapping $I^*_{\mathcal{H}}\colon \mathcal{F} \times \mathcal{M} \to [0, \infty[$ such that*

$$I^*_{\mathcal{H}}(f, \mu) = \bigwedge \left\{ \sum_{i \in J} a_i \mu(A_i)\colon \{A_i\}_{i \in J} \in \mathcal{H}, a_i \geq 0 \ \forall i \in J, \sum_{i \in J} a_i \mathbf{1}_{A_i} \geq f \right\}.$$

Example 2. For decomposition integrals mentioned in Example 1, there is corresponding superdecomposition integral. Observe that in the case of the decomposition system $\mathcal{H}_3$, in both cases the Choquet integral is obtained.

3 New Types of Integrals

In this section we introduce two new types of decomposition integrals called the min-max and the chain integral. Some properties of these integrals are proved. Dual superdecomposition integrals are also defined.

3.1 Chain Integral

If the decomposition system $\mathcal{H}$ consists of only one fixed collection, namely a chain, then we get a new decomposition integral called the chain integral.

Definition 3. *Let B be a chain on X. The operator $\mathrm{ch}_B(f,\mu) = I_{\{B\}}(f,\mu)$ is called the chain integral with respect to chain B.*

The following definition and lemmas will be useful in proving an explicit expression for computing the chain integral.

Definition 4. *A sequence $\{a_i\}_{i=1}^k$ will be called f-$\{A_i\}_{i=1}^k$-feasible if and only if $a_i \geq 0$ for all $i = 1, 2, \ldots, k$ and*

$$\sum_{i=1}^{k} a_i 1_{A_i} \leq f.$$

Lemma 1. *Let $\emptyset \neq A_1 \subsetneq \cdots \subsetneq A_k \subseteq X$. For every f-$\{A_i\}_{i=1}^k$-feasible sequence $\{a_i\}_{i=1}^k$ there exists f-$\{A_i\}_{i=1}^k$-feasible sequence $\{b_i\}_{i=1}^k$ with $b_k = \min f(A_k)$ and*

$$\sum_{i=1}^{k} a_i \mu(A_i) \leq \sum_{i=1}^{k} b_i \mu(A_i).$$

Proof. The proof of this lemma will be divided into two cases.

Case 1: If $\sum_{i=1}^k a_i \leq \min f(A_k)$. Then $\{b_i\}_{i=1}^k$ is given by

$$b_i = \begin{cases} \min f(A_k), & \text{if } i = k, \\ 0, & \text{otherwise} \end{cases}$$

for $i = 1, 2, \ldots, k$. Truly,

$$\sum_{i=1}^{k} a_i \mu(A_i) \leq \sum_{i=1}^{k} a_i \mu(A_k) = \mu(A_k) \sum_{i=1}^{k} a_i \leq \mu(A_k) \min f(A_k) = \sum_{i=1}^{k} b_i \mu(A_i).$$

The fact that $\{b_i\}_{i=1}^k$ is f-$\{A_i\}_{i=1}^k$-feasible can be trivially seen.

Case 2: If $\sum_{i=1}^k a_i > \min f(A_k)$. Then there exists $i^* \in \{1, 2, \ldots, k\}$ such that $\sum_{i=i^*+1}^k a_i < \min f(A_k)$ and $\sum_{i=i^*}^k a_i \geq \min f(A_k)$. Then $\{b_i\}_{i=1}^k$ is given by

$$b_i = \begin{cases} \min f(A_k), & \text{if } i = k, \\ \sum_{i=i^*}^k a_i - \min f(A_k), & \text{if } i = i^*, \\ a_i, & \text{if } i \in \{1, \ldots, i^* - 1\}, \\ 0, & \text{otherwise} \end{cases}$$

for $i = 1, 2, \ldots, n$. Indeed,

$$\begin{aligned}\sum_{i=1}^{k} a_i\mu(A_i) &= \sum_{i=1}^{i^*-1} a_i\mu(A_i) + a_{i^*}\mu(A_{i^*}) + \sum_{i=i^*+1}^{k} a_i\mu(A_i)\\ &\leq \sum_{i=1}^{i^*-1} a_i\mu(A_i) + \left(\sum_{i=i^*}^{k} a_i - \min f(A_k)\right)\mu(A_{i^*}) + \min f(A_k)\mu(A_k)\\ &= \sum_{i=1}^{k} b_i\mu(A_i).\end{aligned}$$

Again, the statement that $\{b_i\}_{i=1}^k$ is f-$\{A_i\}_{i=1}^k$-feasible can be checked trivially using the fact that $\{a_i\}_{i=1}^k$ is f-$\{A_i\}_{i=1}^k$-feasible. □

Definition 5. *A sequence $\{a_i\}_{i=1}^k$ will be called min-f-$\{A_i\}_{i=1}^k$-feasible if and only if it is f-$\{A_i\}_{i=1}^k$-feasible and $a_k = \min f(A_k)$.*

Lemma 2. *Let f be a function, μ a capacity and $B = \{A_i\}_{i=1}^k$ be a chain. Let*

$$\alpha = \bigvee\left\{\sum_{i=1}^{k} a_i\mu(A_i) \colon \{a_i\}_{i=1}^k \textit{ is } f\text{-}\{A_i\}_{i=1}^k\textit{-feasible}\right\}$$

and

$$\beta = \bigvee\left\{\sum_{i=1}^{k} a_i\mu(A_i) \colon \{a_i\}_{i=1}^k \textit{ is min-}f\text{-}\{A_i\}_{i=1}^k\textit{-feasible}\right\}.$$

Then $\alpha = \beta$.

Proof. First of all, it can be seen that

$$\begin{aligned}B^* &= \left\{\{b_i\}_{i=1}^k \text{ is min-}f\text{-}\{A_i\}_{i=1}^k\text{-feasible}\right\}\\ &\subseteq \left\{\{a_i\}_{i=1}^k \text{ is } f\text{-}\{A_i\}_{i=1}^k\text{-feasible}\right\} = A^*\end{aligned}$$

and thus $\beta \leq \alpha$. On the other hand, based on the previous lemma, for every element in $\{a_i\}_{i=1}^k \in A^*$ there exists an element in $\{b_i\}_{i=1}^k \in B^*$ such that $\sum_{i=1}^k a_i\mu(A_i) \leq \sum_{i=1}^k b_i\mu(A_i)$ and thus $\alpha \leq \beta$, which proves that $\alpha = \beta$. □

Let us now consider the computation of the chain integral.

Theorem 1. *Let $B = \{A_i\}_{i=1}^k$ be a chain on X and $\tau = \min f(A_k)$. Then*

$$\mathrm{ch}_B(f, \mu) = \tau\mu(A_k) + \mathrm{ch}_{B\setminus\{A_k\}}(\tilde{f}, \mu\restriction_{\mathrm{Pow}(A_{k-1})}),$$

where $\tilde{f} \colon A_{k-1} \to \mathbb{R} \colon x \mapsto f(x) - \tau$.

Proof. From previous two lemmas we can see that

$$\begin{aligned}\mathrm{ch}_B(f,\mu) &= \bigvee\left\{\sum_{i=1}^{k} a_i\mu(A_i)\colon \{a_i\}_{i=1}^{k} \text{ is } f\text{-}\{A_i\}_{i=1}^{k}\text{-feasible}\right\}\\ &= \bigvee\left\{\sum_{i=1}^{k} b_i\mu(A_i)\colon \{b_i\}_{i=1}^{k} \text{ is min-}f\text{-}\{A_i\}_{i=1}^{k}\text{-feasible}\right\}\\ &= \mu(A_k)\min f(A_k)\\ &\quad + \bigvee\left\{\sum_{i=1}^{k-1} b_i\mu(A_i)\colon \{b_i\}_{i=1}^{k-1} \text{ is min-}f\text{-}\{A_i\}_{i=1}^{k-1}\text{-feasible}\right\}\end{aligned}$$

which proves the theorem. □

Inducting the previous theorem we get the following result.

Corollary 1. *Let* $B = \{A_i\}_{i=1}^{k}$ *be a chain. Then*

$$\mathrm{ch}_B(f,\mu) = \mu(A_k)\min f(A_k) + \sum_{i=1}^{k-1}\mu(A_i)\left(\min f(A_i) - \min f(A_{i+1})\right),$$

or

$$\mathrm{ch}_B(f,\mu) = \sum_{i=1}^{k}\mu(A_i)\left(\min f(A_i) - \min f(A_{i+1})\right)$$

with convention that $\min f(A_{k+1}) = 0$.

Corollary 2. *Let* $B = \{A_i\}_{i=1}^{k}$ *be a chain. Then* $\mathrm{ch}_B(f,\mu) \geq \mu(A_k)\min f(A_k)$.

From the theory of the Choquet integration it is known that

$$\mathrm{Ch}(f,\mu) = \sum_{i=1}^{m}(f_i - f_{i-1})\mu(A_i) \tag{2}$$

where $\{f_i\}_{i=0}^{m}$ is increasing enumeration of $\mathrm{Im}(f)\cup\{0\}$ and

$$A_i = \{x \in X\colon f(x) > f_{i-1}\}$$

for $i = 1, 2, \ldots, m$. Then it can be seen that

$$\mathrm{Ch}(f,\mu) = \sum_{i=1}^{k}\mu(A_i)(\min f(A_i) - \min f(A_{i+1}))$$

with convention that $\min f(A_{k+1}) = 0$.

It can easily be seen that the chain integral and the Choquet integral are related with $\mathrm{ch}_B(f,\mu) \leq \mathrm{Ch}(f,\mu)$ for all chains B, all functions f and all capacities μ. Interesting question arises: What conditions must be satisfied for the equality of these two decomposition integrals? Following theorem gives some insight to the answer to this question.

Definition 6. *A chain $B = \{A_i\}_{i=1}^k$ is called Ch-maximizing for function f if and only if*

$$\{\min f(A_i)\colon i = 1, \dots, k\} \setminus \{0\} = \mathrm{Im}(f) \setminus \{0\}.$$

It seems that the Ch-maximizing property of a chain is enough for the equality of the chain and the Choquet integrals. The following theorem follows from the theory of the Choquet integration.

Theorem 2. *A chain B is a Ch-maximizing for a function f if and only if $\mathrm{ch}_B(f, \mu) = \mathrm{Ch}(f, \mu)$ holds.*

Example 3. Following the original example with workers of Lehrer in [3], let us assume that $X = \{1, 2, 3, 4\}$ represents the set of workers and let $f\colon X \to [0, \infty[\colon i \mapsto 5-i$ denote a positive function where $f(i)$ represents maximum number of working hours for worker $i \in X$. Let us choose a chain $B = \{A_i\}_{i=1}^3$ where $A_1 = \{3\}$, $A_2 = \{2, 3\}$, $A_3 = \{1, 2, 3, 4\}$. This chain can represent the following situation: all four workers start working together, then the first and fourth will leave and then the second one will leave. Let μ represent the number of articles made per hour: $\mu(A_1) = 3$, $\mu(A_2) = 4$ and $\mu(A_3) = 6$, then $\mathrm{ch}_B(f, \mu)$ represents the maximum number of articles made in this situation. From Corollary 1 it is easy to see that

$$\mathrm{ch}_B(f, \mu) = \mu(A_1) + \mu(A_2) + \mu(A_3) = 13.$$

3.2 Min-Max Integral

The definition of the Choquet integral can be rewritten to the form

$$\mathrm{Ch}(f, \mu) = \bigvee_{B=\{A_i\}_{i=1}^n} \bigvee \left\{ \sum_{i \in J} a_i \mu(A_i)\colon a_i \geq 0 \ \forall i \in J, \sum_{i \in J} a_i 1_{A_i} \leq f \right\}$$

where the first supremum operator runs over all full chains B on X. The idea behind the min-max integral is to replace the first supremum by infimum.

Definition 7. *A min-max integral of a function f with respect to capacity μ is defined by*

$$I^{\wedge\vee}(f, \mu) = \bigwedge_{B=\{A_i\}_{i=1}^n} \bigvee \left\{ \sum_{i \in J} a_i \mu(A_i)\colon a_i \geq 0 \ \forall i \in J, \sum_{i \in J} a_i 1_{A_i} \leq f \right\}$$

where the infimum operator runs over all full chains B on X.

Firstly, let us prove that there exists a lower bound on the value of $I^{\wedge\vee}(f, \mu)$.

Lemma 3. *The inequality $I^{\wedge\vee}(f, \mu) \geq \mu(X) \min f(X)$ holds for all functions f and capacities μ.*

Proof. Let $B = \{A_i\}_{i \in J}$ be any full chain. We need to prove that

$$\bigvee \left\{ \sum_{i \in J} a_i \mu(A_i) \colon a_i \geq 0 \ \forall i \in J, \sum_{i \in J} a_i 1_{A_i} \leq f \right\} \geq \mu(X) \min f(X).$$

From the previous part, it is obvious that the expression on the left of the inequality sign is equal to $\mathrm{ch}_B(f, \mu)$. Also, Corollary 2 implies that

$$\mathrm{ch}_B(f, \mu) \geq \mu(X) \min f(X)$$

and thus the result follows. □

Now, let us prove that this value is not only the lower bound but also the value of the min-max integral.

Theorem 3. *It is true that* $I^{\wedge\vee}(f, \mu) = \mu(X) \min f(X)$.

Proof. To simplify writing, notice that $I^{\wedge\vee}(f, \mu) = \bigwedge_B \mathrm{ch}_B(f, \mu)$. Thus we need to find a chain B such that $\mathrm{ch}_B(f, \mu) = \mu(X) \min f(X)$. Let $x^* \in X$ be any element such that $f(x^*) = \min f(X)$. Let $B = \{A_i\}_{i=1}^n$ be any chain such that $A_1 = \{x^*\}$ and thus $x^* \in A_i$ for all $i = 1, \ldots, n$. Also,

$$\begin{aligned} \mathrm{ch}_B(f, \mu) &= \mu(X) \min f(X) + \sum_{i=1}^{n} \mu(A_i) \left(\min f(A_i) - \min f(A_{i+1}) \right) \\ &= \mu(X) \min f(X) + \sum_{i=1}^{n} \mu(A_i) \left(f(x^*) - f(x^*) \right) = \mu(X) \min f(X), \end{aligned}$$

which proves the theorem. □

Example 4. Let f and μ be as in Example 3. From Theorem 3 it follows that $I^{\wedge\vee}(f, \mu)$ represents the maximum number of articles made if only all workers can work together. In this setting, $I^{\wedge\vee}(f, \mu) = \mu(X) = 6$.

3.3 Superdecomposition Duals

For these newly proposed decomposition integrals, namely the chain and the min-max integral, their superdecomposition duals can be considered.

Definition 8. *A super-chain integral of a function f with respect to a capacity μ and a chain $B = \{A_i\}_{i=1}^n$ is*

$$\mathrm{ch}_B^*(f, \mu) = \bigwedge \left\{ \sum_{i=1}^{n} a_i \mu(A_i) \colon a_1, \ldots, a_n \geq 0, \sum_{i=1}^{n} a_i 1_{A_i} \geq f \right\}.$$

The value of a super-chain integral can be computed and proved analogously to its decomposition counterpart.

Theorem 4. *The equality*

$$\mathrm{ch}_B^*(f,\mu) = \mu(X) \max f(X)$$

holds.

The corresponding superdecomposition integral for the min-max integral will be called max-min integral. Note that the min-max and max-min integrals are reasonable bounds for all operators considered to be integrals, e.g. decomposition and superdecomposition integrals.

Definition 9. *A max-min integral of a function f with respect to a capacity μ is*

$$I^{\vee\wedge}(f,\mu) = \bigvee_B \bigwedge \left\{ \sum_{i=1}^n a_i\mu(A_i) \colon a_1,\ldots,a_n \geq 0, \sum_{i=1}^n a_i 1_{A_i} \geq f \right\}$$

where the infimum operator runs through all chains $B = \{A_i\}_{i=1}^n$.

4 Computation Algorithms

Decomposition and superdecomposition integrals show potential in applicability in real-world problems. A natural demand is to find algorithms for their computation. In this section we will formulate basic computational algorithms for their computation and discuss their computational complexity.

4.1 Concave Integration as Linear Optimization Problem

The problem of the concave integration can be trivially rewritten to the following optimization problem:

$$\begin{aligned} \text{maximize} \quad & \sum_{i=1}^{2^n-1} a_i\mu(P_i) \\ \text{subject to} \quad & A\mathbf{a} \leq \mathbf{f} \text{ and } \mathbf{a} \geq 0 \end{aligned}$$

where $\{P_i\}_{i=1}^{2^n-1}$ is any enumeration of $\mathrm{Pow}(X) \setminus \{\emptyset\}$, A is $n \times (2^n - 1)$ matrix with $A_{ij} = 1_{P_j}(x_i)$, $\mathbf{f}$ is n-dimensional vector whose ith element is $f(x_i)$ and $\mathbf{a}$ is unknown n-dimensional vector.

Theorem 5. *The problem of concave integration posed as linear optimization problem is harder than NP [6].*

4.2 Choquet and Chain Integration

The equality (2) can be used to construct algorithm that computes the Choquet integral based on the ordering of $\mathrm{Im}(f)$ that can be done in polynomial time, thus yielding to polynomial time algorithm, more precisely to $O(n \log n)$ algorithm.

Similar approach can be taken for the chain integral defined in this article, yielding also to $O(n \log n)$ algorithm.

4.3 Min-Max Integration

Based on Theorem 3, the computation of the min-max integral is straightforward. The only unknown is to find the minimum of the function f and this can be done using $O(n)$ algorithm. Thus, the computation of min-max integral can be accomplished using $O(n)$ algorithm.

4.4 Brute Force Algorithms

With other types of decomposition integrals, namely the Shilkret and the Pan integral, the situation doesn't seem so easy. Brute force algorithms, i.e. algorithms that check all possible combinations, are discussed.

Theorem 6. *Computation of the Shilkret and the Pan integrals belongs to at most NP class.*

Proof. We need to find polynomial verifiers for both integrals. The solution of Shilkret integral is identified with a set from $\mathrm{Pow}(X) \setminus \{\emptyset\}$. Given such set, A, the minimum $\min f(A)$ can be computed in polynomial time, and also the product $\mu(A) \min f(A)$ which gives us a polynomial verifier for the Shilkret integral. For the Pan integral, the solution is identified with a partition $\{A_i\}_{i \in J}$ of X. Such partition has at most n elements and thus $\min f(A_i)$ for $i \in J$ can be computed using polynomial algorithm. Also, the sum $\sum_{i \in J} \mu(A_i) \min f(A_i)$ can be computed in polynomial time yielding to polynomial verifier. Thus, the computation of the Shilkret and the Pan integral belongs to at most NP class of computational problems. □

Brute force algorithm for the Shilkret integral goes as follows. For every set $A \in \mathrm{Pow}(X) \setminus \{\emptyset\}$ compute $\min f(A)$ and find a minimum of $\mu(A) \min f(A)$. The computation of $\mu(A) \min f(A)$ for every A takes at most $O(n)$ operations. The set $\mathrm{Pow}(X) \setminus \{\emptyset\}$ has $2^n - 1$ elements, thus yielding to $O(2^n n)$ algorithm.

For the Pan integral, we need to check all permutations. The number of permutations of set with n elements is bounded by Catalan numbers, i.e. to generate all permutations we need $O(3^n)$ operations. For each partition, we need to compute at most n minimums, which yields to $O(n^2)$ operations per permutation and thus the brute force algorithm for the Pan integral takes at least $O(3^n n^2)$ operations.

4.5 Special Classes of Capacities

If we restrict ourselves to a special class of capacities then the computation of decomposition integrals is simplified. The first considered such class is the set of measures, i.e. additive capacities.

Theorem 7. *If μ is a measure, then*

$$\mathrm{Ch}(f, \mu) = \mathrm{Pan}(f, \mu) = \mathrm{con}(f, \mu).$$

Also, the value of these integrals is equal to the value of Lebesgue integral of function f with respect to a measure μ.

The same holds if μ is subadditive capacity. On the other hand, if μ is superadditive capacity, we get the following theorem.

Theorem 8. *If* μ *is superadditive capacity, then* $\mathrm{con}(f,\mu) = \mathrm{Ch}(f,\mu)$.

5 Concluding Remarks

In this paper we defined two new types of decomposition integrals, namely the chain and the min-max integral and proved some of their properties.

A similarity between the chain and the Choquet integral was discussed and the equality condition was established. We have showed that the min-max integral can be viewed as a natural lower bound for every operator named integral. Also, superdecomposition variants of these integrals, namely the super-chain and the max-min integrals are proposed.

In the last section of this article, basic computational algorithms for computing some decomposition integrals, namely the concave, the Choquet, the Shilkret and the Pan integrals, are examined. Also we proposed algorithms for newly established decomposition integrals–the chain and the min-max integrals. Also, the computational complexity of these algorithms is discussed.

Acknowledgement. This work was supported by the grants APVV-14-0013 and VEGA 1/0682/16.

References

1. Choquet, G.: Theory of capacities. Annales de l'Institut Fourier **5**, 131–295 (1954). https://doi.org/10.5802/aif.53
2. Even, Y., Lehrer, E.: Decomposition-integral: unifying Choquet and the concave integrals. Econ. Theory **56**, 33–58 (2014). https://doi.org/10.1007/s00199-013-0780-0
3. Lehrer, E.: A new integral for capacities. Econ. Theory **39**, 157–176 (2009). https://doi.org/10.1007/s00199-007-0302-z
4. Mesiar, R., Li, J., Pap, E.: Superdecomposition integrals. Fuzzy Sets Syst. **259**, 3–11 (2015). https://doi.org/10.1016/j.fss.2014.05.003
5. Mesiar, R., Stupňanová, A.: Decomposition integrals. Int. J. Approx. Reason. **54**, 1252–1259 (2013). https://doi.org/10.1016/j.ijar.2013.02.001
6. Šeliga, A.: A note on the computational complexity of Lehrer integral. In: Advances in Architectural, Civil and Environmental Engineering: 27th Annual PhD Student Conference on Applied Mathematics, Applied Mechanics, Geodesy and Cartography, Landscaping, Building Technology, Theory and Structures of Buildings, Theory and Structures of Civil Engineering Works, Theory and Environmental Technology of Buildings, Water Resources Engineering, pp. 62–65 (2017). ISBN 978-80-227-4751-6
7. Šeliga, A.: New types of decomposition integrals and computational algorithms. In: Kulczycki, P., Kowalski, P.A., Łukasik, S. (eds.) Contemporary Computational Science, p. 93. AGH-UST Press, Cracow (2018)

8. Shilkret, N.: Maxitive measure and integration. Indagationes Mathematicae **33**, 109–116 (1971). https://doi.org/10.1016/S1385-7258(71)80017-3
9. Sugeno, M.: Theory of fuzzy integrals and its applications. Doctoral thesis, Tokyo Institute of Technology (1974)
10. Yang, Q.: The Pan-integral on the fuzzy measure space. Fuzzy Math. **3**, 107–114 (1985)

Trend Analysis and Detection of Change-Points of Selected Financial and Market Indices

Dominika Ballová(✉)

Slovak University of Technology, 81005 Bratislava, Slovakia
dominika.ballova@stuba.sk

Abstract. From the macroeconomic point of view, the stock index is the best indicator of the behavior of the stock market. Stock indices fulfill different functions. One of heir most important function is to observe developments of the stock market situation. Therefore, it is crucial to describe the long-term development of indices and also to find moments of abrupt changes. Another interesting aspect is to find those indices that have evolved in a similar way over time. In this article, using trend analysis, we will uncover the long-term evolution of selected indices. Other goal is to detect the moments in which this development suddenly changed using the change point analysis. By means of cluster analysis, we find those indices that are most similar in long-term development. In each analysis, we select the most appropriate methods and compare their results.

Keywords: Trend analysis · Change-point analysis · Cluster analysis

1 Introduction

Stock market indices express the value of a section of the stock market. By observing the historical development of market indices, we can determine the trend of their long term development. It can be useful in construing predictions of the future development of the valuation process. Locating change-points is also essential factor in the analysis of the development of indices.

Understanding the long term development and abrupt changes in the prices of indices is a key factor for the investor in the decision making about where to invest. Therefore, our aim is to reveal the presence of the trend and identify its nature in the time series of 11 selected indices. We use non-parametric tests based on the fact that data follows non-normal distribution. Comparing the results of the Cox-Stuart test, Mann-Kendall test, and Spearman's rho test we seek to find trend and its character. The power of the trend will be expressed by Sen's slope. This will be compared with the values of Kendal's tau and Spearman's rho.

Another important goal is to find change-points in the series. First we obtain single change-points for each time series using the Pettitt's test. Next we use multiple change-point analysis using divisive and agglomerative estimation. We compare the results of all three procedures and we will look for common change-points.

P. Kulczycki et al. (Eds.): ITSRCP 2018, AISC 945, pp. 372–381, 2020.
https://doi.org/10.1007/978-3-030-18058-4_30

In the last part indices, which development in long term are similar, and which are the most different from the development of others will be found. For this purpose agglomerative techniques of cluster analysis will be used. The primary version of this paper was presented at the 3rd Conference on Information Technology, Systems Research and Computational Physics, 2–5 July 2018, Cracow, Poland [1].

2 Statistical Methods

2.1 Trend Analysis

The trend analysis in time series of stock market indices has been evaluated using the following nonparametric tests: Cox-Stuart test, Mann-Kendall test and Spearman's rho test. We will denote $X_1, X_2, \ldots, X_n$ as a sample of n independent variables. The above tests are testing the null hypothesis that there is no trend in the data, against the alternative hypothesis that there is a statistically significant increasing/decreasing trend. Positive/negative values of the statistics implies increasing/decreasing trend.

Cox-Stuart Test. The Cox-Stuart test is based on the signs of the differences

$$\begin{aligned} y_1 &= x_{1+d} - x_1 \\ y_2 &= x_{2+d} - x_2 \\ &\vdots \\ y_d &= x_n - x_{n-d} \end{aligned} \tag{1}$$

where $d = n/2$, if the size n is odd, otherwise $d = (n+1)/2$. Assign $y_1, y_2, \ldots, y_m$ the sample of positive differences. The test statistic of Cox-Stuart test is

$$T = \sum\nolimits_{i=1}^{m} sign(y_i). \tag{2}$$

On the significance level α we reject the null hypothesis if $|T| > t_\alpha$, t_α is the quantile of binomial distribution. For $m > 20$, we can approximate t_α with the α-quantile w_α of the standard normal distribution

$$t_\alpha = \frac{1}{2}[m + w(\alpha)]\sqrt{m}. \text{ [2]} \tag{3}$$

Mann-Kendall Test. The Mann-Kendall test statistic is defined as

$$S = \sum\nolimits_{i=1}^{n} \sum\nolimits_{j=1}^{i-1} sign(x_i - x_j). \tag{4}$$

For $n > 8$, S can be approximate by normal distribution, thus the standardized test statistic is given:

$$Z = \begin{cases} \frac{S-1}{\sqrt{D(S)}} & S > 0 \\ 0, & S = 0 \\ \frac{S+1}{\sqrt{D(S)}} & S < 0 \end{cases} \quad (5)$$

We reject the null hypothesis, if $Z > Z_{1-\alpha/2}$, and that means there is increasing trend in the series, or if $Z < -Z_{\alpha/2}$ what means decreasing trend. $Z_{1-\alpha/2}$ and $Z_{\alpha/2}$ are the critical value of the standard normal distribution [3, 4].

Spearman's Rho Test. The test statistic of Spearman's rho test is given

$$D = 1 - \frac{6\sum_{i=1}^{n}(R_i - i)^2}{n(n^2 - 1)}, \quad (6)$$

where R_i is the rank of the i-th observation in time series. The standardized statistic is given

$$Z_{SR} = D\sqrt{\frac{n-2}{1-D^2}}. \quad (7)$$

If $|Z_{SR}| > t_{(n-2,1-\alpha/2)}$ the trend exists. $t_{(n-2,1-\alpha/2)}$ is the critical value of Student's t distribution [5].

2.2 Change-Point Analysis

First we checked the homogeneity of our time series using Wald-Wolfowitz test and then we used Pettitt's test for single change-point detection and multiple change-point detection.

Wald-Wolfowitz Test. It is a nonparametric test for verifying homogeneity in time series. The null hypothesis says that a time series is homogenous between two given times. The test statistic is given

$$R = \sum_{i=1}^{n-1} x_i x_{i+1} + x_1 x_n, \quad (8)$$

where $x_1, x_2, \ldots, x_n$ are the sampled data. For $n > 10$ we can make an approximation

$$Z = \frac{R - E(R)}{\sqrt{D(R)}}. \text{ [6]} \quad (9)$$

Pettitt's Test. The null hypothesis of this test is that there is no change in the series against the alternative hypothesis there is change. The test statistic of Pettitt's test is

$$\widehat{U} = max|U_k|, \tag{10}$$

$$U_k = 2\sum\nolimits_{i=1}^{k} r_i - k(n+1), \tag{11}$$

where $k = 1, 2, \ldots, n$ and r_i are the ranks of X_i. The most probably change-point is located where $\widehat{U}$ reaches maximum value [7].

Hierarchical Divisive Estimation E-divisive. This method applies single change-point detection iteratively. Details on the estimation of change point locations can be found in [8].

Hierarchical Agglomerative Estimation E-agglo. This method assumes an initial segmentation of the data. If there are no initial segmentations defined, then each observation can be considered as a separated segment. In this method bordering segments are sequentially merged to maximize the goodness-of-fit statistic. The estimated locations of change points are assigned by the iteration which maximized the penalized goodness-of-fit statistic. More details about this method can be found in [8].

2.3 Cluster Analysis

Cluster analysis belongs to multidimensional statistical methods used to seek out similar objects and grouping them into clusters. Clusters contain objects with the highest degree of similarity, while high dissimilarity among each cluster is desirable. Results of cluster analysis can be the best shown by *dendrogram* which represent each object and the linkage distance of these objects. It is a figure which arranges the analyzed objects so that individual joining of objects to clusters can be observed.

Since there are several aggregation methods, each of which generally yields different results, it is necessary to determine the most appropriate method of aggregation. Such measure is the *cophonetic correlation coefficient CC*. The cophonetic correlation coefficient is defined as the Pearson coefficient of correlation between actual and predicted distance. For the most suitable agglomeration method, we choose the one for which the cophonetic correlation coefficient is the highest [9].

In hierarchical clustering, we can choose the appropriate number of clusters from the dendrogram by cutting through its branches at the selected distance level on the corresponding axis. For this several indices has been developed as a criteria. Detailed criteria used to select the number of clusters can be found in [10].

3 Analysis of the Development of Selected Stock Indices

3.1 Trend Analysis

Trend analysis plays vital role in various fields of study since researcher are often interested in the long term development of processes. Describing the long term character of the stock indices can reflect the progress of market efficiency. For this purpose we analyzed the presence and the character of the trend of 11 stock market indices: *SPX, CCMP, INDU, DAX, UKX, CAC, NKY, HSI, LEGATRUU, SPGSCITR* and *CCI* index. We analyzed monthly time series since January 1995 to January 2018.

Since the multidimensional normality was rejected by testing, for this purpose Mann-Kendall, Cox-Stuart test and Spearman's rho test was used. These tests was realized in *R* software using the packages: *trend* [10] for the Mann Kendall test and Cox and Stuart test and *pastecs* [11] for Spearman's rho test. The results of the three trend tests are listed in Table 1. From this table it is found that the Mann-Kendall test indicates the presence of a statistically significant monotonic trend in all the observed indices. Cox and Stuart test and Spearman's rho test rejected the presence of trend at the level 0.05 only for the *NKY index*.

The character (increasing/decreasing) of the trend was obtained by Sen's slope using the R package *TTAinterfaceTrendAnalysis* [12]. All of the indices except *JPY* have a long term increasing tendency. *JPY index* indicated decreasing trend. According to the magnitude of Sen's slope *61.1* the *HSI index* has the highest rising tendency among all the analyzed indices. Other high level of increase was observed in the *INDU Index* with the value of *45.05* and *DAX Index* with the value of slope *29.4*. The only decreasing trend in *NKY Index* reached the value of *−6.73*. P-values of Sen's slope magnitudes indicate that all of the magnitudes are statistically significant at the 0.05 level. We obtained the same results considering the signs of the Z statistics of Mann-Kendall test. Other aspect can be the value of Mann-Kendall's tau and Spearman's rho which all indicate decreasing trend for *NKY* and increasing trend for all the other indices. Remarkably, according to these last three criteria, *LEGATRUU index* shows the highest level of increasing trend and on the other hand *SPGSCITR* and *CAC* the lowest increasing level.

Table 1. Results of trend analysis

Index	Cox-Stuart test		Mann-Kendall test			Spearman's test		Sen's slope	
	T	p-value	Z	p-value	Tau	Rho	p-value	Magnitude	p-value
SPX	9.75	$<10^{-6}$	15.30	$<10^{-6}$	0.62	0.75	$<10^{-6}$	4.98	$<10^{-6}$
CCMP	9.75	$<10^{-6}$	16.58	$<10^{-6}$	0.67	0.78	$<10^{-6}$	12.25	$<10^{-6}$
INDU	9.75	$<10^{-6}$	17.97	$<10^{-6}$	0.72	0.86	$<10^{-6}$	45.05	$<10^{-6}$
DAX	9.75	$<10^{-6}$	16.44	$<10^{-6}$	0.66	0.81	$<10^{-6}$	29.40	$<10^{-6}$
UKX	8.29	$<10^{-6}$	12.02	$<10^{-6}$	0.48	0.62	$<10^{-6}$	9.91	$<10^{-6}$
CAC	4.75	$<10^{-6}$	6.95	$<10^{-6}$	0.28	0.35	$<10^{-6}$	6.51	$<10^{-6}$
NKY	0.38	0.70	−2.06	0.04	−0.08	−0.11	0.06	−6.74	0.04
HSI	9.75	$<10^{-6}$	16.01	$<10^{-6}$	0.64	0.84	$<10^{-6}$	61.10	$<10^{-6}$
LEGATRUU	9.75	$<10^{-6}$	22.37	$<10^{-6}$	0.90	0.98	$<10^{-6}$	1.30	$<10^{-6}$
SPGSCITR	3.30	$<10^{-4}$	4.17	$<10^{-4}$	0.17	0.20	$<10^{-3}$	6.43	$<10^{-4}$
CCI	9.75	$<10^{-6}$	13.67	$<10^{-4}$	0.55	0.79	$<10^{-6}$	1.26	$<10^{-6}$

3.2 Change-Point Analysis

Presence of change-point in time series is a vital question in the development of processes. Our aim was to find abrupt changes in the time series of each index. First we used Pettitt's test from package *trend* for single change-point detection. After finding the single change-points we carried out multiple change-point analysis using divisive estimation and agglomerative estimation of change-points from the *ecp* [13] package. Results of these analysis are in Table 2 and Fig. 1. According to Pettitt's test three indices- *SPX*, *CCMP* and *UKX* have a significant change-point in 2010. Other common significant change was detected in *INDU, DAX, HSI, LEGATRUU* and *CCI* indices in from October 2005 to August 2006. *CAC index* has a significant change in December 1998. *NKY index* in September 2000, *SPGSCITR* in December 2012.

Next, the results of multiple change-point analysis was compared. As we can see in general we obtained more results using the divisive estimation. Some of the results are similar to the agglomerative estimation, although there are differences. All the detected change-points obtained by divisive estimation are statistically significant. We can see some pattern in the positions of the change-points. Most of the indices has the first abrupt change from September 1997 to April 1998. Other significant period can be considered from December 2000 to September 2001. *DAX, UKX* and *CAC* changed abruptly in June 2002. Further common changes was observed in the period from June 2005 to August 2006. Another signifficant changes in most of the indices was from March 2008 to July 2009. *LEGATRUU, SPGSCITR, CCI* and *CCMP* has significant changes from August 2010 to April 2011. Other changes was found from the end of 2012 to the beginning of 2014. For most of the indices it was the last change-point. For *INDU, DAX* and *UKX*, the last change-points was detected in August of 2015. For *CCI* and *SPGSCITR* indices it was in October and December 2014. As we can see in Fig. 1, for most of the indices, divisive estimation allocated one of the multiple change-points near to the ones found by Pettitt's test. Also in most case the results of agglomerative estimation are close to the ones gained by divisive estimation.

Table 2. Results of single and multiple change-point analysis

Index	Pettitt's test		Divisive estimation		Agglomerative estimation
	Position	p-value	Position	p-value	Position
SPX	191	$<10^{-6}$	34, 80, 133, 163, 223	0.01	31, 226
CCMP	188	$<10^{-6}$	37, 74, 127, 159, 190, 223	0.01	47, 74, 119, 225
INDU	130	$<10^{-6}$	37, 75, 106, 136, 166, 218, 248	0.01	230
DAX	140	$<10^{-6}$	37, 90, 129, 162, 218, 248	0.01	60, 80, 131, 227
UKX	191	$<10^{-6}$	33, 90, 126, 162, 218, 248	0.01	25, 91, 117, 165, 175, 217
CAC	48	$<10^{-6}$	38, 90, 126, 165, 225	0.01	39, 91, 125, 165, 230
NKY	69	$<10^{-6}$	34, 72, 129, 165, 197, 227	0.01	79, 129, 157, 217
HSI	138	$<10^{-6}$	51, 81, 140, 175, 216	0.01	144
LEGATRUU	139	$<10^{-6}$	40, 97, 155, 188, 229	0.01	91, 192
SPGSCITR	96	$<10^{-6}$	63, 111, 166, 196, 240	0.01	110, 239
CCI	135	$<10^{-6}$	40, 102, 132, 189, 238	0.01	136

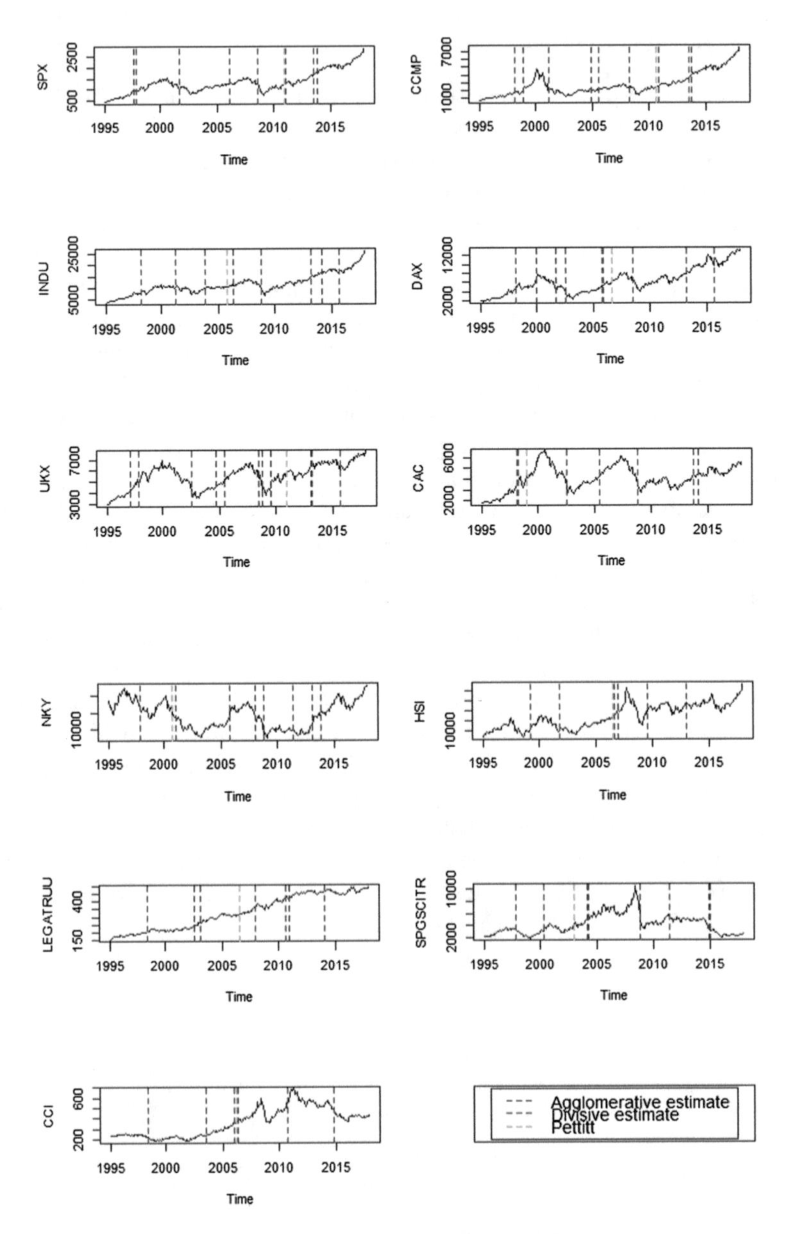

Fig. 1. Change-point analysis results

3.3 Cluster Analysis

In practice it is common to seek for similar objects and explain the relationship between these objects. Thus our other goal was to identify the indices which development are similar in time and can be separated into clusters. We used hierarchical methods from the *stats* package [14], because of the low number of the clustering objects. Since indices are measured in various currencies, first we standardized the values of the series. To determine the best cluster analysis method we calculated the cophonetic correlation coefficient from package *stats*, for each method using Euclidean distance. We chose the method which contains the highest cophonetic correlation coefficient. According to this coefficient the best clustering methods to use are the average linkage method and Ward's method. In average linkage method the average distance between objects of each cluster is used as distance between clusters. Ward's method is based on minimization of the within-cluster variance. The number of clusters was determined using the *NbClust* package based on 30 indices as criteria. Most of the indices proposed three cluster solution. The results of this analysis was creating using *dendextended* package [15] and can be found in Fig. 2.

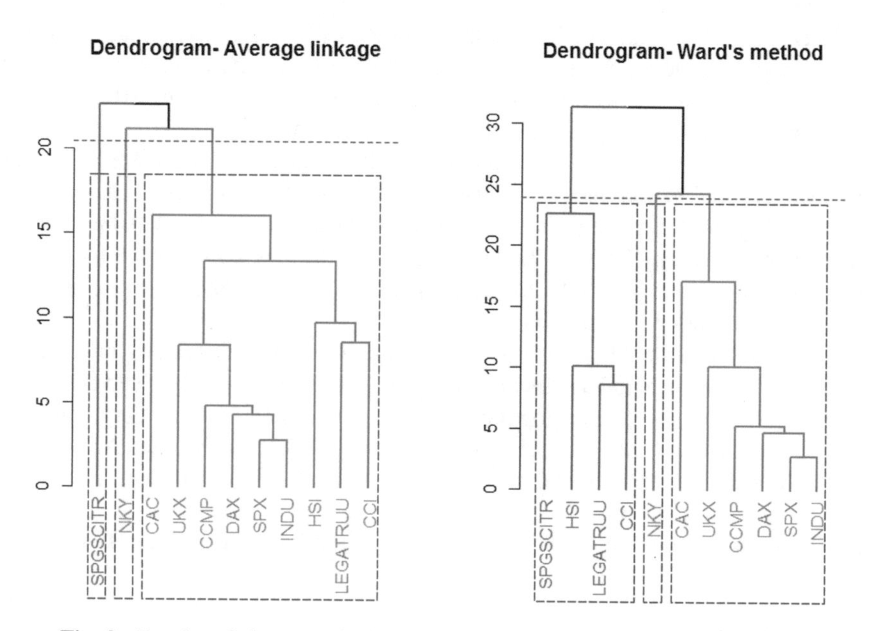

Fig. 2. Results of cluster analysis- average linkage method and Ward's method

As we can see the results have something in common. On the lowest linkage distance *SPX* and *INDU indices* are joined into a cluster. *DAX* and *CCMP* are very similar to them. The most different from other indices are *NKY* index and *SPGSCITR* index. *NKY* index was the only index showing decreasing trend. It also has a unique change-point found with Pettitt's test. In *SPGSCITR* some different change-point can be found from the change-points of other indices. We can also observate the basic behaviour of Ward's method clustering, that objects are joined into existing clusters and new cluster is created just in case of high dissimilarity.

4 Conclusions

The aim of this paper was to analyze the development of 11 market indices. Indices are very important indicator of market development. Our first goal was to reveal the trend in time series of indices. We compared the results of three nonparametric methods to determine the trend-Cox-Stuart test, Mann-Kendall test and Spearman's rho test. The Cox-Stuart test showed a significant increasing trend for all of the indices except *NKY index*. Mann-Kendall test and Spearman's test showed a statistically significant trend for all of the indices. Except *NKY index* it was a statistically significant increasing trend. The magnitude of the trend was calculated by Sen's slope. According to this statistic *HSI index* has the highest increasing tendency. All magnitudes are statistically significant on the level of 0.05.

Other important point of view on the development of indices is finding change-points. Single change-point detection was carried out by Pettitt's test. Single change-points was found in 1998, 2000, from 2005 to 2006, in 2010 and 2012. Multiple change-point analysis was performed by using divisive and agglomerative estimation. The results of these methods return similar results although a little biased. Also the agglomerative estimation proposes less change-points then divisive methods. The results of divisive estimation are statistically significant on the level 0.05. Also the change-points found by Pettitt's test are located near to the ones obtained by divisive estimation. These change-points can be caused by changes in the components of the indices or by economic depression.

In the third part of this paper we found indices which development is similar in time. Results of average linkage method and Ward's method basically give very similar clusters. The most similar development has *INDU* and *SPX indices*. Very similar to them are *DAX, CCMP* and *UKX*. The lowest level of similarity is between *NKY* and the other indices. We determined the three cluster solution as the most appropriate.

Acknowledgement. The support of the grant VEGA 1/0420/15 is kindly announced.

References

1. Kulczycki, P., Kowalski, P.A., Łukasik, S. (eds.) Contemporary Computational Science, p. 94. AGH-UST Press, Cracow (2018)
2. Lehtinen, E., Pulkkinen, U., Pörn, K.: Statistical Trend Analysis Methods for Temporal Phenomena. SKi (1997)
3. Kendall, M.G.: Rank Correlation Methods. Griffin, London (1975)
4. Mann, H.B.: Nonparametric tests against trend. Econometrica **13**, 245–259 (1945)
5. Sneyers, R.: On the statistical analysis of series of observations. World Meteorological Organization, Technical Note no. 143, WMO no. 415 (1990)
6. Wald, A., Wolfowitz, J.: On a test wether two samples are from the same distribution. Ann. Math. Stat. **11**, 147–162 (1940)
7. Pettitt, A.N.: A non-parametric approach to the change-point problem. J. Appl. Stat. **28**(2), 126–135 (1979)
8. Matteson, D.S., James, N.A.: A non-parametric approach for multiple change point analysis of multivariate data. J. Am. Stat. Assoc. **109**(505), 334–345 (2014)

9. Gilmore, B.T.: Dynamic Time and Price Analysis of Market Trends. Bryce Gilmore & Associates, Helensvale (1998)
10. Edwards, R.D., Magee, J., Basetti, W.H.C.: Technical Analysis of Stock Trends, 8th edn. CRC Press LLC, Boca Raton (2001)
11. Brodsky, B.: Change-Point Analysis in Nonstationary Stochastic Models. CRC Press Taylor & Francis Group, Boca Raton (2017)
12. Everitt, B.S., Landau, S., Leese, M., Stahl, D.: Cluster Analysis, 5th edn. Wiley, Hoboken (2011)
13. Kassambara, A.: Practical Guide To Cluster Analysis in R. STHDA, Alboukadel Kassambara (2017)
14. Chen, J., Gupta, A.K.: Parametric Statistical Change Point Analysis With Applications to Genetics, Medicine, and Finance, 2nd edn. Birkhäuser, Boston (2010)
15. Racine, J.S.: Nonparametric Econometrics. Now Publisher Inc., Hanover (2008)
16. Mills, T.C.: Modelling Trends and Cycles in Economic Time Series. Palgrave Macmillan, Basingstoke (2003)
17. Palma, W.: Time Series Analysis. Wiley, Hoboken (2016)
18. Mathworks Statistics Toolbox. http://www.mathworks.com/help/stats/cophenet.html. Accessed 7 Mar 2018
19. https://cran.r-project.org/web/packages/NbClust/NbClust.pdf. Accessed 5 Mar 2018
20. https://cran.r-project.org/web/packages/trend/trend.pdf. Accessed 9 Mar 2018
21. https://cran.r-project.org/web/packages/pastecs/pastecs.pdf. Accessed 6 Mar 2018
22. https://cran.r-project.org/web/packages/TTAinterfaceTrendAnalysis/vignettes/TTAVignette.pdf. Accessed 8 Mar 2018
23. https://cran.r-project.org/web/packages/ecp/vignettes/ecp.pdf. Accessed 10 Mar 2018
24. https://stat.ethz.ch/R-manual/R-devel/library/stats/html/00Index.html. Accessed 1 Mar 2018
25. https://cran.r-project.org/web/packages/dendextend/dendextend.pdf. Accessed 12 Mar 2018
26. https://www.r-project.org/other-docs.html

Author Index

P. Kulczycki et al. (Eds.): ITSRCP 2018, AISC 945, pp. 383–384, 2020.
https://doi.org/10.1007/978-3-030-18058-4

Printed in the United States
By Bookmasters